Y0-BXR-693

ROOTS OF THE EARTH

This book has been published with the help of a generous
subvention from the Publications Committee, La Trobe University

PAUL SILLITOE

Roots of the earth

CROPS IN THE HIGHLANDS OF PAPUA NEW GUINEA

MANCHESTER UNIVERSITY PRESS

First published 1983 by

MANCHESTER UNIVERSITY PRESS
Oxford Road, Manchester M13 9PL
and
51 Washington Street, Dover, N.H. 03820, USA

British Library cataloguing in publication data

Sillitoe, Paul
Roots of the earth : crops in the highlands of Papua New Guinea.
1. Food habits — Papua New Guinea
2. Man, Primitive — Food
I. Title
306'.3 GN671.N5

ISBN 0-7190-0874-3

Library of Congress Number 82-62247

Library of Congress cataloguing in publication data applied for

Printed in Great Britain

FOR JACKIE

CONTENTS

MAPS, FIGURES, TABLES AND PLATES

MAPS

FIGURES

TABLES

PLATES

PREFACE

The Wola, living in the cordillera that runs like a backbone through the centre of Papua New Guinea, subsist largely on a vegetarian diet, not through any dislike of meat nor belief that killing other creatures is wrong but because their environment and lifestyle offer them little choice. Their existence depends on the plants that are the subject of this book, which they cultivate in gardens cleared on the slopes of the mountains that make up their homeland.

On one level this is an ethnographic flora,[1] devised to facilitate the identification of the plants discussed; written for those unable to comprehend the technical-telegraphese of botanical descriptions. The floristic accounts, although they pay due regard to the relevant findings of scientific botany, have a Wola perspective: they present the plants as they appear to them — so far as this is knowable to an outsider and conveyable on paper.

On another level, this study explores the factors that condition and circumscribe the perceptions those socialised in alien and pre-literate cultural traditions have of natural phenomena. It uses the floristic accounts as a map and compass in these explorations, giving in detail the evidence on which the argument stands.

Currently, it is fashionable in anthropology to consider what others think, how they order, classify and symbolise with flora and fauna, yet give scant regard to the occurrence of the biota concerned elsewhere in their lives. This book contends that comprehensive consideration is necessary: experience influences, and is in turn influenced by, ideas relating to nature's products. The eighteenth-century clergyman and celebrated amateur naturalist Gilbert White put this point long ago in his timeless and ever-popular book, *The Natural History and Antiquities of Selbourne*, when he wrote in June 1778:

[1] I have deposited pressed specimens of the plants discussed in this book in the herbariums attached to the University Museum in Manchester and the Botany School in Cambridge.

'The standing objection to botany has always been, that it is a pursuit that amuses the fancy and exercises the memory, without improving the mind or advancing any real knowledge: and where the science is carried no further than a mere systematic classification, the charge is but too true. But the botanist that is desirous of wiping off this aspersion should be by no means content with a list of names; he should study plants philosophically, should investigate the laws of vegetation, should examine the powers and virtues of efficacious herbs, should promote their cultivation; and graft the gardener, the planter, and the husbandman, on the phytologist.' Subsequent Victorian scholars, portending the intellectual break throughs of Darwin and Mendel, criticised museum or closet naturalists engaged in arranging organisms in families, genera and so on, for not thinking beyond their groupings of dried and stuffed specimens.

It would appear that in some regards, anthropological understanding of certain issues pertaining to other cultures stands on a similar threshold: it needs to proceed from documenting others' classifications to excogitating the understanding they achieve with them. This in no way discounts the importance of the former, as White went on in the letter quoted above: 'Not that system is by any means to be thrown aside; without system the field of nature would be a pathless wilderness: but system should be subservient to, not the main object of, pursuit.' There is a tendency for the pendulum of academic debate to swing from one polemical position to another, when the mid-point of its arc gives optimum insight; I am not advocating a swing from cognition to functional praxis but a judicious consideration of both. While some Victorian critics may have considered taxonomy as a juvenile stage through which any science must initially pass before reaching explanatory maturity, subsequent events have proved them wrong — classification remains central to scientific endeavour. Indeed, the constant revision of named groups is essential for reflection and communication. The confusion of life demands ordering into classes — a place for everything and everything in its place.

Some refer to this advocacy of a wide approach as holism, others will recognise a long-established functionalist dictum, to wit: any culture is a tightly interrelated whole and to study any feature in isolation is to give at best a partial account and at worst a distorted one. Unlike the botanist who can specialise in a narrow field without fear of missing crucial material, the anthropologist is obliged to bear in mind the whole world inhabited by the people studied, both their environmental and man-made cultural worlds. But there is a logical absurdity in this requirement, which demands either the knowledge of everything or nothing: clearly some boundaries need to be fixed, as is customary anthropological practice.

The paralogism to guard against when restricting the field of enquiry is tunnel-vision reasoning, not becoming myopic as moles digging along narrowly defined burrows overlooking the significance of factors that fall without the arbitrarily drawn boundaries, outside the field of vision. A workable balance is needed between two opposed and contradictory requirements: to produce a detailed and sophisticated study which, by definition, demands narrowing the field of enquiry, without becoming too restricted and missing related issues and distorting the overall view: to maintain perspective while working in depth.

The ethnographic material presented in this book I collected in the course of two extended periods of fieldwork with the Wola, from July 1973 to September 1974 and from December 1976 to July 1978. Throughout this time I lived in the west of their region, in the settlement of Haelaelinja situated in a sweeping bend on the Was river (see Map 2). From the first I endeavoured to work in Wola, like other Non-Austronesian languages a difficult verb-dominated tongue. I refer the reader to my previous book *Give and Take* (1979: xi–xiv) for a brief note on this language, the orthography I use in writing it, and my approach to its translation, often using literal transcriptions.

It is my agreeable obligation to close this prefatory note by thanking all those who have helped in this work.

To my wife Jackie I am unable to express the extent of my gratitude. She has accompanied me during my fieldwork and on our return continued to give invaluable assistance: she helped weigh food in my surveys of yields and consumption, she typed up drafts of this work, only to see me deface them as I struggled to express myself, and she has constantly given sensible criticism and needed encouragement. To her I dedicate this book as a token of my thanks. I also thank my mother, who as always, has assited in numerous little ways.

To my friends in the Was valley I owe a debt of gratitude which I am also unable to reciprocate in any measure. They patiently endured my enquiries and good humouredly answered my, to them, sometimes ridiculous and naïve, questions. Slowly they taught me about their plant lore. All those who live in the Haelaelinja region, too many to name, have helped me in one way or another with my research; although in particular I benefited from the assistance of Maenget Saendaep, Wenja Neleb, Maenget Kem, Mayka Kuwliy and Maenget Waebay, and those, mentioned elsewhere in this book, who co-operated in the yield and consumption surveys. They would, I know, like me to record for posterity our sorrow at the passing of old Kem, great friend and sage, 'narrator' in the film *Bird of the Thunderwoman.*

On the academic side, I particularly wish to thank the botanists who have advised me, notably by supplying authoritative scientific identifications of my tentatively named plant specimens: Barbara Croxall of the Cambridge Botany School Herbarium, who has proved a sterling ally in this work; William Erskine, researching into tropical legumes, and Nigel Banks, working on bananas, both formerly of the Applied Biology Department in Cambridge; David Frodin of the University of Papua New Guinea and Bob Johns of the

University in Lae; and the staff of the Royal Botanic Gardens of Kew and Edinburgh and the Rijksherbarium, Leiden. Throughout my research I have received help from several anthropologists, notably Marilyn and Andrew Strathern, whose extensive experience and knowledge of the New Guinea Highlands it has been my good fortune to draw upon, and my colleagues here in Manchester, both in informal discussions and in seminar criticism.

In Nipa I thank the staff of the administration for assistance during my stays, especially John Gordon-Kirkby, a reliable and good friend. I also thank for their help the missionaries of Tiliba: Jan and Ed Staich. And Father Sam Driscoll at St Fidelis.

Finally, I record my thanks to the several institutions and bodies that financed the research culminating in this book, which I started during my tenure of a Senior Rouse Ball Scholarship at Trinity College, Cambridge and completed here in Manchester as a Simon Fellow. For sponsoring the fieldwork I am grateful to the Social Science Research Council for a Project Grant and Studentship; the Leverhulme Foundation for an Overseas Scholarship tenable in Papua New Guinea; Trinity College, Cambridge for an award from the William Wyse Studentship Fund; the University of Cambridge for grants from the Smuts Memorial Fund; the Royal Anthropological Institute for an Emslie Horniman Anthropological Scholarship; and the Museum of Ethnology and Archaeology in Cambridge for a grant from the Crowther—Benyon Fund. I am particularly grateful to the Publications Committee of La Trobe University for a most generous grant towards the cost of publication.

PAUL SILLITOE

His own language was so much part of the earth and animals that abstract ideas in the European sense either did not exist, or the translation had to be too roundabout, or made abstruse beyond meaning. There were many senses ... that had no expression in English, and many shadings of the mind not to be understood except by intuition.

(Richard Llewellyn, 1961, *A Man in a Mirror*)

Chapter 1
WOLA CROPS, COUNTRY AND CULTURE

THE CROPS

This book is a study of the crops grown by the Wola people who live in the Highlands of Papua New Guinea. These people subsist primarily by gardening, as so-called shifting cultivators, and it is on the plants described here that their food quest largely depends. Indeed, measured by the weights consumed, they depend only on a few of them, and on one tuber in particular — the sweet potato. For them the 'fruits of the earth' are largely roots, this has a marked influence on their lives, in a number of ways.

The aim is to give a Wola-centric account of their crops: to describe how they see and classify them, how they cultivate them, how they think they grow, how they use them, and so on. Thus this work is intended as a contribution to the study of ethno-botany, that field of anthropological endeavour concerned with the documentation of exotic flora in the indigenous categories of those who live in the region where it grows.

The study of plants in anthropology has a respectable history stretching back over fifty or more years (for example Smith 1923, 1928, Roys 1931, Wyman and Harris 1941). Early on however, two distinctive approaches developed, which have become marked with time. On the one hand there are those who have concerned themselves with the use of plants, and on the other there are those who have concentrated on documenting their autochthonous classification.

The former approach held sway for the functionalists with their interests in practical issues relating to the current existence of cultures and provisioning of their members. They documented the use of plants; that is, their consumption, place in the diet, uses in other aspects of live, and so on (for example see Richards's (1932, 1939) now classic studies along these lines). This interest has passed today, within anthropology, to the so-called neo-functionalists who are primarily concerned to trace causal connections between peoples' life-styles and their environment (for such a study in a New Guinea context, see Rappaport 1968). Outside anthropology others, notably nutritionists and geographers, continue to pursue research into the use, largely the consumption of plants, although in a restricted way on the whole, ignoring relevant social and cultural issues (for examples of such studies in New Guinea see Hipsley and Clements 1947, Venkatachalam 1962, Waddell 1972).

The other approach to the study of plants in anthropology, concerning itself with their conceptual ordering, has burgeoned in recent years. This parallels developments elsewhere in the subject, stimulated in some measure by the advent of structuralism with its emphasis on human thought. Here there are two intellectual lines of enquiry, between which it is necessary to maintain an unambiguous distinction. One concerns the classification of plants themselves, which, with its more empirical concern with natural phenomena, is more closely associated with the foregoing functionally orientated approach. These studies vary markedly in sophistication from those that draw unconvincing parallels with the taxonomic thinking of Western science, to those that try to reveal the unique principles and logic lying behind others' botanical classifications (compare for example Berlin *et al.* 1973 and Brown *et al.* 1976 with Friedberg 1968 and Strathern 1969). The other line of enquiry concerning people's conception of plants investigates their use in symbolic contexts, and is sociologically sophisticated while biologically naive (see Needham 1979).

Recent studies in anthropological botany have tended to concentrate on what people think of their flora, to the virtual exclusion of how they use it and how it otherwise figures in their lives. An attempt is made here to redress this imbalance and demonstrate that the integration of the above functional and cognitive approaches can benefit both (for an early study along these lines see Conklin 1954). It is argued that what people do with plants influences what they think of them, and *vice versa.* To keep this attempted comprehensive ethno-natural history within manageable limits, I have restricted it to a narrow range of plants, namely those cultivated by the Wola in their gardens. These are important plants, essential for their subsistence, although they amount to only a fraction of their region's total plant life.[1]

Others have recently pointed out that the current exclusive concern with matters of classification results in biased and incomplete studies of the flora, fauna or whatever, and sometimes in the reification of inappropriate models to explain their ordering (Johnson 1974, Ellen 1975: 221–2, Hardesty 1977: 17). After all, the functionalists' dictum, put forcefully by Malinowski, that cultures ought to be studied as relatively integrated wholes, is the accepted

[1] I hope to deal with the remaining plants in other contexts when my network of contacts with botanists is developed enough to facilitate it (I currently have several hundred specimens in various herbariums, although sadly I lost many when the herbarium at the University of Papua New Guinea burnt down in 1978).

watchword of field orientated ethnographic research. It is agreed that the social and environmental context should be taken into account as far as possible, that considering anything in relative isolation is to risk distortion. Yet current anthropological studies of natural history do not always heed this. Plants and animals have a practical side concerned with subsistence, environmental exploitation and so on, and to consider only what people say about them, and extrapolate from this what they possibly think of them, is to present a partial, maybe cock-eyed view.

Cultivating ideas relating to crops in a culturally holistic bed, in Malinowski's cultural context, is shown to be a fertile approach resulting in a worthwhile intellectual harvest, yielding insights into Wola thinking and behaviour. There are many ramifying and mutually modifying connections between these plants, the food quest of the Wola and their society in general. To mention a few briefly, discussed in detail later: the elaborateness with which the Wola classify their crops relates in a number of ways to their cultivation and consumption, as does their occurrence in gardens at locations where they flourish; their proportionally uniform occurrence in different men's gardens reflects features of the egalitarian social order; and the gender ascribed to them, their association with one sex or the other, relates to the social order too, signifying an important distinction upon which this exchange polity stands.

As these remarks suggest, this study attempts to be more than a contribution to the narrow field of ethno-botany. It proceeds on three broad fronts, only one of which is ethno-botanical. The other two are ethnographic and floristic: to make a contribution to Highland New Guinea ethnography in general and Wola subsistence practices in particular, and to serve as a simplified agronomic flora to the crops found growing in this part of the world.

On the ethnographic front, this book is part of an ongoing study into the food quest of the Wola; indeed it grew out of this work. Previous studies of subsistence practices in the Highlands of New Guinea go little further than listing the crops cultivated by the people and giving their botanical identification (see Rappaport 1968: 44–6; Clarke 1971: 225–40; Waddell 1972: 226–8; Pospisil 1963: 108–15 – the studies of Powell *et al.* 1975, Fischer 1968: 258–88 and Hays 1981 are something of exceptions). I made extensive enquiries into the crops cultivated by the Wola with a view to pushing on beyond this, although I did not anticipate a work of this size, which started life as a chapter in a study of Wola gardening practices. It proved, however, to be something of a dehydrated vegetable, expanding to the present work when I watered my field material with theoretical ideas and started to cultivate the quantitative data which I had collected!

The perspective of this anthropological botany switches between the plants and the people who cultivate them. In places it gives prominence to the crops themselves and resembles an elementary botanical study. But botanists' interests remain fixed on the plants; unlike this Wola-centred study, they treat the ethnographic situation (when they pay heed to it) as little more than a setting, and not something to be integrated into their accounts (Massal and Barrau 1956, Yen 1974, Powell 1976). When these scholars do venture into the anthropological realm it is usually to speculate on cultural evolution (Powell *et al.* 1975, Yen and Wheeler 1968). The study of plant morphology and genetics, palaeontology and pollen, can supply valuable evidence about the origin and movement of plants, from which (together with evidence from other fields such as linguistics, archaeology and physical anthropology) suggestions can be made, and hypotheses developed, for answering questions about cultural evolution and change in the distant past. Other than noting the possible origin of the crops discussed, and giving Wola views on the subject, this issue is not taken up in this study. Having no data nor qualifications to make constructive suggestions, the focus is on present-day Wola plant cultivation practices and thoughts.

The botanical component in this work, where it gives primacy to the plants discussed, is intended to serve as a flora for amateurs uninitiated into the complicated technical terminology of botanical taxonomy. Unlike many anthropological natural histories, which assume some scientific knowledge on the part of the reader of the flora, fauna or whatever under study, this study describes the plants concerned. These botanical facts, while presented as the Wola see them, are put in layman's language with the intention of helping others to identify and come to know the crops discussed. Throughout this part of the world at least some of these plants are important in peoples' quest for food, but non-specialists frequently experience difficulty in identifying them or finding out anything about them which might be relevant to their work. This is notably so for anthropologists, whose wide-ranging interests often lead them into other fields. It is hoped that this book will help to introduce such non-specialists to this field, and so serve a practical puspose as a manual (especially for fieldworkers)[2] and save others the trouble experienced by the author in identifying and understanding the botany of the crops grown by the Wola. Prior to arriving in the Highlands of Papua New Guinea, for example, I had no idea of the difference between a sweet potato and a yam, and it was some time before I knew what a kudzu was (and talking to others I do not think I was an ignorant exception, but the ignorant rule).

Table 1 introduces the crops grown by the Wola in their gardens.[3] It is designed as a simple floristic key which leads into the straightforward botanical descriptions contained in the early chapters. It gives a brief description of the crops, sufficient to distinguish them from others, their common and botanical names, and the page on which they are described in more detail. It is intended that this key, together with the later descriptions, will enable others to identify the plants. As a further aid to fieldworkers

[2] Not only anthropologists, but fieldworkers in other disciplines too.

[3] This book deals only with those plants cultivated by the Wola and not wild ones which they also collect and eat (see Triede 1967).

trying to do this, a list is given in Appendix 1 of the Pidgin or Neo-Melanesian names for each.

To use this book as a botanical manual refer firstly to this key. Find out which part of the plant is eaten and look under the relevant section (A, B or C etc.). Then follow down the list of questions, which have positive or negative answers that guide you until you identify the plant. For example, suppose people eat the tuberous roots of the plant before you, then refer to the alternatives under A2. If the plant has an erect stem, then refer to A4. If the plant has single unbranched stems, the key guides you to A5. If it has stout spongy stalks with large, heart-shaped leaves, then you go on to A6, where you are told that if your plant specimen has spherical to oval tubers and smooth silky leaves which are edible, then you are looking at a taro (*Colocasia esculenta*).

If you wish to know more about this plant then turn to the page number given in the right-hand column, where you will find a detailed account of it.[4] This is written for laymen by a layman (helped by specialists),[5] eschewing the unfamiliar technical terms, which make the accurate use of botanical flora by the uninitiated difficult. Some common technical terms remain, such as simple and compound leaf, but every effort has been made to make their meaning clear.[6]

The accounts of each crop follow the same pattern. They start by giving the Wola, common English, and Latin botanical names for the plant in question, and also its botanical family designation. Then they note the plant's origin, as far as it is known by botanists and also according to the Wola. There follows a description of the crop in question, accompanied, in the case of plants unfamiliar to those from temperate regions, by a drawing on which the botanical parts are identified in Wola.[7] Next comes a discussion of the cultivars[8] distinguished by the Wola, in which the criteria they use to differentiate between plants are outlined, supported by tables listing cultivars and their identifying features.

[4] For further information on the plants described here, and tropical crops not dealt with, the reader is recommended to consult Purseglove (1968 and 1972) and Cobley (1976), in addition to the more specialist works listed in the bibliography. Other useful general agronomic works include French and Bridle (1978), and Herklots (1972); Womersley (1972) gives a useful summary.

[5] And here I thank especially W. Eskine of the Applied Biology Department in Cambridge and B. Croxall of Cambridge University Herbarium.

[6] In their exhaustive ethno-botanical study of some Mayan Indians, Berlin *et al.* (1974: 156) do otherwise, their use of technical botanical terms making their descriptions difficult for non-specialists to comprehend.

[7] The reader is referred to the glossary of Wola botanical terms given in Appendix II.

[8] This book follows the accepted convention of agricultural science (as recommended in the *International Code of Nomenclature for Cultivated Plants*) and refers to named sub-types of crops as cultivars and not varieties. This avoids confusion with botanical usage, where variety can refer to a morphological variation within a species, something which the term cultivar does not imply. Throughout this account cultivar names are given in quotation marks (e.g. '*konoma*', a sweet potato cultivar).

Table 1: Phytological key to crops

		KEY			COMMON NAME	GENUS AND SPECIES	SEE PAGE NO.
A	1	*Tuberous roots eaten*	see	A2			
		Roots *not* eaten		B1			
	2	Plant stem/stalks: creeping and vine-like		A3			
		erect		A4			
		twining and climbing		A8			
	3	Leaves and tubers variably shaped (tubers long, spindle-like to spherical, usually with tapered ends; sweet taste)			Sweet potato	*Ipomoea batatas*	29
	4	Single stem/stalks		A5			
		Branched stems		A7			
	5	Stout spongy stalks with large heart-shaped leaves		A6			
		Spindly grass-like stems with alternate, narrow and pointed leaves, and knobbly pinkish rhizome			Ginger	*Zingiber officinale*	49
	6	Tuber spherical to oval; leaves smooth & silky (& eaten)			Taro	*Colocasia esculenta*	37
		Clump of sausage-shaped tubers growing from parent; larger plant than taro with coarser leaves more pointed at three corners			Tannia	*Xanthosoma sagittifolium*	42
	7	Clump of small spherical tubers; compound leaves			Irish potato	*Solanum tuberosum*	51
	8	Clump of variably shaped tubers (often elongated and sausage-like) joined at apices; small heart-shaped leaves			Greater yam	*Dioscorea alata*	44
		Clump of elongated fibrous tubers, tapered at ends and joined at apices; bean-like winged flowers and trifoliate leaves			Kudzu	*Pueraria lobata*	46
B	1	*Leaves eaten as greens*	see	B2			
		Leaves *not* eaten		C1			
	2	Plant stem: erect		B3			
		creeping		B12			
		woody		B13			
	3	Plants grow: singly		B4			
		in clumps		B11			
	4	Stems long, thin and branching		B5			
		Stems short, stout and single		B10			
	5	Single leaves: simple and oval		B6			
		wavy edged and lobed		B9			
		Compound leaves of small opposed feather-like leaflets		B8			

		Description		Key	Common name	Scientific name	Page
	6	Small chaffy flowers arranged in dense spikes		B7			
		Clusters of flowers up stem, with white petals flanked by violet leaflets; leaves taper to long points			Acanth meat spinach	*Dicliptera papuana*	62
	7	Egg-shaped leaves have tiny notch in apex, plant often tinged red and yellow			Indian spinach	*Amaranthus tricolor*	53
		Similar to *A. tricolor* but taller plant and leaves have tiny point on apex			Prince's feather spinach	*Amaranthus cruentus*	56
		Similar to *A. cruentus* but spikes drooping not erect			Love lies bleeding spinach	*Amaranthus caudatus*	56
	8	White flowered, parsley-like herb			Javanese dropwort	*Oenanthe javanica*	64
	9	Annual herb with yellow flowers and fruits small, spherical translucent capsules in terminal clusters			Crucifer spinach	*Rorippa* sp.	57
	10	Dense round head of smooth, thick, overlapping leaves			Cabbage	*Brassica oleracea*	66
		Elongated cabbage-like leaves with pale stalks and ribs, growing up stem in loose cluster			Chinese cabbage	*Brassica chinensis*	67
	11	Perennial; leaves simple, egg-shaped, thick and somewhat crinkly			Acanth spinach	*Rungia klossii*	60
		Hollow, cylindrical pointed leaves with bulb; familiar onion odour when crushed			Onion	*Allium cepa* var *aggregatum*	68
	12	Grass-like herb with jointed stem and alternate, spear-like leaves joined at nodes with sheaths (not stalks)			Spiderwort	*Commelina diffusa*	70
		Aquatic perennial herb with compound leaves composed of leaflet lobes; characteristic 'hot' taste			Watercress	*Nasturtium officinale*	60
	13	Trees		B14			
		Perennial shrub with large, varyingly lobed, palm-like leaves and large, bell-shaped, yellow flowers			Hibiscus spinach	*Hibiscus manihot*	71
	14	Oval leaves and spherical green fruits (also eaten)			Fig	*Ficus wassa*	73
		Large, crinkly, oval leaves and small cabbage-like fruits			Highland breadfruit	*Ficus dammaropsis*	75
C	1	Shoots/stems eaten	see	C2			
		Not eaten		D1			
	2	Shoots eaten		C3			
		Stems chewed		C4			
	3	Perennial grass growing in clumps; broad spear-like leaves taper to long point and are longitudinally corrugated			Highland *pitpit*	*Setaria palmifolia*	78
		Hollow cane growing in clumps; tall, erect, drooping at ends; large, bullet-shaped, sucker shoots eaten			Bamboo	*Nastus elatus*	81
	4	Cane with fibrous pith chewed for sweet juice; grow in clumps; leaves long, pointed and joined alternately at nodes with sheath round cane			Sugar cane	*Saccharum officinarum*	84

Table 1 (continued)

		KEY		COMMON NAME	GENUS AND SPECIES	SEE PAGE NO.
D	1	Fruits eaten	see D2			
		Not eaten	E1			
	2	Pulses (pod containing beans)	D3			
		Cucurbits (fleshy, variably shaped, rounded fruits)	D4			
		Syncarps (mass of nuts)	D7			
		Globular fruits with many seeds in juicy flesh	D8			
		Others	D9			
	3	(*a*) Climbing stems with trifoliate leaves and winged flowers producing: (i) Flat, broad pods		Hyacinth bean	*Lablab niger*	89
		(ii) Slender long pods (sometimes dwarf plants too)		Common bean	*Phaseolus vulgaris*	92
		(iii) Long square pods with four prominent wavy wings		Winged bean	*Psophocarpus tetragonolobus*	94
		(*b*) Short plant with compound leaves terminating in tendril (medium pod and round seeds)		Pea	*Pisum sativum*	93
	4	Plant stem: climbing	D5			
		spreading and trailing	D6			
	5	Large heart-shaped leaves, tendrils, and oblong, pendent, red fruits		Climbing cucurbit	*Trichosanthes pulleana*	101
	6	White flowers, produce variably sized and shaped fruits with moist, pale green flesh; if left develop durable rind		Bottle gourd	*Lagenaria siceraria*	96
		Yellow flowers, produce large, variably shaped fruits with orange flesh (new leaves also eaten)		Pumpkin	*Cucurbita maxima*	98
		Yellow flowers, but smaller plant than above, produces short, dumpy, yellow/orange fruits		Cucumber	*Cucumis sativus*	100
	7	Trees with aerial prop-roots and crowns of long, stiff, sword-like leaves armed with barbs:				
		Tall trees producing large, spherical, pale green, pendent fruit composed of many nuts		Screw-pine (*karuga*)	*Pandanus brosimos* & *Pandanus julianetti*	103

		Low trees producing elongated red fruit composed of many drupes			Screw-pine (*marita*)	*Pandanus conoideus*	112
	8	Climbing perennial with three-lobed leaves and large, showy, white and purple flowers; fruits purple with black seeds in yellowish, tart-flavoured pulp			Passion fruit	*Passiflora edulis*	113
		Trailing, branched, hairy plants with compound leaves; fruits red and smooth			Tomato	*Lycopersicon esculentum*	115
	9	Large plants with long, oblong leaves producing many curved, yellow fruits on a single pendent stem			Banana	*Musa* hort var.	116
		Erect plant terminating in grass-like tassel; long, wavy edged leaves alternately clasp stem with sheaths, and cobs grow from their point of attachment			Maize	*Zea mays*	120
E	1	*Inedible crops*	see	E2			
	2	Plant stems: erect and herbaceous		E3			
		woody		E5			
	3	Leaves simple and egg-shaped		E4			
		Leafless, aquatic perennial with round, hollow stems tapering to a point			Sedge	*Eleocharis cf. dubia*	129
	4	Large leaves dried and smoked			Tobacco	*Nicotiana tabacum*	122
		Purple perennial with soft, velvety leaves and small blue flowers			Dye plant	*Plectranthus scutellarioides*	130
	5	Trees		E6			
		Shrubs with long, narrow, bluntly pointed leaves growing in terminal heads			Palm lily	*Cordyline fruticosa*	125
	6	Variably shaped, woolly leaves; catkins and spherical round fruits with bristles			Paper mulberry	*Broussonetia papyrifera*	131
		Long drooping branches resembling horse-tails; leaves reduced to pointed scales and fruit a small globular cone			She-oak	*Casuarina oligodon*	133

The focus of each account then shifts from the way the Wola classify their crops and distinguish between them, to a discussion of their cultivation. This covers their propagation and husbandry; that is, how they are planted, who plants them, the kinds of garden in which they are cultivated, their place in the sequence of planting, and so on. Where possible Wola reasons for these practices are given, relating to how they think the plants grow, the sort of location they understand them to prefer, and how these ideas influence their agricultural practices (for example, prompting them to plant certain crops in specific areas within gardens where they think conditions are particularly suitable for them). An associated issue is the relation between plants and the natural conditions of the Wola region; that is, the kind of soil they prefer, their tolerance to climatic variations, which natural resources they draw upon when growing, and so on. Each account concludes with a discussion of the crop's consumption. The time plants take to reach maturity and yield food is indicated, as is the length of time for which people can harvest from them. The parts of the plant eaten, and their preparation are then described, and finally the importance of the crop in the diet noted.

Table 2 lists the crops of the Wola and gives a summary of the points discussed in each account, as outlined above. This listing of the salient points is intended as an *aide de memoire*, as a source of ready reference to help guide those unfamiliar with the plants through the coming accounts.

This outline of the structure of each crop account points to the Wola orientation of this study, which is only a simplified agronomic flora as a sort of by-product of its ethnographic aims. Later chapters proceed to consolidate these, moving away from any floristic pretensions: they deal with topics in the same order as the crop accounts though, starting with classification, moving on to cultivation and concluding with consumption. Throughout, plants are presented in a way that would be meaningful to the Wola: their ordering for instance according to part, or parts eaten. One result is that they are grouped together, and consequently described, in a way that differs from straightforward agronomic accounts. These would order and describe plants according to their distinguishing botanical features (some of which would be as obscure to the Wola as the layman), whereas here these are not given prominence; for example beans are dealt with in a different chapter to kudzu, although these plants are all legumes and so in the same botanical family.

An important methodological point, taken up in detail later in the discussion of classification, is that this singular ordering of crops, and the drawing up of the floristic key, is the author's doing. It is founded on monothetic principles (Needham 1975, Ellen 1979), that is the plants are grouped together by their possession of a single common feature taken as diagnostic for the purpose — in this case the part eaten. While the Wola could appreciate this ordering, it is not an expressed artifact of their thinking.[9]

[9] This relates to the notion of 'covert categories' (Berlin *et al.* 1968, 1974: 59—61; Hays 1976), that is the existence of implicit levels of ordering within folk classifications, the implications of which are discussed in Chapter 8.

Table 2: A compendium of facts pertaining to Wola crops

COMMON NAME	WOLA NAME	BOTANICAL NAME	NO. CULTIVARS	REGION OF ORIGIN	PRE-/POST-CONTACT	KIND OF PLANT	PROPAGATION	PLACE IN PLANTING SEQUENCE	MALE/FEMALE CROP	TYPE OF GARDENS PLANTED IN *	PARTS EATEN/USED	IMPORTANCE IN DIET
Acanth greens	omok	Dicliptera papuana	1	Papuasia	Pre	erect herb	cuttings	3rd	M,F	1,2,3,5	leaves	<0.001
" "	shombay	Rungia klossi	3	Papuasia	Pre	erect clump herb	cuttings	3rd	F	1,2,3,5	leaves	0.962
Amaranth greens	komb	Amaranthus tricolor	5	Papuasia	Pre	erect herb	seed	5th	M	1,3,5,6	leaves, flowers	0.015
" "	mbolin komb	Amaranthus caudatus	1	S.America	Post	erect herb	seed	5th	M	1,3,5,6	leaves, flowers	0.001
" "	paluw	Amaranthus cruentus	2	C.America	Post	erect herb	seed	5th	M	1,3,5,6	leaves, flowers	0.382
Bamboo	taembok	Nastus elatus	1	Asia	Pre	erect large canes	lateral shoots	6th	M	1,2,3,5,8	sucker shoots	0.010
Banana	diyr	Musa hort. var.	10	Papuasia	Pre	giant large leafed herb	suckers	6th	M	1,2,8	fruit, stem heart	1.250 0.059
Beans : common	taeshaen pebway	Phaseolus vulgaris	4	S.America	Post	climbing/erect herb	seed	4th	M,F	1,3,5	seeds	0.805
: hyacinth	sokol	Lablab niger	4	Asia	Pre	climbing herb with pods	seed	4th	M,F	1,3,5	seeds	
: winged	wolapat	Psophocarpus tetragonolobus	1	Asia	Pre	climbing herb with pods	seed	4th	M,F	1,3,5	pod, seeds, tuber	0[2]
Cabbage	cobaj	Brassica oleracea	5	Mediterranean	Post	stout,'headed' herb	lateral buds	3rd	M,F	1,2,3,5,6	leaves	0.315
Chinese cabbage	kwa	Brassica chinensis	2	E.Asia	Post	erect herb	seed	5th	M	1,3,5,6	leaves	0.013
Climbing cucurbit	tat	Trichosanthes sp.	1	Papuasia	Pre	climbing cucurbit	seed, seedling	6th	M	5,8	fruit	0.009
Crucifer greens	taguwt	Rorippa sp.	6	Papuasia	Pre	erect herb	seed	5th	M	1,3,5,6	leaves	0.216
Cucumber	laek	Cucumis sativus	2	India	Pre	trailing cucurbit	seed	4th	F	1,3,5,6	fruit	0.231
Dye plant	komnol	Plectranthus scutellariodes	1	Old World Tropics	Pre	erect herb	cutting	6th	F	1,2,5	leaves for dye	n.a.
Fig	poiz	Ficus wassa	1	Papuasia	Pre	tree	cutting, seedling	6th	M	1,2,3,5,8	leaves, figs	0.006
Ginger	shombiy	Zingiber officinale	3	Asia	Pre	erect tuberous herb	budded rootstock	4th	M	1,3,5	rhizome	0.036
Gourd	senem	Lagenaria siceraria	3	Africa	Pre	trailing cucurbit	seed	4th	M,F	1,3,5,6	fruit	0.330
Hibiscus greens	huwshiy	Hibiscus manihot	5	China	Pre	bushy shrub	woody cutting	3rd	M	1,3,5,6	leaves	0.243
Highland breadfruit	shuwat	Ficus dammaropsis	1	Papuasia	Pre	tree	seedlings	6th	M	1,2,3,5,8	new leaves	<0.001
Highland pitpit	kot	Setaria palmifolia	9	Papuasia	Pre	robust grass	lateral shoots	2nd	F	1,2,3,5	stem hearts	10.032
Irish potato	aspus	Solanum tuberosum	2	S.America	Post	erect tuberous herb	budding tuber	4th	F	1,2,3,5	tuber	0.013
Kudzu	horon	Pueraria lobata	2	Papuasia	Pre	climbing tuberous herb	cuttings	4th	M	1,2	tuber	<0.001
Maize	kwaliyl	Zea mays	2	C.America	Post	tall stout grass	seed	4th	M,F	1,2,3,5	seeds	1.324
Onion	ᵉnyun	Allium cepa	1	C.Asia	Post	erect bulbed herb	lateral shoots	3rd	F	1,2,3,5	leaves, bulb	0.031
Palm lily	aegop	Cordyline fruticosa	25	Asia	Pre	shrub/tree	cuttings	6th	M	1,2,3,4,	wear leaves	n.a.
Paper mulberry	korael	Broussonetia papyrifera	1	Papuasia	Pre	tree	cuttings	6th	M	8	use bark fibre	n.a.
Parsley (dropwort)	taziy	Oenanthe javanica	2	Papuasia	Pre	erect herb	cuttings	3rd	F	1,2,3,4	leaves	0.299
Passion fruit	ya iyl	Passiflora edulis	1	S.America	Post	woody climber	seed	6th	M,F	1,2,3,5,8	fruit	0.003
Pea	mbin	Pisum sativum	1	S.W.Asia	Post	climbing herb with pods	seed	4th	M,F	1,3,5	seeds	0[2]
Pumpkin	pompkin	Cucurbita maxima	1	S.America	Post	trailing cucurbit	seed, cutting	3rd	F	1,2,3,5,6	fruit, new leaves	13.804
Screw-pine	aenk	Pandanus brosimos & P. julianetti	45	New Guinea	Pre	sword-leafed tree	crowns, seedlings	6th	M	1,2,3,5,7,8	fruit	0.882
" "	wabel	Pandanus conoideus	4	New Guinea	Pre	sword-leafed tree	crowns, seedlings	6th	M	1,2,3,5,7,8	fruit	0.104
Sedge	hurinj	Eleocharis cf. dubia	1	Papuasia	Pre	erect aquatic herb	budded rootstock	6th	F	4,7	wear stems	n.a.
She-oak	naep	Casuarina oligodon	1	Papuasia	Pre	tree	seedling	6th	M	1,2,3,5,8	n.a.	n.a.
Spiderwort	hombiyhaem	Commelina diffusa	1	Pan-tropical	Pre	creeping herb	cuttings	3rd	F	1,2,3	leaves, shoots	0.003
Sugar cane	wol	Saccharum officinarum	12	New Guinea	Pre	large cane-like grass	apical cutting	6th	M	1,2,3,5	stems, tips	1.643
Sweet potato	hokay	Ipomoea batatas	64	S.America	Pre	creeping tuberous herb	cuttings	1st	F	1,2,3,5	tubers	100
Tannia	mbolin ma	Xanthosoma sagittifolium	1	S.America	Post	erect tuberous herb	cuttings	1st - 2nd	M,F	1,2,3,4,5	tubers	<0.001
Taro	ma	Colocasia esculenta	43	Papuasia	Pre	erect tuberous herb	cuttings	1st - 2nd	M,F	1,2,3,4,5	tubers, leaves	1.279 0.276
Tobacco	miyt	Nicotiana tobacum	6	S.America	Pre	erect herb	seed, cutting	6th	M	5,6,8	smoke leaves	n.a.
Tomato	tomasow	Lycopersicon esculentum	1	S.America	Post	erect herb	seed	4th	M,F	1,3,5,6	fruit	0.024
Watercress	kuwmba	Nasturtium officinale	1	Europe	Post	trailing aquatic herb	cuttings	n.a.	M	4,7	leaves	0.074
Yam	bet	Dioscorea alata	1	S.E.Asia	Pre	climbing tuberous herb	cuttings	2nd	M	1,2,3	tubers, bulbils	<0.001

Notes

* Garden types = (1) new sweet potato garden; (2) established sweet potato garden; (3) dry taro garden; (4) wet taro garden; (5) mixed vegetable garden; (6) garden on abandonned house site; (7) outside gardens; (8) around houseyards.

[1] This ratio was computed as a ratio of all other crops to sweet potato. Sweet potato total weight was taken to be 100 and all other crop weights harvested converted to a ratio of 100. E.g. weight of sweet potato = 6757.6 kg and this was taken to be 100, the weight of harvested Highland *pitpit* = 677.9 kg and this was converted to a ratio by multiplying 677.9 by 100/6757.6 = 10.032. These figures are calculated from the raw unprepared weights of the harvested crops (see Chapter 12 for further details).

[2] These crops score 0 because they are absent in the Haelaelinja region where the survey from which these diet ratios are calculated was conducted.

The absence of any direct correspondence between Wola ideas and the English categories in which they are presented raises difficult methodological problems. After all, the lasting value of any ethnography depends on the integrity and accuracy of its field material, not on the intellectual games played with it by the anthropologist. The difficulty is not to foul and present a gross distortion in making some theoretical point (Lewis 1980). The problems I encountered when trying to extrapolate from Wola comments how they conceive of their crops indicate the magnitude of the conceptual difficulties faced.

They thought ridiculous questions like 'What is a sweet potato?' or 'How do you know a sweet potato from a yam?'. Although such questions can be framed in their language, they bring forth either embarrassed giggles, laughter or looks of amazement. In answer people said things like 'a sweet potato's a sweet potato, everyone knows that' or 'look' (pointing to a plant), 'this is a sweet potato' (implying that only a fool could confuse it with anything else). Given these responses, how could an outsider 'enter' a Wola head and see plants their way? Clearly some artificial, and so probably distorting, technique was required. I resorted to prodding informants, who were familiar with the kind of unusual answers I expected in other contexts, until they pointed to some features as diagnostic in the recognition of the plants before them. There were no standard responses. Different people pointed to varying features, indicating that when the Wola identify a plant they do not look for specific cues as criteria for naming, but see it as a familiar whole.[10] According to Strathern (1969: 192), the Hageners tend to discriminate between plants principally by leaf type (speaking of plants of one leaf), which contrasts with the apparently simultaneous consideration of all cues observed by the Wola, who see a crop plant as a distinct entity rather than as consisting of a certain number of distinctive features.[11]

The characteristics which appear to figure prominently in the configuration seen by the Wola, as the coming accounts make clear, are the size and shape of a plant's organs and their colouring.[12] Smell is not important and neither

[10] Frake (1961) distinguished between 'cues' and 'criteria' in his discussion of disease among the Subanum. Criteria are the cues (or observable attributes) used explicitly to define categories, something which the Wola do not apparently do systematically with their crops.

[11] This is not to say that some plants do not have distinctive leaves by which the Wola can identify them. See Ellen (1979: 12); also Bulmer and Tyler (1968: 353), who point out that the Karam look at the overall configuration of features when identifying frogs.

[12] In the coming accounts these features are recorded in English categories, some of which vary notably from those of the Wola. For instance, I have recorded in English the colour which I saw on the plants and not a gloss of the colour category stated by the Wola (in this situation Panoff 1972: 385 gives the Maenge colour terms in her descriptions of their cultivars, followed in brackets by an English gloss; I have decided against this procedure because to make any sense it would require an extensive discussion of Wola colour perception, something not germane to the issues under consideration). I assume that physiologically my informants saw the same colours as I did, although they categorise them differently (Berlin and Kay 1969; Turton 1980), in a way that would confuse a European and make using them in the

is habitat, all plants occurring in gardens. Taste is not always significant either, people can identify plants without eating them; indeed once cooked identification sometimes becomes uncertain.[13] When making identifications at the cultivar level, the points looked for become narrower, micro-morphological variations and colour changes becoming significant. These cues are looked for more systematically because of their fineness, such that informants could more readily appreciate demands for a list of the criteria by which they distinguished between cultivars and could point to them (as the tables in the following chapters show). The problem with cultivars was the disagreement encountered about the use of these diagnostic criteria to name any plant specimen (an issue taken up later).

It is pertinent here to note how I learnt about the idiosyncratic way the Wola 'see' and identify their food plants. My first lessons in crop identification and cultivation came during the surveying of gardens, when I noted the plants growing in each area measured. Later I made more detailed investigations by collecting specimens of all the crops, and those cultivars that could be found (which amounted to the majority in the end), and asking informants to describe each one to me and point out its diagnostic features.[14] When they had finished, I pressed the specimens between newspaper in a plant press to preserve them for future reference (some of these feature in the photographs illustrating this book).

This method of learning was foreign to the Wola, who normally pass on knowledge in a piecemeal and casual way. Asking them what they looked for when identifying plants was, as noted, an artificial question demanding a contrived answer; as shown when those questioned disagreed over the naming of some plants and pointed to different features as the criteria they use for identification. Yet the principles they used came over clearly, and it is on these that the following plant descriptions stand. In this sense they are Wola-centred.

Even so, they inevitably distort Wola ideas. Indeed the way these people order their crops, and their disagreement over this, questions and challenges the assumptions made by many writers in their discussions of folk classifications. The evidence suggests that Wola crop classification is of a different order to that of Western thought; the general concepts of which others have accepted, in line with the philosophical arguments for their universal

coming descriptions difficult, rendering the use of these accounts for identification purposes ambiguous and subject to error.

[13] See Berlin *et al.* (1974: ch. 4) for a description of the features looked for by Tzeltal Mayan Indians when identifying plants.

[14] Maenget Pes, Mayka Kuwliy, Wenja Neleb and Maenget Kem acted as my principal informants throughout and helped me collect the plant specimens. We referred to Wenja Leda and Ind Orliyn, two knowledgeable older women, when we had female crops of which we were uncertain. Also, there were invariably half a dozen or so other men and women around my house during this work and they also offered comments and advice. To them all I express my thanks for patiently instructing me in Wola plant lore.

applicability advanced by Lévi-Strauss (1966) and others. Equating their ideas, so far as we understand them, with the categories and thinking of Western science would be inadequate. What they think and how they order things demands appreciation in its own right, without implying some expected orderly scheme.

This understanding will come not only from attempts to appreciate what people think of natural phenomena, which will inevitably be partially successful only, but also from what they do with them. A consideration of functional issues (such as gardening methods and dietary intakes) is equally relevant. After all people do not engage in concrete or symbolic classification in isolation. Nor do they engage in practical pursuits without thinking about the things involved; in the case of crops relating them to cultivation practices, etiquette surrounding their consumption, and so on. Practical and intellectual issues, the functional and cognitive approaches, are intimately connected, and considering anything from only one point of view is distorting and stultifying. They spark off one another when considered together, each casting further light on the other.

This is so for issues other than classification too. The rationale behind the planting of crops relates to ideas about the locations in which they are thought to flourish and the extent to which they are considered to draw on the soil and deplete it of nutrients such that gardens have to be abandoned; which, contrary to the accepted cycle of slash and burn cultivation, some Wola crops (notably their staples) do not necessarily do rapidly (the reverse being true for a while, yields increasing as a garden matures). Similarly, the consumption of crops relates to both functional and cognitive issues. Edibles are good to chew over and digest both literally and figuratively. The analysis of quantitative data relating to the food eaten in a number of households, like that of similar studies in the Highlands, demonstrates the overriding importance of the single root crop of sweet potato in the Wola diet, something which not surprisingly influences their thinking, and also results in nutritional deficiencies and vulnerability to environmental, notably climatic, perturbations.

WOLA COUNTRY

Any discussion of crop cultivation requires some consideration of the ecology of the region concerned. The two aspects of the environment that are particularly relevant here are the soil and the climate because they determine largely the range of plants that may be grown.

The Wola people occupy five river valleys in the rugged Southern Highlands Province of Papua New Guinea. These valleys trend from the northwest to southeast and the following rivers flow along them: the Augu, Was, Nembi, Lai and Mendi.[15] The stippled area on Map 1 is that occupied by the Wola

[15] See Sillitoe (1979: 13) for a note on the inclusion of the Mendi valley within the Wola region.

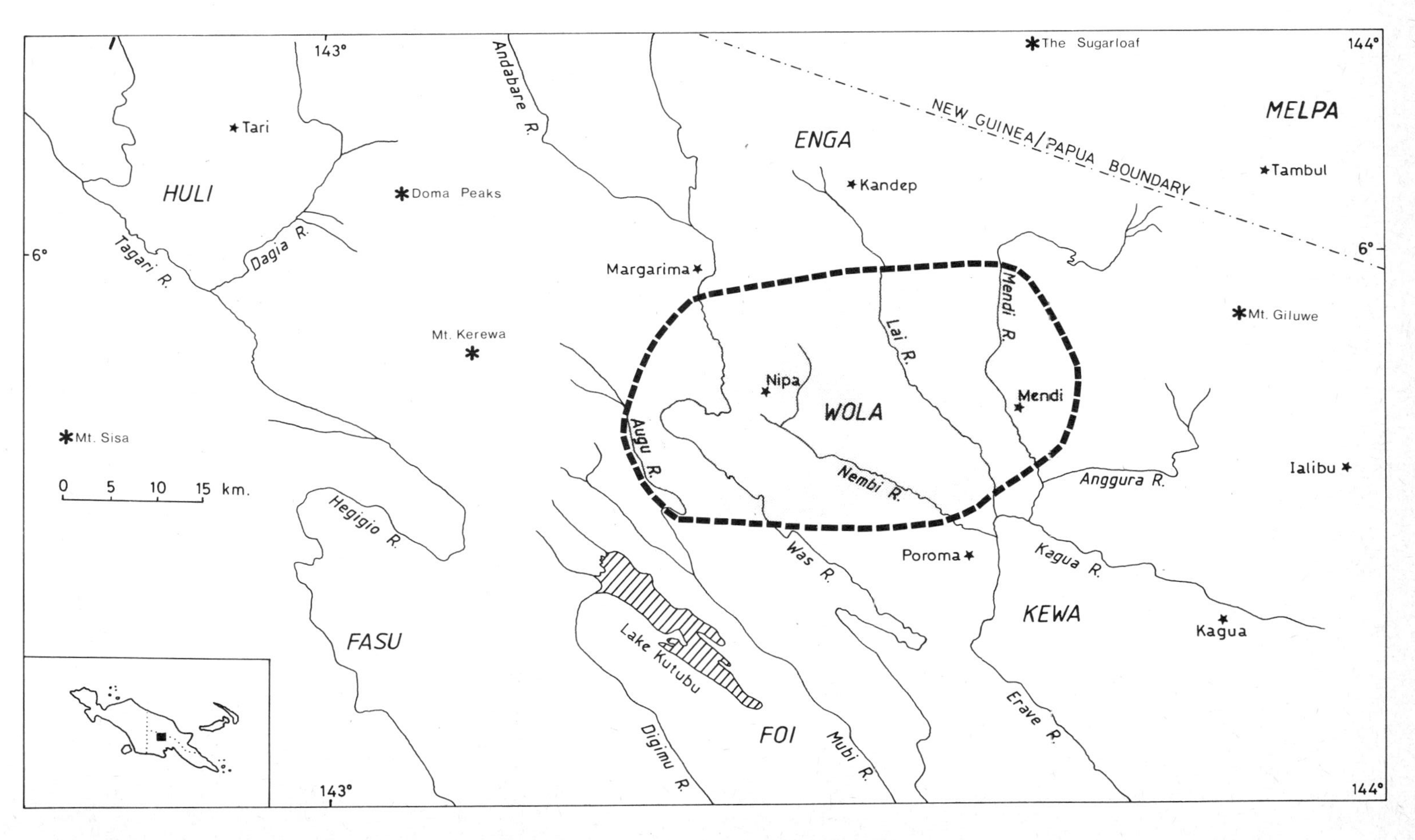

Map 1: The country of the Wola and their neighbours

people; this map also shows the people who occupy the regions surrounding them.

The country of the Wola is mountainous, rugged and precipitous, with turbulent and fast-flowing rivers running along the valley floors. They live on the sides of the valleys, leaving the intervening mountainous watersheds largely unpopulated. In the valleys, where they have cultivated extensively, there are areas of dense *Miscanthus* cane interspersed with the grassy clearings of recently abandoned gardens and the brown earth and dark green foliage of current ones. On the mountains, and in the unpopulated areas of river valleys, lower montane forest occurs in which the southern beech (*Nothofagus* spp.) predominates.

Map 2 outlines the typography and terrain of the region (the locality in the west called Haelaelinja, situated within a sweeping bend of the Was river, is the place where I lived and collected the ethnographic data recorded in this book). The majority of gardens and most settlement occur between 1600 and 2000 m (in the white valley areas on the map), although occasionally gardens are located up to 2400 m above sea-level (notably in the Lake Egari region and the headwaters of the Kolpa river). Considerable areas are left unoccupied, some of them substantial mountain ranges; notably in the east with Mount Giluwe (rising to an inhospitable 4367 m) and in the west with Mounts Kerewa and Imila (3235 and 2998 m respectively). This indicates that the Wola are cultivating some crops up to their altitudinal limits, beyond which they will not grow in this part of the world.

This region of the Southern Highlands consists mainly of limestone, a marine sedimentary rock which the people call *hat haen.* On the western and eastern margins there are igneous and metamorphic rocks of more recent volcanic origin (the large peaks in both these regions, named above, are former volcanoes). The region was uplifted, and folded and faulted in the recent geological past, and frequent earth tremors indicate that these earth movements are continuing now. This folding gave rise to the north-west/south-east orientation of today's landscape, and its relatively recent occurrence explains its rugged and sharp relief. Current geomorphological processes are continuing to change the area rapidly, so maintaining its raw and youthful topology. Weathering on the unresistant limestone, largely through solution, proceeds rapidly in the damp, cool conditions that suit it; the rivers and streams are vigorously cutting into their beds so maintaining steep slopes; and on almost all of these there is constant downward creep and the occasional large scale slump or landslide.[16]

This vigorous erosion of the underlying calcareous bedrock has played an important part in determining the nature of the soil found in the region. This relatively soft and readily eroded parent rock has allowed relatively deep soils to develop, except on the steepest slopes where constant creep keeps them

[16] For further details on the region's geology and geomorphology see Perry (1965).

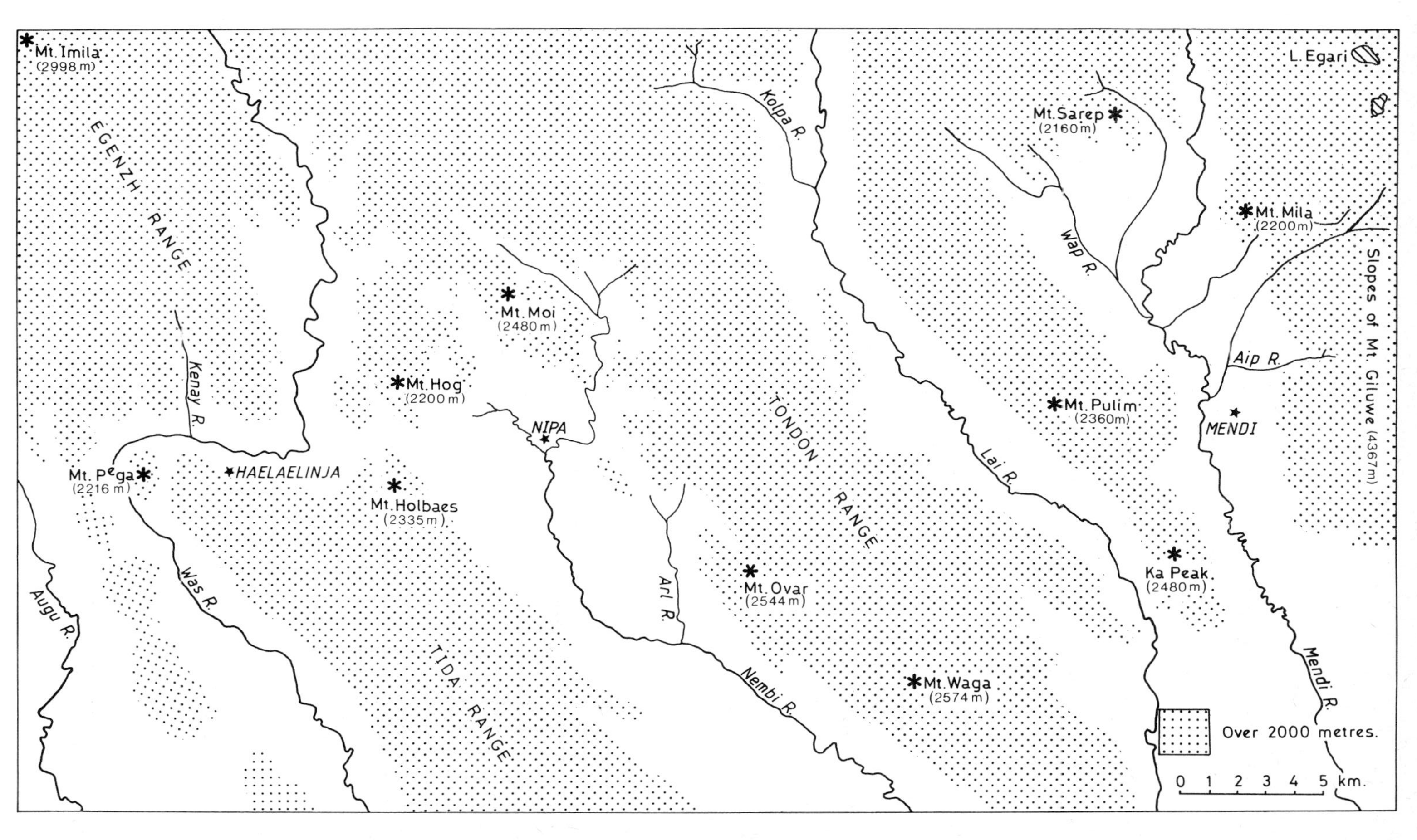

Map 2: The topography of the Wola region

shallow. The top-soil ranges in colour from dark brown to black, and the Wola call it *suw pombray* (lit: soil black). It rests on a heavy clay sub-soil which they call *suw hundbiy* (lit: soil orange).[17] The top-soil varies in both texture and depth; from relatively dry, friable soils to damp, sticky ones, ranging from less than one to over one hundred centimetres in depth (with an average of about thirty centimetres).

Table 3 gives the average results of a series of chemical tests conducted on a number of top-soil samples collected from gardens in different locations throughout the area of the Was valley where fieldwork was conducted.[18] These show soils that are markedly acid, although without toxic levels of soluble salts. They have variable nutrient levels (notably of calcium), but owing to severe leaching they are generally low to deficient in phosphorus, calcium, magnesium and potassium. Their organic content is variable, although this variation, and that in the nutrient levels, correlate to some extent with the varying ages of the gardens from which the samples were taken.[19]

The general picture is of a region with relatively poor soils which, combined with the steep slopes and precipitous nature of the terrain, does not give very promising garden land. Yet the Wola, with their intimate knowledge of the soils of their home territories and canny awareness of which crops will flourish on which sites, contrive not only to survive, but also to support a vigorous and expanding population. So the soil, its poor nature notwithstanding, is adequate to support the crops grown by the Wola, and notably their staple, the root sweet potato. Table 4 shows that this is so, giving the chemical composition and type of soil preferred by many of their crops. Comparing this table with the previous one of the chemical composition of soils indicates that they meet adequately the requirements of the principal crops grown (notably the first three on the list, which make up the greater part of the Wola diet, as the final column of Table 2 shows).

Another important influence on soil formation, indeed a dominant factor, is the climate; notably the moderate temperatures and relatively high rainfall of the region. The latter gives rise to markedly wet soils in poorly drained areas.

There are no notable seasons in the region which influence crop cultivation. Although the Wola divide the year into two named parts, which they call *ebenjip* (October to March) and *bulhenjip* (April to September), they say that the difference between them (the former being somewhat drier and finer than the latter) is not marked, nor is the weather pattern predictable according to

[17] This clay ranges from a pale brown to deep orange colour, and in places is white (called *suw tongom* by the Wola).

[18] For conducting these tests on my behalf I am grateful to F. Fahmy of the Soil Laboratory of the Department of Primary Industry in Port Moresby and M. Clarke of the Geography Department, Manchester University.

[19] For further information on the soils of this region see Rutherford and Haantjens (1965).

Table 3: Soil composition in Haelaelinja area

	pH	SPECIFIC CONDUCTIVITY (mho X 10)	TOTAL SOLUBLE SALTS	OLSEN P (ppm)	EXCHANGEABLE CATIONS (meq %)					CATIONIC EXCHANGE CAPACITY (meq %)	BASE SATURATION (%)	CARBON (%)	NITROGEN (%)	RATIO C/N	MOISTURE (%)	LOSS ON IGNITION AT 375°C (%)
					Ca^{2+}	Mg^{2+}	K^{+}	Na^{+}	Sum							
AVERAGE RESULT	5.14	0.1	0.065	6.29	13.01	2.36	0.87	0.82	16.81	20	69.64	14.06	0.8	16.59	15.85	39.7
NO. OF SAMPLES	25	15	15	14	25	25	25	25	25	15	15	23	27	23	12	10
STANDARD DEVIATION	0.7	0.15	0.147	6.6	10.43	1.59	0.73	1.35	11.52	8.63	29.31	7.3	0.22	5.32	7.51	4.26

Table 4: Soil preferences of crops

CROP	pH	PREFERRED CHEMICAL	PREFERRED SOIL TYPE
Sweet potato	5.25	If N too high, vine flourishes at tuber's expense	Tolerates wide range. Ideal is light sandy top-soil and clay sub-soil. Intolerant to waterlogging
Pumpkin	5.25		Tolerates wide range of well-drained soil
Highland *pitpit*		Yields reasonably on infertile, inferior soils	Intolerant of waterlogging
Taro		K_2O and organic matter are important	Prefers heavy, wet soil; tolerates waterlogging
Tannia			Intolerant of waterlogging
Yam			Deep, friable and well drained
Kudzu	5—6		
Ginger		Organic matter important i.e. good humus supply	Medium loam. Intolerant of waterlogging
Irish potato	5.25		Friable loam. Intolerant of waterlogging
Crucifer greens	7.5	Need good soil	Sandy loam
Onion	6—7	Requires good supplies of P, K_2O and N	Friable, damp loam
Common bean	6—6.75	Sensitive to concentrates of Mn, B and Al	Tolerates wide range (from sandy to clay)
Hyacinth bean	6—6.75	Yields reasonably even on poor soils	Tolerates wide range of well drained soils
Winged bean	6—6.75		Friable loam best. Intolerant of waterlogging
Pea	5.5—6.5	Cannot tolerate very acid soil	Intolerant of waterlogging
Cucumber	6.5—7.5		Well drained sandy loam
Sugar cane		Tolerates acid soils. Requires good levels of N, P and K_2O	Heavy soils, but must be deep and well drained.
Banana		Requires good levels of Ca, N, P and K_2O	Well drained loam with high humus content
Passion fruit		Tolerant of variety of soils	Medium and well drained
Tomato	5—7	Tolerant of wide variation	Light, well drained loam
Maize	5—8	Requires high level of N	Deep well drained soils
Tobacco	5—6.5	Requires high organic matter, Ca and K	Variety of well drained soils (sandy or clay)

the season. Hence the same climatic conditions largely prevail throughout the year, and the planting and harvesting of crops follows a constant and regular daily pattern, unless there is an adverse perturbation, notably an extended period of cold clear nights giving rise to frosts which may "burn" crops severely and lead to food shortages and hunger.

The weather is usually equable though, and follows a relatively predictable daily pattern. The mornings normally start bright and clear, with some low cloud settled in the bottoms of the valleys. This clears by about 8.00 a.m. and the temperature then rises quickly as the sun, generally unobscured by any cloud, rises to its zenith. In the early part of the afternoon clouds generally build up and the sky becomes overcast with a consequent fall in the temperature. It usually rains in the latter part of the day, if not earlier. The nights can be clear and cold.

The graphs in Fig. 1 show the monthly rainfall and temperatures for Haelaelinja during the course of a year. Both fluctuate from month to month, following no noticeable annual pattern (there is the same randomness in the daily fluctuations). In the course of the year 286.2 cm of rain fell. This compares with the annual average of 280 cm at Mendi Provincial Headquarters, calculated from thirteen years' readings; although it is somewhat less than Nipa's annual average of 318.2 cm, calculated from nine years' readings. Given the annual, monthly and daily fluctuations at the same place,[20] such variation between different places is only to be expected. The Mendi Mission annual average for instance, based on nine years' readings, is 15.6 cm lower than the above recorded at the Provincial Headquarters only about one kilometre away; this illustrates clearly the unpredictable and variable rainfall pattern of the region — raining heavily at one place it may be clear nearby.[21] These figures show that although rainfall varies from one period and place to another, it is heavy overall; that the region has a damp climate.

Over the year the daily temperature at Haelaelinja reached an average maximum of 23.8°C and fell to a minimum of 11.7°C. These figures compare with those recorded over a considerably longer time span in Mendi, of 23.5°C maximum and 12.7°C minimum.[22] During the day when the sun is at its height and unobscured by cloud it is uncomfortably hot out of the shade; whereas a clear night following such a day may bring temperatures down to uncomfortably low levels, even, on rare occasions, so low as to precipitate a devastating frost. An important factor in this temperature pattern, and the weather in general, is the altitude, variations in which, together with the mountains dividing up the region, give rise to numerous micro-climates.[23]

[20] For example, expressed as a percentage of the mean, the standard deviation of the above average annual rainfall figures is 10 per cent for Mendi and 12 per cent for Nipa.

[21] Rainfall figures at Mendi and Nipa taken from McAlpine, Keig and Short (1975).

[22] Figures taken from McAlpine, Keig and Short (1975).

[23] For further information on the climate of this region see Fitzpatrick (1965), and McAlpine, Keig and Short (1975).

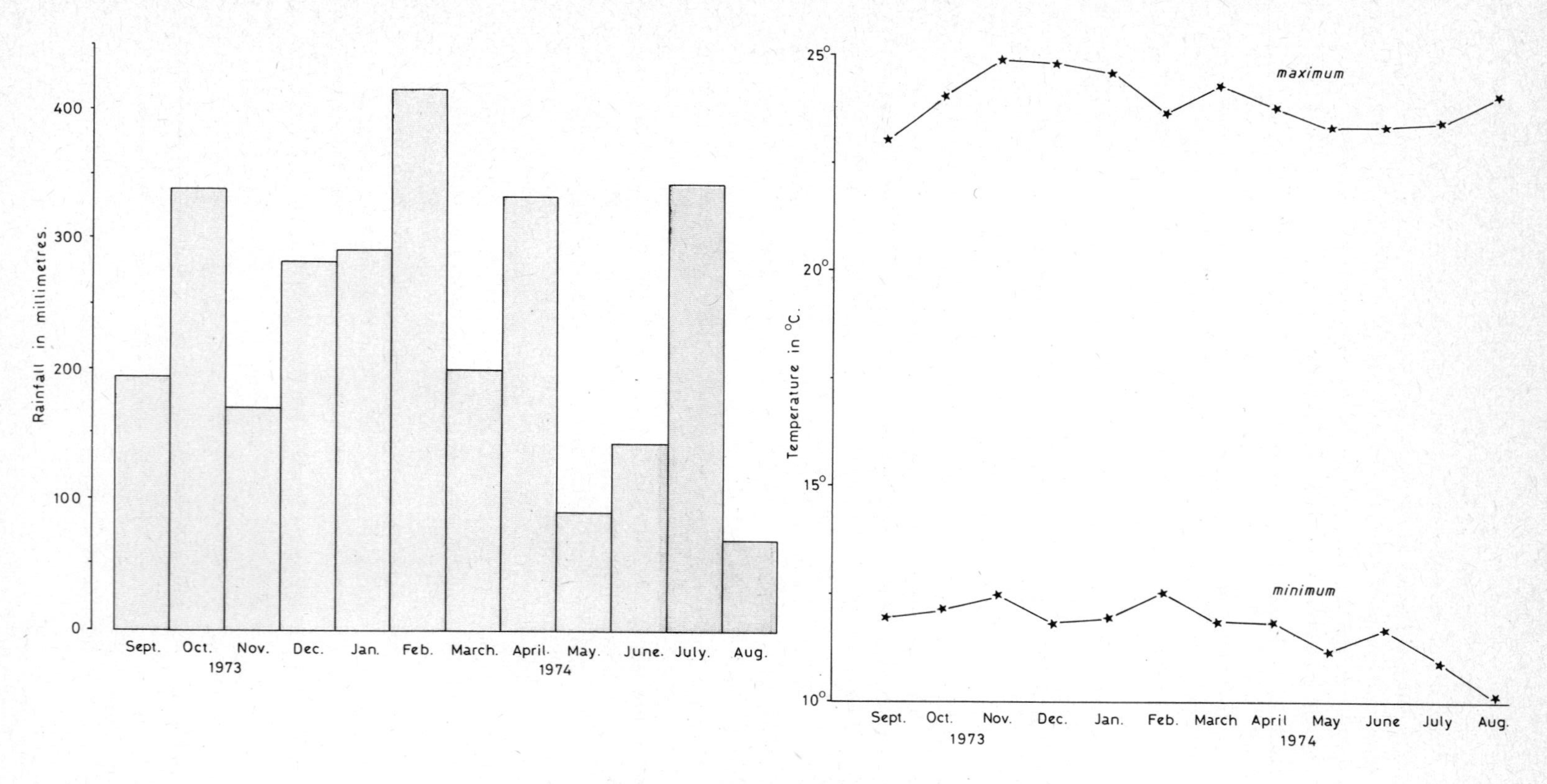

Fig. 1: Monthly rainfall and temperature at Haelaelinja

Regardless of its variability and unpredictability, the climate of the region generally suits the crops grown by the Wola, as shown by Table 5 which notes the climatic conditions preferred. The annual rainfall is more than most of the plants like, but the steep slopes gardened by the Wola ensure rapid drainage and so probably serve to ameliorate this. The other notable climatic shortcoming is frost, which devastates crops susceptible to it (especially those in

Table 5: Climatic preferences of crops

	RAINFALL (cm)		TEMPERATURE (°C)		
CROPS	RANGE	OPTIMUM	RANGE	OPTIMUM	REMARKS
Sweet potato	76—127	100		24	Susceptible to frosts and drought
Pumpkin	Medium		Warm		
Highland *pitpit*					Susceptible to drought
Taro	min.—175	250			Tolerates shade
Tannia	100—150				Tolerates shade
Yam	100—300	150	25—30	30	Frost kills; prefers a dry season
Ginger		150		21	Prefers short dry season
Crucifer greens		89	10—24		
Onion	Not excessive		Mild — no extremes		Needs long daylight hours
Common bean	Medium				Susceptible to drought and frost
Hyacinth bean		60			Drought resistant
Winged bean	High		Hot		
Pea				16	
Cucumber			18—29		Susceptible to mildew when humidity is high
Sugar cane	min.—150				Prefers brief dry season
Banana	min.—200		min.—21	25—30	Susceptible to high winds; prefers not humid conditions
Passion fruit		100			Heavy rain at flowering prevents pollination
Tomato	Light		10—20		Susceptible to frost. If too cloudy and wet the fruit is poor
Maize	40—90		20—30		Susceptible to frost
Tobacco	min.—25	50		24	Susceptible to frost and strong winds
Bamboo			Vulnerable to cold		

low lying and shaded situations). Indeed the vulnerability of sweet potato to frost puts the Wola in a precarious position and can, on rare occasions, lead to famine and starvation.[24]

WOLA CULTURE

The Wola live in an acephalous society. Indeed, the importance placed on the equality and sovereignty of all individuals poses, at first sight, problems for the maintenance of the social order. These are solved and, to use Evans-Pritchard's (1940) aptly enigmatic aphorism, 'ordered anarchy' permitted, through the elaborate institutionalisation of the ceremonial exchange of wealth (see Sillitoe 1979a).

These people live scattered along the valleys, in homesteads consisting of variably composed family groups; ranging from a man and his wife or wives and their children, to three or four related men together with their families and other relatives (such as widowed mothers, unmarried adolescent sisters and so on). Each family is recognised as belonging to its father, which is one of many reflections of the marked jural superiority of males.[25] Indeed men carefully observe a surprising number of conventions which keep them apart from their womenfolk, one of which is their invariable occupation of separate houses within the homestead (see Sillitoe 1979b).

The Wola call a family a *sem*, and it is on the basis of the aggregation of several variably related families that the permanent groups of their society are built. Several families, which are genealogically related by male and female links and reside and garden in the same area, constitute small named groups called *semg*^{e}*nk* (lit: family-small). Varying numbers of these small groups, which people think share a genealogical connection in the distant past (the precise nature of which they may not remember) and which occupy specific territories, constitute named communities called *semonda* (lit: family-large). The members of these communities ought to observe a rule of exogamy because they are related, although in reality they do marry sometimes.

These permanent and named groups result from the collection of several related families on territories where they have rights to claim garden land, on which to grow the crops discussed in this book. It is on these plants that their subsistence depends, hence access to land on which to grow them plays an important part in the constitution of their society. Men can claim rights to land through a wide variety of connections with a place (both kin and situational factors playing a part) and given their eclectic exercise of these, the empirical constitution of groups in bilateral — regardless of men's spoken preference for residence on their father's land (where, in all probability, they

[24] The last such famine (in which many people starved to death) occurred in the Wola area within living memory sometime in the mid-1920s.

[25] This is not to imply that women are mere pawns without any influence; for instance see Strathern (1972) on their position in a nearby Highland society.

may well wish to remain though, having lived there prior to marriage and being in a position to remain, having the right to insist that their wives move to them).

The related members of these permanent territorial groups do not recognise any obligations to support them in certain situations, they do not unite for action. The action groups of their society are temporary coalitions which come together to pursue some activity (such as to stage a pig kill or perform a dance) as the result of individuals deciding that it is in their interests to join together and participate in the planned event. A crucial point is that nobody participates to support the group, they all take part for self-interested reasons. While some of the resident members of a permanent territorial group will make up the nucleus of any such *ad hoc* action group, their relatives and friends living elsewhere may join it. This pattern of recruitment results directly from the surprising freedom extended to individuals to do as they wish and to pursue their own ends within the normative framework of Wola society (which is such that it allows for a considerable degree of apparent freedom).

One result of this is that the Wola behave in ways that nullify the assumptions of corporate theory (modified New Guinea versions of the segmentary one or otherwise) that obligations to groups maintain social order. The social principle which encourages the maintenance of social order among these densely populated people is exchange, something which pervades all human social interaction and is elaborated here into a complex institution to facilitate co-operation and harmony in social life. Exchange, unlike some other institutionalised mechanisms for the maintenance of social order, does not overtly infringe on the freedom of the individual, and for this reason the Wola value it highly. Indeed, they accord high status to those who excel at it, and it is sometimes even possible for these more able men, by virtue of their success, to exert a marginal degree of influence over the decisions reached by those united for some activity. But their success does not earn them the status of leader, as it does in some other Melanesian societies; they are unable to exert any control over the actions of others, which would openly offend against their society's ethos of equality.

The valuables which the Wola exchange with one another, on such prescribed occasions as marriages and funerals, include pearl and cowrie shells, crude salt, birds' feathers, large bladed stone axes, a thick cosmetic oil extracted from trees at Lake Kutubu, various marsupials, and the flightless cassowary bird. Pigs are another important source of wealth; herded by women, they consume a considerable proportion of the crops grown, especially the sweet potato tubers.

This root crop is the staple of pigs and man alike, and is supplemented in the case of human beings with the various other crops described in this study, which include pumpkin, taro, bananas, sugar cane, and various kinds of greens and shoots. The Wola grow these crops under what is generally known as a regime of swidden or shifting cultivation, although with many gardens this is somethings of a misnomer because they remain under almost continuous cultivation for several years, even decades. This careful and prolonged cultivation of plots is at odds with the stereotype of a swidden as a rudely cleared area planted once and then abandoned to regenerate under secondary regrowth while the gardeners shift their cultivation elsewhere (see Spencer 1966). The agricultural practices of the Wola result in two broad categories of gardens: those cleared and planted once (the classic shifting pattern) and those established for a number of years and planted over and over again.

When looking for a new garden site, men's choices are constrained by a number of factors. An important one, already mentioned, is the land tenure system which restricts their choice to those areas where they have rights, traced through either consanguines or affines, to lay claim to land. Other considerations concern the nature of the site; such as its slope, its soil, its aspect (they prefer sites facing north-east), its distance from the homestead, the ease with which they can enclose it, and so on.

Once a man has chosen a site, his first task is to clear it. This is men's work. The vegetation they will have to clear will be either secondary regrowth, of trees and *Miscanthus* cane grass, or primary montane forest, with its associated tangle of understory vegetation. Today the Wola clear this with machetes, using axes on the larger material, and it takes them 250 man-hours of work to clear a hectare.[26] Once they have cleared the required area, men have to enclose it. Often they situate their gardens in localities where they can use natural features (such as rivers, limestone cliffs and steep sided gullies) for this purpose. Otherwise they have to erect a fence to keep pigs out, which they make from stakes split from hardwood trees (particularly beech), sharpened and driven into the ground, with a sapling beading lashed along the top edge to give added strength. It takes them 342 man-hours to enclose a hectare in this fashion.

Following its enclosure, they prepare the site for planting. This involves firstly the pollarding of any trees left standing and the pulling up of as many roots as possible. Men usually arrange work bees to execute quickly this tedious and hard work, which involves the use of heavy pointed sticks to lever out the roots of trees and cane clumps. It takes 749 man-hours to clear a hectare ready for burning off.

Men and women work together to burn the cut vegetation when it has dried out sufficiently in the sun, men handling the large pieces of wood and women burning the smaller bits of vegetable refuse. This firing is important because it releases the nutrients that are locked up in the natural vegetation that covered the site, and returns them to the soil in a state which the crops

[26] For further details on the work involved in gardening, including the time taken to do the above task using stone axes (which, until recently, were the tools employed by the Wola) see Sillitoe 1979c.

to be planted can readily take up. It is time-consuming work, taking 1014 work-hours to burn the vegetation cleared from a hectare.

Once the burning off is completed, the garden is ready for planting. This is largely women's work, as they plant the staples which predominate in most gardens. Later chapters detail the work involved in planting and husbanding the various crops grown. It takes 678 work-hours to plant a hectare of garden, which brings the overall total for establishing a new garden of this area to 3033 work-hours.

By comparison, the total time spent preparing and planting an established garden for a second or subsequent time is 2545 work-hours. This involves firstly the clearing of the area to be replanted of the grass and weedy herbs that have established themselves there. Women do this largely, it taking them 371 hours to clear a hectare. They generally use the uprooted vegetation as compost in the centre of the earth mounds which they invariably heap up in rows across these gardens for the growing of sweet potato. The other 2174 hours spent reworking a hectare of established garden goes on the time consuming tasks of breaking up of the soil with digging sticks, heaping it up into these mounds and planting them (largely with sweet potato vines). Women largely tend gardens and harvest the crops, as described in later chapters.

This skeletal outline of Wola culture has touched on three of the topics which constitute the accepted anthropological quadrivium and concludes on the fourth, namely supernatural or religious beliefs. These rest on a belief in the malevolent powers of the ghosts of ancestors to return and bring sickness, or even death, to their descendants. The Wola think that these ghosts lead a nomadic existence, residing temporarily at any one of a number of defined places, such as in ancient stone mortars and pestles, in the skulls of ancestors and in certain deep pools of water. If someone falls seriously ill, a man with the required knowledge divines the current residence of the attacking ghost and the victim's relatives make an offering of a pig at that place to placate the offending ancestor spirit and, as the Wola say, persuade it to stop 'eating' the sick person. On other occasions several men combine to perform large rituals for their, and their community's collective benefit. They believe that such rituals in some way please their ancestor spirits and encourage them to both lessen their malevolent attacks on their descendents and bring about a period of general well-being.

Section I: Crops described

Chapter 2
TUBERS

SWEET POTATO

Ipomoea batatas (family: Convolvulaceae)
Wola name: *hokay* (No. of cultivars = 64)

This is the most important crop grown by the Wola. It originated in tropical America, but has been in the Wola region for as long as anyone can remember and they think it indigenous. It is so important to them that they do not believe there could have been a time when their ancestors did not cultivate it. They have no myths accounting for its arrival, it has always been with them.[1]

It is a variable plant in the size, shape and colouring of leaves and tubers, and this diversity has led to the identification of numerous cultivars by the Wola. But regardless of this variety, the sweet potato is readily identifiable.

It is a perennial herb, which means the same plant can be harvested from several times, so long as its stems are earthed up each time. It has a white and sticky latex in all its parts when broken, which the Wola call *paenjay*. The stem, the *iysha*, is vine-like and trailing (1—5 m long) and it grows roots at nodes along its length, such that when planted it spreads, rooting itself at intervals, and produces a dense ground cover. The stems are fairly thin (3—10 mm in diameter) and vary in colour from light to dark green, sometimes with purple; under the soil they are yellow or brown.

The leaves, the *shor*, are simple and vary enormously in size and shape

[1] Considerable debate goes on over the arrival of sweet potato in the Pacific. Some botanists, as already mentioned, speculate about cultural evolution on the basis of evidence extrapolated on the origin and movement of this and other crops (Keleny 1962; Yen 1963, 1971; Nishiyama 1963); while human scientists argue over the impact this root crop had on its arrival and the pace of resulting agricultural demographic and other changes (Watson 1965; Brookfield and White 1968; Waddell 1972: 206—14; Golson 1976).

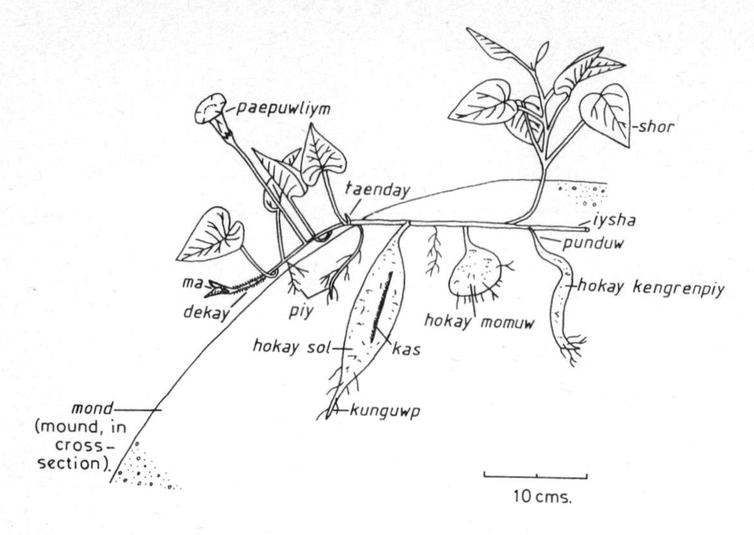

Fig. 2: Sweet potato (*Ipomoea batatas*)

(even on the same plant, depending on its age), from heart-shaped and entire to deeply dissected and lobed. They grow spirally round the stem and like it vary in colour, through shades of green to purple. The leaves have slender stems (5—30 cm long), those growing on the underside of the vine developing a characteristic bend in them to carry their leaves to the same level as the others, so producing a dense carpet of foliage.

The flowers, the *paepuwliym*, of the sweet potato are purple, a deep colour at the throat and lighter at the margin. They are funnel-shaped (5 cm long by 4 cm wide), with five united petals, and they open early in the mornings and wither within a few hours. They are cross pollinated by hymenopterous insects, and develop fruits which are globular capsules that burst open to release up to four small angular, black seeds with hard shells.

The important edible part of the plant, the tubers, which the Wola call *hokay*, develop from the secondary thickening of some roots in the top 25 cm of soil. A single plant produces several tubers (maybe ten or more). They vary greatly in size, shape and colour, from long and spindle-shaped to almost spherical, and from white to yellow through orange, pink and red to purple. The Wola distinguish three shapes of tuber:[2] long spindle-shaped ones called *hokay sol*, round ones called *hokay momuw*, and crooked ones called *hokay kengrenpiy*. They explain that these differences are due to variations in the hardness of the soil in which the tubers grow; in the best crumbly soft soil they grow straight and *sol*, whereas in more compact soil they are unable to push down and so become *momuw*, and in soil which is crumbly in part and compact in others, they twist about to follow the softer soil and crooked

[2] See Yen 1974: 206—11 on the shape of tubers.

kengrenpiy tubers result. Thin stringy roots which fail to develop into real tubers are called *kunguwp*. The Wola say that the hardness of the soil, and the difficulty tubers have penetrating it, explains why vines planted in mounds, where the soil is broken up and crumbly, grow tubers more quickly and are ready to harvest before those planted *suwl*, straight into the unmounded soil which has not been broken up.

When tubers begin to age their skins turn a dark grey colour and somewhat flaky. The Wola call this *hokay horok* (they use the same word *horok* for the soot from the fire which accumulates in the rooves of their houses, something which *hokay horok* resembles). The *horok* develops firstly at the top of the tuber where it joins the stem and then extends down it, the growing end always remaining a fresh white, pink or whatever colour for the cultivar. The Wola say that *horok* develops from the soil, and that in newly cleared gardens the abundance of rotting vegetation inhibits its development.

Some tubers grow with smooth skins, while others develop fissures called *kas*. The flesh of a tuber is called *hokay hobor*, and varies considerably in firmness and texture, from dry and powdery, to moist and gelatinous, through to coarse, fibrous and unpalatable (fed to pigs). Some tubers, notably older ones, are diseased. The Wola distinguish two principal kinds of blight, one called *dimbuw* where the tuber is watery and soft, and the other called *kol* where it has dark yellow discolourations in the flesh.

Sweet potato plants grow in two ways according to the Wola. Newly planted cuttings grow a single trailing vine and are called *sol*, whereas established ones develop buds called *taenday* (where the leaf stems join the vine), which grow into secondary vines, giving branching plants called *shoba*. The growing end and new leaves of any vine is its *ma*.

When distinguishing between sweet potato cultivars, the Wola look firstly at the size, shape and colour of the leaf, which is often enough for identification, and then, if necessary they look at the skin and flesh colour of the tubers. Another variable feature is the presence or absence of hairs on the stems, called *dekay*. Table 6 lists the sweet potato cultivars distinguished by the Wola and the features by which they identify them.[3] These features, as stressed in the opening chapter, are those that Wola informants pointed out as diacritical for each cultivar, what they look for in naming them (as pointed out, some of these vary from what a botanist would see: for example, the Wola classed some stems with a fine down on them as glaborous, saying this did not qualify as *dekay* or hairs).

This list of features is somewhat gross: it does not indicate all the subtle variations in colouring and shape which might be seen. The former for example varies through numerous shades from pale to dark green, and various reds through to deep maroon. Colours even vary on different parts of the same plant.[4] The size and shape of leaves vary to some extent on a single vine

[3] For further details on sweet potato variation see Yen (1974: 174—222).
[4] See Yen (1974: 212) for colour chart of tuber colours.

Table 6: Sweet potato cultivars

CULTIVAR NAME	PRE-/POST CONTACT	COLOURING								LEAF SHAPE	LEAF SIZE (mm)		TUBERS		
		PLANT STEM	LEAF STALK	WHERE STALK JOINS STEM	WHERE LEAF JOINS STALK	NEW UNFOLDED LEAF	MATURE LEAF	LEAF RIBS	HAIRS		LENGTH	BREADTH	SKIN COLOUR	FLESH COLOUR	TEXTURE
'Aton'*	Pre	green	green	green	green	maroonish green	green	green		F			red	white	
'Baegai'	Pre	pale green	green	maroon	maroon	maroonish green	green	maroon	slight	B	55	56	pale brown	white	dry
'Balay'	Post	green	green	green	maroon	dark maroon	green	maroon	present	G	66	75	red	white	
'Bizuw'	Post	green & maroon	maroon	maroon	maroon	maroon	green	maroon	slight	G-H	62	66	pale brown	pale cream	powdery dry
'Boraynonk'	Pre	green	green	green	green	green	green	green	present	D-F	65	61	white	white	damp soft
'Bordorwiy'	Pre	green	green	green	green	green - red edge	green	faint red	present	B-C	52	57	white	white	damp hard
'Daebayda'	Pre	green	green	red	red	reddish green	green	green	present	A	62	58	red	cream	moist firm
'Denshor'	Pre	pale green	green	green	green	green	green	green	slight	C-E	51	49	white	white	
'Gaentuwla'*	Pre	green	green	green	green	red	green	green	absent	A			red	white	
'Hagen'	Post	green & maroon	green	maroon	maroon	dark maroon	green - edge maroon	maroon	absent	D-F	56	45	dark red	pale yellow	moist
'Hulma'	Pre	green	green	green	maroon	green	green	green	slight	A-B	56	56	red	white	
'Hungabuw'	Pre	green	green	green	green	maroonish green	green	green	slight	A-B	34	41	red	white	
'Iybob'	Pre	green	green	green	red	green	green	faint red	slight	B-C	70	62	white	white	moist
'Kauwa'*	Pre	green	green	green	green	green	green	green		D			red	white	
'Keret'	Post	green	green	maroon	maroon	maroonish green	green	maroon	absent	B-C	47	39	pale brown	pinkish orange	damp
'Kinjuwp'	Pre	green	green	green	green	green	green	green	present	D	58	42	off white	white	dry stringy
'Kirael'	Pre	green	green	green	green	maroonish green	green	green	present	F	65	64	red	white	dry stringy
'Kiyga'	Post	green	green	green	green	reddish green	green	red	absent	C	54	53	pale brown	pale cream	dry or moist
'Kiygahaez'	Post	green	green	green	green	maroonish green	green	green	absent	C-E	61	60	white	white	
'Kogba'	Post	green	green	maroon	maroon	green - ribs maroon	green	maroon	absent	E	44	41	white & brown blotches	purple & white veins	very moist
'Kolwep'	Pre	red & green	green	red	green	reddish green	green	red	slight	E	60	56	red	purple & yellow veins	
'Komiya'	Pre	red & green	green	red	green	reddish green	green	red	slight	E	60	56	red	white	
'Konoma'	Post	red & green	green	red	green	red	green	red	absent	G-H	58	41	bright red	white	moist
'Kuwl'	Pre	green & maroon	green	maroon	green	maroon	green	maroon	absent	C	61	57	red	pale cream	dry stringy
'Kuwliy'	Pre	green & maroon	green	maroon	maroon	green - ribs maroon	green	green	slight	B	40	50	cream	pale yellow & orange veins	damp
'Laegaben'*	Pre	green	green	green	green	green	green	green		D			red	white	stringy
'Ma'	Pre	green & maroon	green	maroon	maroon	green - edge maroon	green - edge maroon	pale maroon	present	B-C	46	45	pale brown	white	dry firm
'Maegai'	Pre	green	green	green	green	maroonish green	green	green	absent	A	88	76	red	white	
'Maezayzuwla'	Post	green	green	maroon	maroon	green - ribs maroon	green	maroon	present	I	70	78	pale brown & pink blotches	white	dryish
'Mapopo'	Post	green	green	green	red	reddish green	green	green	absent	B-C	47	56	pale brown	white	dry
'Mayow'	Post	red & green	green	green	green	green - edges red	green	green	present	A-E	71	86	red	cream	hard
'Mbaduw'	Post	pale green	pale green	maroon	maroon	dark maroon	green	maroon	absent	A-B	70	78	pink	pale cream	very moist
'Mendkauwa'	Post	green	reddish	green	reddish	green - edge red	green	reddish	absent	D-E	57	56	red	white	hard
'Miytbiyp'	Post	green	green	green	green	green	green	green	present	D-E	61	55	white	white	
'Monay'*	Post	green	green	green	green	green	green	green	present	G-H			white	white	dry powdery
'Mormuwn'	Post	red & green	green	green	green	green	green - edge red	green	absent	A-B	92	68	pink	pinkish orange	damp firm
'Mukba'	Pre	green	green	green	green	maroon	green - maroon underside	maroon	absent	B	35	33	pale brown	white	powdery dry
'Mundiyaem'	Pre	green	green	green	green	green - edge reddish	green	green	present	A	60	72	pinkish	white	dry powdery
'Muwtuwpiy'	Pre	green & maroon	green & maroon	maroon	maroon	maroonish green	green	dark maroon	absent	E-F	43	46	maroon	pale yellow	
'Ngorow'*	Pre	green	green	green	green	pale maroon	green	green	present	A-E			red	white	
'Nolim'	Pre	green & maroon	maroon	maroon	maroon	green	green	maroon		B-C			red	white	
'Oba'	Pre	green	green	green	green	maroon	green - edge maroon	maroon	slight	A-B	73	68	red	white	
'Olshor'	Post	red & green	reddish	reddish	reddish	green - ribs & edge red	green	green	absent	B	27	36	pale brown	cream	moist soft
'Oma'	Post	pale green	pale green	maroon	green	green-ribs & edge maroon	green	maroon	absent	H	91	89	maroon	white	
'Ombuwabay'	Pre	green & maroon	green	green	green	maroonish green	green	pale maroon	present	B-E	55	59	red	orangish yellow	
'Paeng'	Pre	green	green	green	green	red	green	green	slight	A	53	58	red	white	
'Shorhiy'	Pre	green & maroon	green & maroon	maroon	maroon	maroonish green	green	maroon	absent	C-E	64	66	red	white	
'Shorwat'	Post	green & grey	green	green	green	green - edge reddish	green	green	present	G-H	79	80	off white & pink blotches	white	moist
'Showmay'	Pre	pale green	green	green	green	maroon	green - edge maroon	green	slight	A	84	75	dark red	white	
'Sigiliy'	Pre	green	green & maroon	maroon	maroon	pale maroon	green	maroon	slight	C-D	47	49	red	white	moist
'Simbil'	Post	green	green	red	green	reddish	green	red	absent	I	91	104	white	off white	damp firm
'Sokol'	Post	red	red	red	red	green - ribs reddish	green	reddish	absent	B	88	100	off white	yellow orange	damp
'Tat'	Pre	green & maroon	maroon	maroon	maroon	green - ribs maroon	green	maroon	present	B-C	45	44	red	white	stringy
'Taemboliy'*	Pre	green	green	green	green	maroonish green	green	green		A-B			red	white	stringy
'Taengeltwem'	Pre	pale green	green	maroon	green	maroonish green	green	maroon at ends	present	B	68	58	maroon	pale yellow	
'Taengiyael'*	Pre	green	green	green	green	green	green	green	slight	C-F			white	white	
'Taenguwl'	Pre	green	green	green	green	green	green	green	present	C	57	58	pale brown	white	
'Tiriy'	Pre	green	green	green	maroon	green	green	green		A-B			red	white	stringy
'Toriya'	Pre	green	green	green	green	green	green-pale maroon under	maroon at ends	slight	B-C	53	46	white	off white	damp
'Tuwgiy'	Pre	green	green	green	green	green	green	green	slight	B	82	75	red	white	damp stringy
'Uwgaimkat'	Pre	green & maroon	maroon	maroon	maroon	maroonish green	green	pale maroon	present	E	50	44	red	white	stringy
'Waipgənk'	Pre	green	green	green	green	green - edge faint red	green	green	slight	C-F	58	48	red	white	moist
'Wənmuwn'	Post	green	green	green	green	reddish	green	green	present	H	80	60	off white	white	dry
'Yaebkil'	Pre	pale green	pale green	pale green	pale green	maroonish green	green	pale green	absent	F-G	46	55	pale pink - red patches	white	very moist
No name		green	green	green	green	maroonish green	green	green	absent	A-B	52	51	white	white	
No name		pale green	green	green	maroon	green	green	green	slight	G	67	74	off white	off white	stringy
No name		green	green	maroon	green	maroonish green	green	maroon	absent	A	66	64	red	white	
No name		green	green	green	green	reddish	green	green	absent	G-H	66	70	off white	off white	moist

* = no examples observed, based on informants' descriptions only.

too, and the data on the table represents an 'average' leaf for each cultivar. The various leaf shapes identified, and the dimensions measured, are shown on Fig. 3; where a cultivar falls between two shapes it is recorded as a hybrid (for example B—C is a leaf which falls between shapes B and C). The leaves are arranged according to the pattern of their lobes, which the Wola call *dink dink bay*, becoming progressively more serrated.[5]

The data on the texture of tubers are particularly variable. Although sometimes marginally diacritical, texture varies considerably for the same cultivar and depends to a large extent on the soil in which the tuber grows (in much the same way as shape does). So tubers cultivated in soft crumbly soil have a powdery flesh whereas those grown in compact 'strong' soil have a firm hard one, dry soil produces sapless tubers and wet soil moist ones.

This classification of cultivars covers those grown in the settlement on the Was valley where fieldwork was conducted, or known to people living there as growing elsewhere (such as the '*monay*' cultivar, which occurs only in the gardens of Huli people living in the forests towards Lake Kutubu, or the '*oma*' one which is only found towards Porsera, where the Wola give way to the Huli).[6] The list is undoubtedly incomplete. Even if at the time of field work all the cultivars present were documented and specimens collected, it is partial in the sense that over time the stock of cultivars changes. This change has been dramatic since Europeans contacted the Highlands and introduced new cultivars; as Table 6 shows 34 per cent of the sweet potato cultivars currently recognised by the Wola have arrived since contact (and as pointed out in discussion of classification in Chapter 8, some of these now figure prominently in their diet).

Trying out and experimenting with new plants is something which the Wola do with alacrity, as their rapid adoption of several new crops shows.[7] The rate at which new cultivars arrive and are adopted is unexpectedly swift; for example, in about seven years (from 1970 to 1977) six new named cultivars arrived in the Haelaelinja area and are now cultivated widely (these are the '*hagen*', '*keret*', '*mbaduw*', '*mormuwn*', '*shorwat*' and '*sokol*' sweet potato cultivars). Sometimes people are aware of the direction from which new cultivars arrived, as some of their names testify; for example, '*hagen*' from the Western Highlands, '*mendkauwa*' from Mendi, '*hulma*' from the Huli

[5] See Yen (1974): 195—200) on the shape and size of leaves (the dimensions used here are those he outlines on p. 199). The above typology omits the size of the two lobes on either side of the stalk though, which on some leaves are non-existent and on others are markedly rounded. See Yen's fig. 66 where they are marked as +0 and —0.

[6] For comparison, see Fischer's (1968: 259—63) discussion of the sweet potato cultivars identified by a group of Kukukuku people (including a table of the diacritical features they look for when naming — pp. 260—1), and Powell *et al.* (1975: 16—19) on those identified by the Melpa.

[7] Friends always showed keen interest in seeds we took to the field and fought for the privilege of planting them in their gardens, where they would insist, 'conditions are the very best' — they even tried orange pips, and were particularly disappointed when a sprouting avocado stone was rooted up and eaten by a pig.

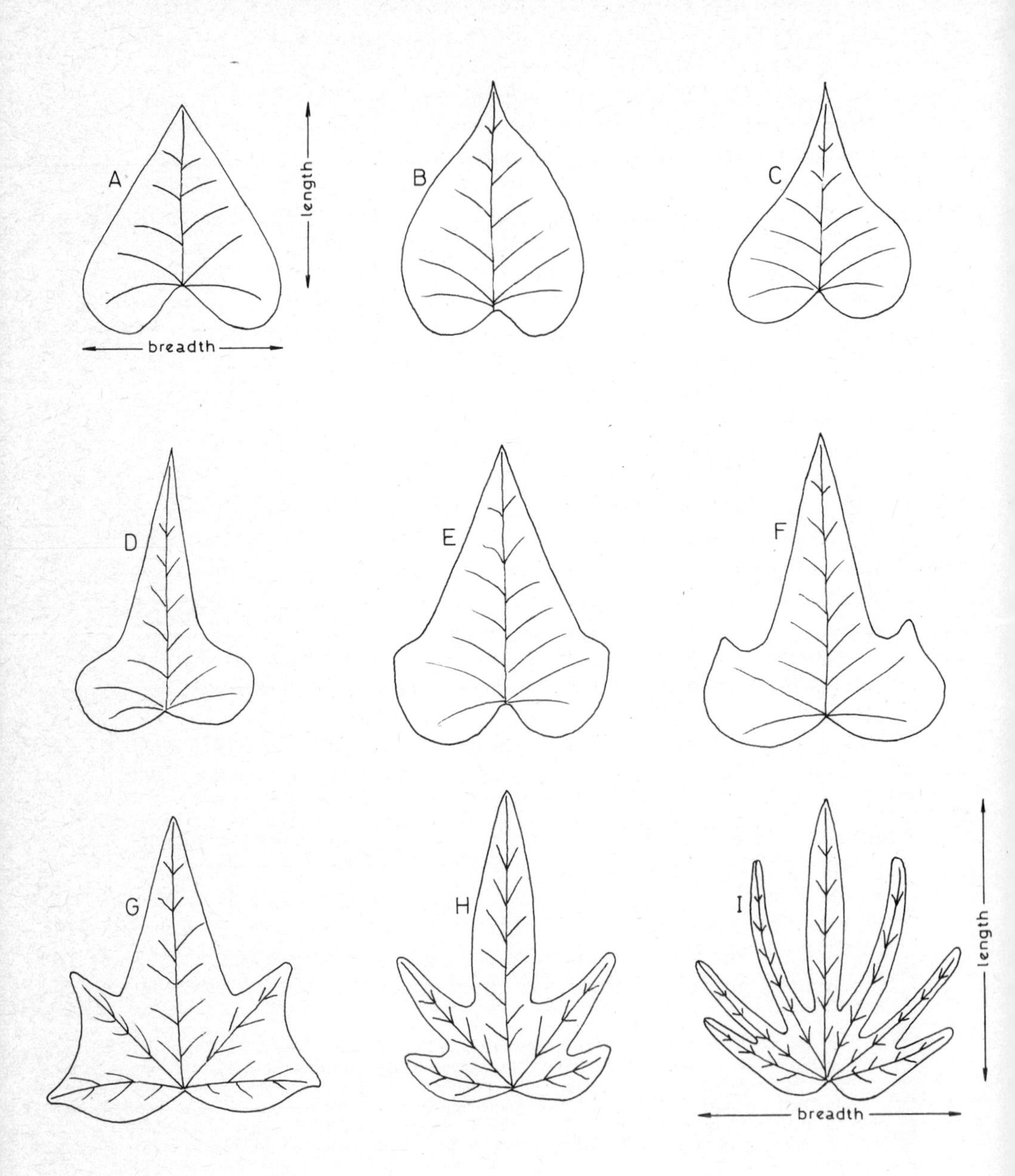

Fig. 3: Sweet potato leaf types

people, and '*keret*', which is also called '*tari*', from Tari. The four cultivars '*bisuw*', '*hulma*', '*keret*' and '*mbaduw*' they say come from a westerly direction and the two '*hagen*' and '*mendkauwa*' from an easterly one.

The arrival of these new improved cultivars, which grow more quickly and yield more and larger tubers, has resulted in the loss from cultivation of some indigenous ones. For example nine of the cultivars in Table 6, those with asterisks (excepting '*monay*'), are now only names remembered by older people, and the Wola say that they are *bumhaez* (lit: lost). Thorough searching of gardens and questioning of women could not locate any examples of these lost cultivars, which people said were inferior because they invariably grew spindly and stringy tubers (some of them had other drawbacks too, like '*nolim*' which took an inordinately long time to grow any tubers and '*taemboliy*' which only yielded a few tubers before dying back).

This search of gardens for cultivars also produced four plants which had no names, which women said were *imbiy na wiy* (lit: name not has). This suggests that the change in the cultivar stock not only results from the introduction of new plants from outside, but that genetic changes and plant mutations within the existing population are also partially responsible for it. Prior to European contact this was probably a more significant source of change and variety than it is with today's tidal wave of newly arriving cultivars. Another term for these individually unnamed cultivars is *diyr tay hokay* (lit: banana base sweet-potato). This is the name given to vines which grow round banana plants, where they root themselves by chance from cuttings tossed aside by women after planting or weeding. Any cultivar may take root in this way and become *diyr tay hokay*, and such largely undisturbed areas of semi-wild vines are the kind of places where genetic changes might take place and establish themselves. The Wola are aware that such changes in leaf and tuber take place, and with two of the unnamed examples found said that one was like the brother (*haemay non biy*) of the cultivars '*mayow*' or '*maegai*' and the other of '*kinjuwp*' or '*kiygahaez*'. They were not certain though why such changes occur, one person suggested that they happen when an old tuber is left in a newly worked and planted mound where it sprouts roots and a stem which might be different to the parent.[8] They said that once such changes occur they remain constant if the plant is brought into cultivation.

Sweet potato is a short-day plant, and only flowers freely where the hours of daylight are below eleven. Above this flowering decreases until there is none at all where the photoperiod exceeds 13½ hours. So it is doubtful whether flowering ever reaches the optimum in the Wola region, where the hours of daylight do not fall below eleven. This probably influences the rate

[8] For an intriguing account of how some Highlanders think sweet potato cultivars are propagated through the agency of birds and their droppings, see Bulmer (1965, 1966); Panoff (1970: 240—1) also records that the Maenge of New Britain have similar beliefs.

of cross-pollination and hence genetic change and establishment of chance seedlings. But the sweet potato is also an out-breeding crop, that is most cultivars are self-incompatible, which favours the breeding of diverse off-spring when germination does occur, hence there is wide scope for change. It is the vegetative method of propagation employed by the Wola that maintains the integrity of clones under cultivation.

Women plant the crop from stem cuttings called *hokay way*, which they take from the apical growth of mature vines.[9] They take these cuttings just prior to the harvesting of any tubers, following this the foliage thins out and dies back somewhat. The stem cuttings are 30 to 60 cm long. They are planted in one of two ways: either straight into the otherwise unprepared soil of a new garden or into mounds of earth worked to a fine tilth. In the former method, called *suwl* planting, a woman drives a sharpened stake into the soil, to a depth of about 30 cm at an angle of about 30 degrees to the horizontal, and into this hole thrusts two or three stem cuttings, around which she firms up the earth with her fingers. In the latter method of planting, called *mond kolay* (lit: mound heap-up), a woman breaks up the soil with a digging stick and heaps it up into a mound, usually over a pile of rotting vegetation, and into the soft and crumbly soil pushes two or three cuttings, up to the middle of her forearm in depth. Sometimes, in new gardens, women employ a variation of the mound and plant vines into raised irregularly shaped beds of broken up soil called *paen*.

The Wola explain that they can plant straight into the soil of a newly cleared garden without breaking it up into a fine tilth, because previously shaded by vegetation it is soft and damp. Whereas, when they come to replant an older established garden, exposure to the sun will have dried out the soil and made it hard; women must break this up into a soft and crumbly loam so that tubers can penetrate it (compare above points on the shape and texture of tubers). Two or three cuttings are always planted together because, women say, sometimes one dies without taking root; they are thus ensuring that at least one takes.

Sweet potato is the first crop planted. It is usually cultivated throughout a garden and not restricted to certain locations. It is a plant that requires a well drained, light soil (making slopes good sites); in a wet one its foliage turns brown and it does not develop tubers. It is obviously the crop that predominates in sweet potato gardens (which constitute most of the area under cultivation), to the extent that in established gardens replanted for a second or subsequent time it may be the only crop. It is rarely planted in taro gardens, which are often located on wet sites, and only to a limited extent in the small mixed vegetable gardens near houses.

When growing the sweet potato requires little attention. It establishes itself quickly and grows a dense foliage which, covering the soil, both protects

[9] See Yen 1974: 60—70 for details of planting and soil preparation.

it against erosion and inhibits the growth of weeds. Once or twice before this dense cover develops women may weed the crop. It takes about six months for the crop to reach maturity, although this varies somewhat according to the weather and the site; tubers maturing more quickly in dry soil than in moist. Also, impatient for new tubers, people sometimes harvest from plants before they are fully mature, called *pindbiy nay* (lit: immature eat).

Women judge maturity in two ways. They watch for the tubers growing near the surface to push up little humps in the soil, and for the leaves at the base of the vine to turn yellow and drop off (leaving a largely leafless stem with a brown bark-like skin similar to the *horok* of an old tuber).[10] The first harvest of tubers from a plant is called *waeniy* (lit: new), and subsequent ones from it are *puw* or *taembel*. By earthing up sweet potato plants each time they harvest from them, women can keep gardens in production without replanting for two years or more. The yields at each harvest decline though, and it is more usual after a few months of *puw* harvesting for women to rework and replant gardens.

The Wola eat only the tubers of the sweet potato. These are overwhelmingly important in their diet, both in terms of the weights eaten and their nutritional contribution. They bake them mostly under the hot ashes and embers of a fire, although they also cook them in earth ovens, by roasting them over a fire in the flames and today, by boiling them in water. If desperately hungry they will eat them raw. They also feed them raw to their pigs.

TARO

Colocasia esculenta (family: Araceae)
Wola name: *ma* (No. of cultivars = 43)

This crop possibly evolved in Melanesia or in nearby south-east Asia.[11] As far as the Wola are concerned they have always grown it, and the spells and taboos which surround its cultivation suggest that it is an ancient crop.

It is a herb which may grow a metre or so high, and it produces a tuber which varies from a globular to an elongated cylindrical shape (up to 30 cm long and 15 cm in diameter). The tuber is marked with growth rings and has a flaky new potato-like outside skin which varies in colour from orangish brown (when grown in a dry soil) to black (when grown in a waterlogged one). The flesh of tubers varies in colour from white through yellow to purple.

Established parent corms, which the Wola call *ma injiy* (lit: taro mother), grow lateral cormels, called *pora*, that send up their own leaves. They call this

[10] The Wola explain that this occurs because there are no leaves to shade it from the baking sun.
[11] See Yen and Wheeler (1968).

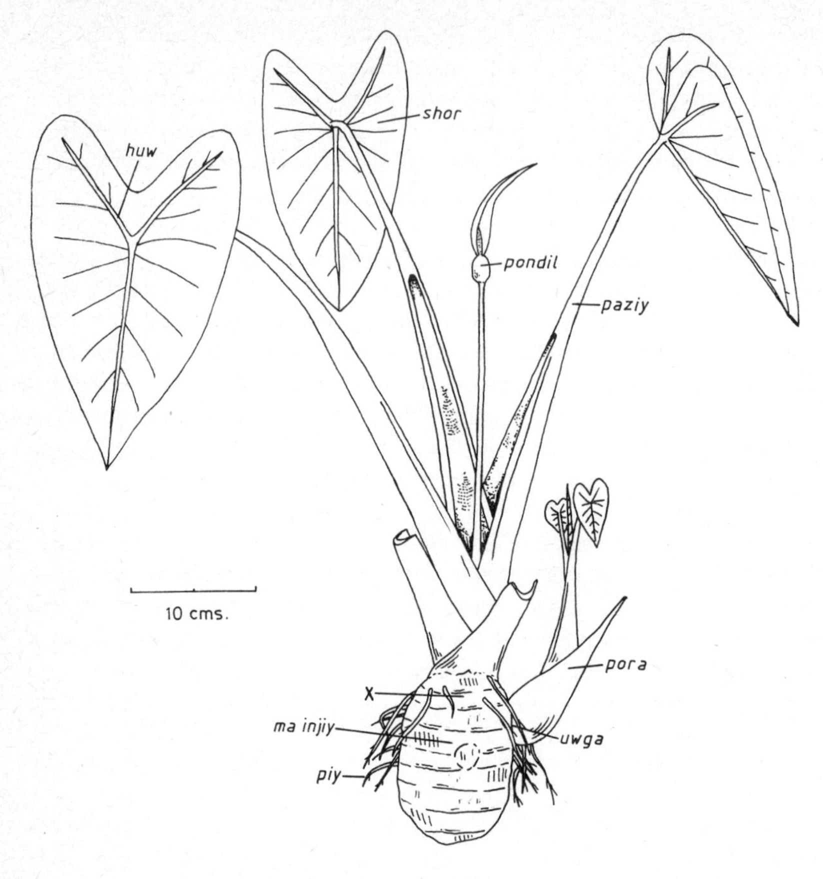

Fig. 4: Taro (*Colocasia esculenta*)

process *sem biy* (lit: family make), and the tuberous connection between a *ma injiy* and a *pora*, which is sometimes quite long and spindly, they call the *uwga*. These suckers are always smaller than the parent tuber and often bitter and inedible, in which case they are fed to pigs. If their tops are planted, they grow to become *ma injiy* themselves, in turn sprouting *pora*.

From the apex of a taro tuber grows a whorl of stout spongy stalks which the Wola call *paziy*; each new stalk growing up, with its leaf furled, through a split in the previous one. The stalks vary considerably in colour from white through various shades of green to red and dark maroon, and they are often veined with two colours.

These stalks support large heart-shaped leaves (30 cm or so long) which are thick, hairless and silky smooth. All cultivars have the same sized and shaped leaves, with a dark green upper surface and paler lower one; some of them have maroon ribs. In some cultivars (see Table 7) a single tuber produces more than one whorl of stalks and leaves, and the Wola call these multiple-topped corms *habuwp* (when harvested each clump of stalks may be replanted separately to grow independent plants).

Some plants, when mature, send up a pale green central stem topped by a spike shaped inflorescence called *pondil.* A yellow spathe encloses these spikes which carry both male and female flowers (before terminating in a long pointed sterile section). It is rare for a plant to produce seeds.

Taro exhibits considerable variation and many edible clones are known.[12] The Wola distinguish several cultivars, largely by the colour of the tuber's flesh and the colouration of the leaf stalk (see Table 7).[13]

In pre-contact times the numbers of taro and sweet potato cultivars were about the same, but unlike the sweet potato no new taro cultivars have been introduced since contact. Yet there are a number of cultivars described here for which a search of many gardens failed to find any examples. Whether these are lost, like some sweet potato cultivars, is not clear, and neither is the reason for their disappearance; for example no one said they were poorer, all taro are the same to eat. Also there are two cultivars ('*degelesh*' and '*kwilow*') which are remembered as only names; nobody can remember ever seeing these plants, today their names occur only in spells. It is not clear whether they were ever cultivated or have always been spell names only, the origin and meaning of which have been forgotten.

Taro is propagated vegetatively, and both men and women plant it. They plant the leaf bearing top from a mature corm, called the *ma way.* They cut the tops off (at 'X' on Fig. 4) and then trim all the stalks down (like the two outside ones on the figure). They then plant these with club-like digging sticks called *ma way g^{e}mb*, which have a long point one end called the *wil* (lit: nose) and a blunt club the other called the *huruw*. They firstly drive the long *wil* point into the ground, and then turn the tool round to enlarge this hole into a bowl shape with the rounded blunt end. They place the tuber top in this bowl and firm the earth from round the edge over it, patting it down with the blunt *huruw* end such that only the trimmed tops of the stalks protrude from the ground. The Wola say that planting taro in this way ensures the growth of good large tubers because it surrounds them with soft soil. Like sweet potato, taro cannot grow large tubers in compact or 'strong' earth. The bowl-shaped hollow of crumbly soil they say, 'shows' the growing corm how to develop large and round, encouraging it to expand and displace the soft earth. It is usual to plant tops in pairs, about 10 cm apart, to ensure

[12] See Whitney *et al.* (1939) for example, they discuss the numerous taro cultivars grown in Hawaii.

[13] The previous comments on the variation between sweet potato plants, and the somewhat gross nature of any listing of the diacritical features seen by the Wola, apply here too, as they do to all the forthcoming accounts of crop cultivars.

For comparative material on taro cultivars see Fischer (1968: 263—5) for an account of the 41 clones distinguished by the Kukukuku, Panoff (1972) on the 129 identified by the Maenge of New Britain, and Powell *et al.* (1975: 19—21) on the 39 named by the Melpa. See also Malinowski (1935 (vol. 2): 104—7) on Trobriand cultivation and botany of taro.

CULTIVAR NAME	TUBER FLESH	LEAF RIBS			LEAF BORDER	STALK & LEAF JOINT	PLANT STALKS							OTHER POINTS
							TOP 4 cm					MARGINS OF SPLIT		
		MAIN	LOBE	OTHER			SPLIT SIDE	REAR	MIDDLE	BOTTOM 10 cm	WHERE JOINS TUBER	UPPER HALF	LOWER HALF	
'Borbae'	white	green	green	green	green	pale maroon	pale green	pale green	green & dark maroon veins	green & dark maroon veins	pale green	maroon	maroon	
'Daelma'	white	green	green	green	green	red	green	green	green	green	pink	red	red	
'Goizmayja'	white	maroon	maroon	maroon	maroon	maroon	pale green	pale green	dark maroon	dark maroon	pink	dark maroon	dark maroon	
'Gwiy'	pale yellow	dark maroon	dark maroon	dark maroon	green	dark maroon	dark maroon	dark maroon	maroon	green & maroon veins	pink	dark maroon	green & maroon veins	
'Habuwp'	white	green	green	green	green	green	pale green with a few thin maroon veins	pale green with a few thin maroon veins	pale green with a few thin maroon veins	greenish white -maroon veins	pink	red	red	
'Hael'	white	green	pale maroon	green	green	green	maroon	yellowish green	green	green & dark maroon veins	pink	red	red	
'Hebiyab'	white & purple	green	green	green	green	green	yellowish green	yellowish green	yellowish green	yellowish green	white	brown	red	
'Hiyuw'	pale yellow	green	green	green	green	green	white flecked green	pale green	pale green	pale green	white	white flecked green	white flecked green	
'Holor'	white	green	green	green	green	green	pale green with maroon flecks	pale green with maroon flecks	pale green with maroon flecks	pale green with maroon flecks	white	maroon	maroon	
'Huwshwaeb'	white	green	green	green	green	green	yellowish green	yellowish green	maroon	maroon	pale green	green	green	
'Iybaip'*	white	green	green	green	green	green	yellowish green	yellowish green	maroon	maroon	pale green	green	green	bitter, small
'Iyshma'*	white	green	green	green	green	green	pale green	pale green	green & dark maroon veins	green & dark maroon veins	white	green	green	short stalks
'Kendkend'	white	green	maroon	green	green	green	dark maroon	yellowish green	pale green & maroon veins	pale green & maroon veins	pink	maroon	maroon	
'Keret'	white	green	green	green	green	green	dark maroon	pale green	pale green	pale green	white	pale red	pale red	
'Kogba'	white & purple	green	green	green	green	green	green & maroon veins	green & maroon veins	green & maroon veins	green & maroon veins	pink	greenish white	greenish white	multiple tops
'Kolmaen'*	white	red	red	red	red	red	red	red	red	red	pink	dark maroon	dark maroon	
'Kombol'	white	green	green	green	green	green	green & maroon veins	green & maroon veins	green & maroon veins	green & maroon veins	white	green & maroon veins	green & maroon veins	
'Koronkin'	white	green	green	green	green	maroon	pale green	pale green	pale green	pale green	white	pale green	pale green	multiple tops
'Lab'	white	green	green	green	green	green	green & dark maroon veins	green & dark maroon veins	green & dark maroon veins	green & dark maroon veins	pale maroon	green & dark maroon veins	green & dark maroon veins	waterlogged soil only
'Mamuwn'	white	pale maroon	pale maroon	green	dark maroon	green	dark maroon	pale green	dark maroon	dark maroon	red	dark maroon	dark maroon	
'Mez'	yellow	green	green	green	green	green	red	red	green	green	white	green	green	
'Mond hundbiy'	white	green	green	green	green	red	green	green	green	pale maroon	white	maroon	maroon	multiple tops
'Mond kaeriyl'	white	green	green	green	green	red	green	green	green	green	white	green	black	multiple tops
'Mond kongol'	white	green	green	green	green	green	pale green	pale green	green	green	white	green	green	multiple tops
'Mond pombray'	white	green	green	green	green	green	green	green	green & dark maroon veins	green & dark maroon veins	white	green	green	multiple tops
'Muwt'	yellow & purple	green	green	green	green	green	maroon	maroon	green	green	white	green	green	
'Naenk'*	white	green	green	green	green	green	green	green	green	green	pale green	green	green	never fruits
'Ndobok'	white	green	green	green	green	green	green, maroon & grey veins	maroon	green, grey & maroon veins	green, grey & maroon veins	pink	red	red	
'Nemhaez'	white	green	green	green	green	green	green	green	green	green	white	red	red	
'Paela'	white	green	green	green	green	green	maroon	green	green	green	white	green	green	never fruits
'Paym'	white & purple	green	green	green	green	green	green	green	green	green	pink	pink	pink	new leaf pink
'Pimbadom'	greeny yellow	green	green	green	green	green	green	green	yellowish green	white	white	yellowish green	white	
'P^{e}nk'*	white	green	green	green	green	green	red	red	red	red	pink	dark maroon	dark maroon	short stalks
'Shorhobor'*	yellow	green	green	green	green	green	yellowish green	yellowish green	green	green	white	green	green	short stalks
'Showmow'*	white	green	green	green	green	green	green	green	green	green	white	green	black	
'Sigab'	white	red	red	red	red	red	red	red	red	red	pink	red	red	
'Sizil'	white	green	green	green	green	green	maroon	pale green	maroon	maroon	white	maroon	maroon	
'Taegat'	white & purple	green	green	green	green	green	green & maroon veins	green & maroon veins	green & maroon veins	green & maroon veins	pink	red	red	
'Taeshaen'	white	green	green	green	green	green	red & pale green streaks	red & pale green streaks	red & pale green streaks	white	white	white	white	
'Tolop'	white	green	green	green	green	green	greenish white	greenish white	pale green	pale green	white	maroon	maroon	waterlogged soil only
'Tongay'*	yellow	green	green	green	green	green	dark maroon & green veins	dark maroon & green veins	dark maroon & green veins	dark maroon & green veins	pale maroon	dark maroon & green veins	dark maroon & green veins	

* = no examples observed, based on informant's descriptions only.

against one failing to root or growing into a spindly plant; this way plants will cover the whole garden. The planting of taro also requires the recitation of a spell, and the offering of marsupials to the ancestors in a ritual, to ensure its healthy growth.

The Wola cultivate taro primarily in gardens cleared especially for it, called *ma em* (lit: taro garden). These are usually situated on waterlogged sites or at least in places with wet soils, which the Wola call *suw pa* (lit: ground marshy). They say that the biggest tubers grow in these places because waterlogged soil is very soft and so allows for their growth, although corms take longer to reach maturity growing here than in drier soils. According to them, in waterlogged soils tubers grow long and thin first and then fill out, whereas in drier soils they grow consistently in a round shape, unable to expand into the more compact earth beyond the club-made hollow of soft soil. Consequently, although they sometimes plant small areas of taro in newly-cleared sweet potato gardens, it is more usual to see only one or two plants dotted about here and there in them; they rarely cultivate this crop in subsequent plantings of such gardens. Occasionally they also plant a few taro in the small mixed vegetable gardens located near their houses.

When planted in the drier soils of sweet potato and mixed vegetable gardens, the faster maturing tubers are harvested a few at a time, as needed, like other crops. Whereas in marshy taro gardens the tubers are all harvested at once and distributed in an exchange to relatives and friends. The large numbers of tubers planted in such gardens all reach maturity about the same time and are more than the gardener and his family could eat, so they give them away in a taro exchange. But before they can hold such a harvest and exchange they must clear and fence a new taro garden in which to plant all the corm tops, or else they will lose their stock. This repeated planting of the 'same' tops is a tangible link with the past, they are the 'ones' the ancestors cultivated and handed down, something which gives this archaic crop a special quality that is highlighted by the spells and rituals to ensure its healthy growth, the exchanges of its tubers and the various taboos that surround it.

Taro is the first crop planted in a taro garden whereas in a sweet potato garden it is, together with Highland *pitpit*, the second, planted after sweet potato. Both men and women may plant taro in any garden, although only men say the spells and perform the rituals to promote its development.

The planted tops send up new stalks from the centre of the old trimmed ones, which die away as the new plant establishes itself. Once established, the crop requires little attention or weeding, the large leaves casting a continuous shadow which inhibits the growth of weeds. The stalks of plants cultivated in water-logged conditions grow noticeably longer than those planted in dry soils, although the latter, if planted too close together, will grow tall stalks as they race for the sunlight (the Wola call such crowded tall-stalked plants *huguwp huguwp pay*).

Taro takes nine months to a year to reach maturity in dry soils and eighteen months or longer in waterlogged ones. It is judged to be mature when leaves begin to turn yellowish and the tuber lifts slightly. The Wola call this lifting *bokor nay*, and say it occurs because at this late stage the corm develops a narrow apex that pushes up the few leaf stalks that it supports. If the top of the tuber breaks the surface though and the sun shines on it, they say that this will ruin it, turning it soft and mushy; they call such blasted taro, *ma sinjbiyay*. Something else that ruins tubers is a chafer beetle (*Papuana* sp.) the Wola call *twenj*, which bores into tubers.

The Wola eat the tuber, leaves and inflorescence of taro. They consume both the parent tuber and its cormels; although not small ones, which are bitter and fed only to pigs.[14] They bake the tubers in the ashes of a fire, while they cook leaves and inflorescences over embers in bamboo tubes. They also steam entire plants in earth ovens, when they sometimes leave the leaves attached to the scraped tuber, wrapping round it to make a parcel for cooking. They never eat taro raw because it is bitter and astringent. Although taro cultivars, unlike those of sweet potato, all taste the same, when eating the purple-fleshed ones people may not speak (if they break this taboo, the tops from the tubers will not grow again). Taro figures significantly in the diet of the Wola, although it is not eaten daily, nor even weekly.

TANNIA

Xanthosoma sagittifolium (family: Araceae)
Wola name: *mbolin ma* (No. of cultivars = 1)

This is another edible aroid, similar to taro; called literally the 'whiteman's taro' by the Wola. Tannia originated in tropical America and did not arrive in the Pacific until last century. It reached the Wola after contact with the outside world, coming to them from Lake Kutubu sometime in the 1950s (Maenget Kem claims to be the first man to have cultivated it in the Haelaelinja area).

It is a robust herb which may approach two metres in height. Its tubers differ from those of taro in shape, size and growth pattern. The planted top develops a smaller parent tuber or *ma injiy* than taro, and from this grows a large number of 'light bulb-shaped' cormels. Many of these *pora* do not send up sucker shoots of their own, although some do and these may sprout 50 cm or so from the parent plant.

The leaf stalks, which the Wola call *paziy*, are stout and always green in colour. They support large leaves or *shor*, 60 cm or more long, which are more arrow-shaped and coarser than those of taro and have slightly wavy

[14] Any cultivar may on occasion produce bitter tubers, but the tops of such plants may grow palatable corms next time, and *vice versa*.

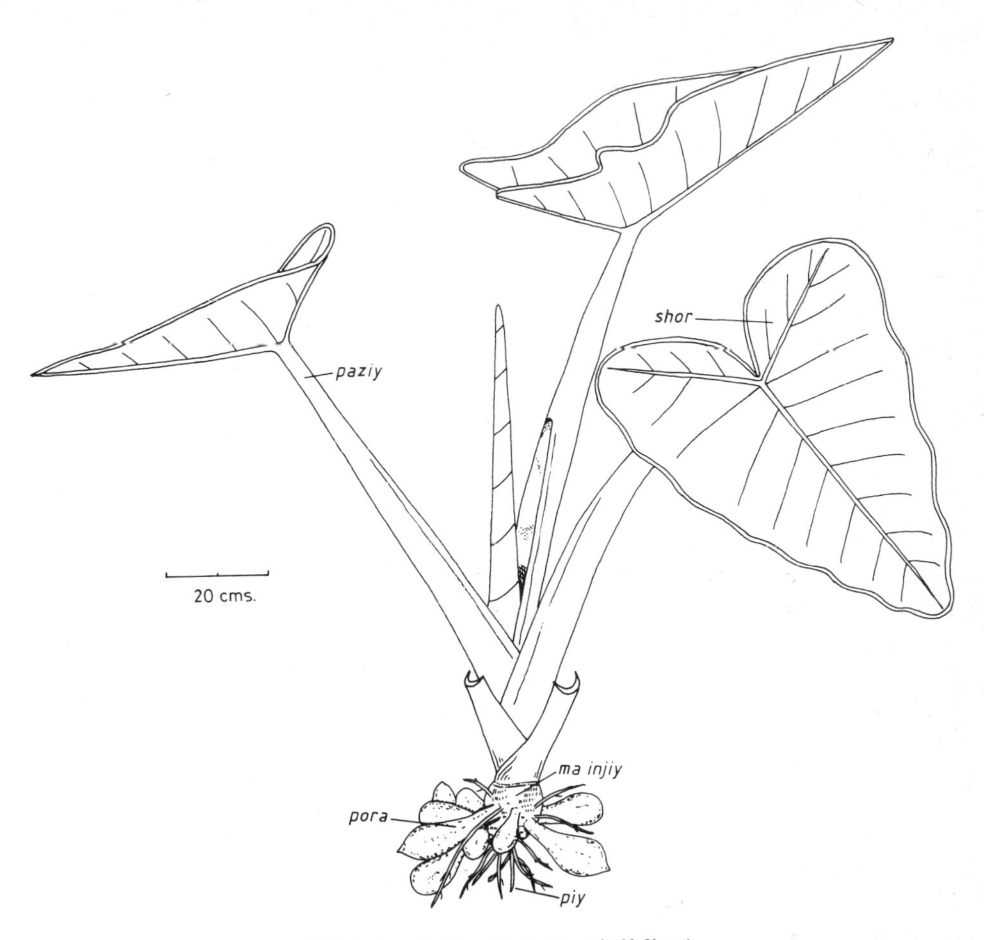

Fig. 5: Tannia (*Xanthosoma sagittifolium*)

edges. It is not usual for mature plants to flower, but when they do, they produce a single spike-like inflorescence wrapped in a pale green spathe which carries flowers of both sexes. As with taro, it is rare for seeds to be produced.

Although several tannia cultivars are recognised in other parts of the world, (this species, like taro being a variable and polymorphic one), the Wola distinguish only one. They propagate it in the same way as taro, planting the tops cut from corms, but unlike taro they do not cultivate it in waterlogged places, where it fares badly. They sometimes plant one or two tannia around the edge of sweet potato gardens, in the same locations as bananas, both having large shade-casting leaves which would inhibit the growth of other crops if planted in with them. Also like bananas, the Wola sometimes plant tannia around the edges of their house yards.

Tannia may be harvested from about nine months onwards by scraping back the soil from round the plant's base and excising some of the mature cormels. By earthing up the parent plant after harvesting, it is possible to keep it in production for some time and crop it again and again. The Wola never harvest the *ma injiy*, which in contrast to taro, is bitter and inedible; they only use its top for replanting. The edible *pora* have a white, firm and moist flesh, similar to that of the Irish potato, whereas taro has a drier more crumbly flesh. The Wola do not eat the leaves or inflorescences of tannia as they do with taro. They cook the tubers in the same ways as taro, but they are not as significant in their diet, being eaten rarely.

YAM

Dioscorea alata (family: Dioscoreaceae)
Wola name: *bet* (No. of cultivars = 1)

There are a number of edible tuber-bearing species of yam (see Massal and Barrau 1956: 12–15), although the Wola cultivate only one: the greater or winged yam.[15] This crop originated in nearby south-east Asia. According to the Wola their ancestors have cultivated it since time immemorial. But unlike some regions of Melanesia, where the yam is an important crop, it is rarely seen in their gardens (although they say that they think it was more popular with their ancestors, and was cultivated more prior to contact).

The tubers of the winged yam vary considerably in size and shape, although those grown by the Wola usually occur in small clumps, joined at their apexes where the vine starts. They are globular to long and 'sausage-shaped' (although cultivars grown in other regions produce branched, lobed and fan-shaped tubers). The tuber skin is pale brown and the flesh is white, moist and firm, like that of an Irish potato.

The stems of this yam species are a green or purplish colour and spineless. Gardeners train them up poles around which they twine in an anti-clockwise direction.[16] This staking is important for good yields because it presents the greatest leaf area to the sun. The stem is square in cross-section, with its four corners extended to flaps or wings running longitudinally, these grip the pole as the plant climbs.

The leaves are variable but roughly heart-shaped, and grow from the stem in opposed pairs. They have distinctive ribs which follow the leaf margin. The plant sometimes produces small round green bulbils, or aerial tubers, where the leaf stalks join the stem. They also produce here male flowers, in clusters on a main stalk, and female flowers in spikes. The fruits are elliptic in shape and three-winged, but it is rare for them to produce fertile seeds.[17]

[15] There is another species of *Dioscorea* that grows wild in the Wola region, which they call *bawiy* and eat in times of famine.

[16] The direction in which a vine twines is important in distinguishing between different yam species (anti-clockwise climbers are in the Enantiophyllum section and clockwise climbers in the Combilium section); the presence or absence of spines is also significant here.

[17] For further botanical information on the yam see Burkhill (1951) and Coursey (1967).

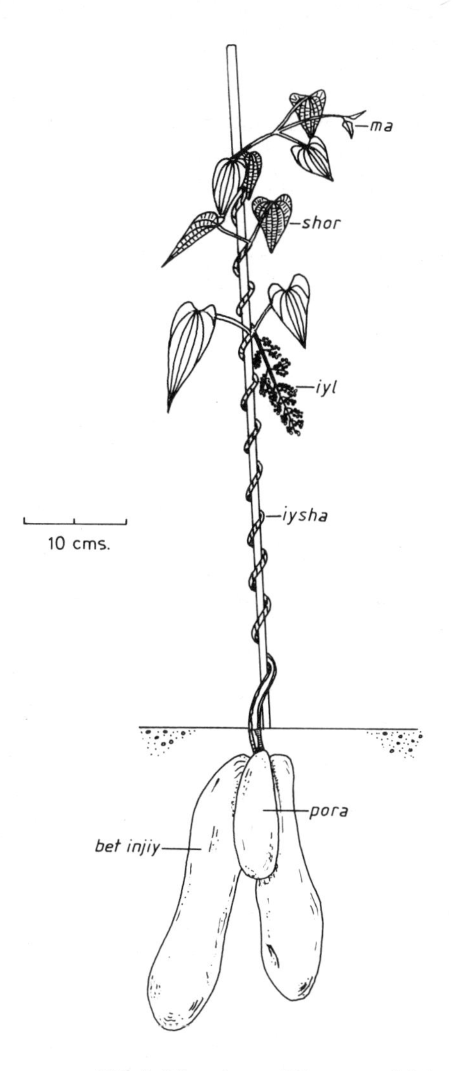

Fig. 6: Winged yam (*Dioscorea alata*)

They are perennial plants and the Wola describe how the older part of the stem sheds its leaves while the tip or *ma* continues to climb and grow. They say that the plant has to shed its leaves in this way about four times before its tubers are mature and ready to harvest; this contrasts with other places where people cultivate the yam as an annual (the probable reason for this is that, given the high altitude of the Wola region, the plant is here approaching the limits of its climatic range).[18] This leaf-shedding is actually part of the plant's adaption to a dry season, when the tubers become dormant.

[18] The higher the temperature the better it grows, and below 20°C it does poorly.

This climbing plant with its tuberous food-storage organs varies considerably, and those living in other parts of Melanesia have distinguished a large number of cultivars according to differences in the shape and colour of leaves, stems and tubers, but the Wola only distinguish the one cultivar of *bet.*[19]

Only men plant yams. They plant the top off a mature tuber with 15 cm or so of stem still attached. To plant, a man simply excavates a shallow hole, inserts the cutting and firms the soil around it with his fingers, taking care to leave the short length of stem protruding above the surface. They cultivate yams in the centre, not on the shaded edge, of newly-cleared sweet potato gardens, because they say they need plenty of sunshine to flourish. They do not cultivate them in gardens planted for a second or subsequent time, the Wola say yams only grow well on the soft, moist soil of newly cleared sites. But the yam, like the sweet potato, cannot tolerate conditions that are too wet: it demands a well-drained location.

Newly-planted cuttings remain dormant for three to four months before they sprout, at which time the vine consumes the food reserves of the planted piece, which shrivels up and dies to be replaced by new tubers. Men plant yams at the same time as taro and Highland *pitpit*, that is following the planting of sweet potato. The plant requires little husbandry once it is established, other than a stout pole up which to climb. It will be surrounded on the ground by sweet potato vines, which inhibit competition from weeds.

The Wola eat both the tubers and the bulbils. They bake them in the ashes or roast them on the top of a fire, steam them in an earth oven or today boil them up in a tin can. They are very rarely eaten by the Wola; indeed we found yams growing in only two Haelaelinja men's gardens in the Was valley.

KUDZU

Pueraria lobata (family: Leguminosae)

Wola name: *horon* (No. of cultivars = 2)

According to the Wola this crop, like the yam, has been cultivated by them and their ancestors for all time. Today, they rarely cultivate it, which is the situation in other Highland regions too (Bowers 1964, Strathern 1969). Some informants conceded though that it might have been more important in the past, while others doubted this because of the tuber's toughness and stringiness. On the grounds that it might have figured more in the pre-Ipomoean diet of Melanesians, some writers have shown interest in this marginal crop.[20]

[19] See Fischer (1968: 266) for an account of the yam cultivars distinguished by the Highland Kukukuku, and Malinowski (1935 (vol. 2): 98—104) for those identified by the Trobriand Islanders (he also discusses the cultivation of yams, which feature prominently in some ceremonial exchanges, in addition to the diet). See also Powell *et al.* (1975: 24) on those cultivated by the Melpa.

[20] See for example Watson 1964, 1968; Bowers 1964, and Barrau 1965: 282—5. This writing is of the kind already mentioned, which proceeds on botanical evidence to speculate on prehistoric situations.

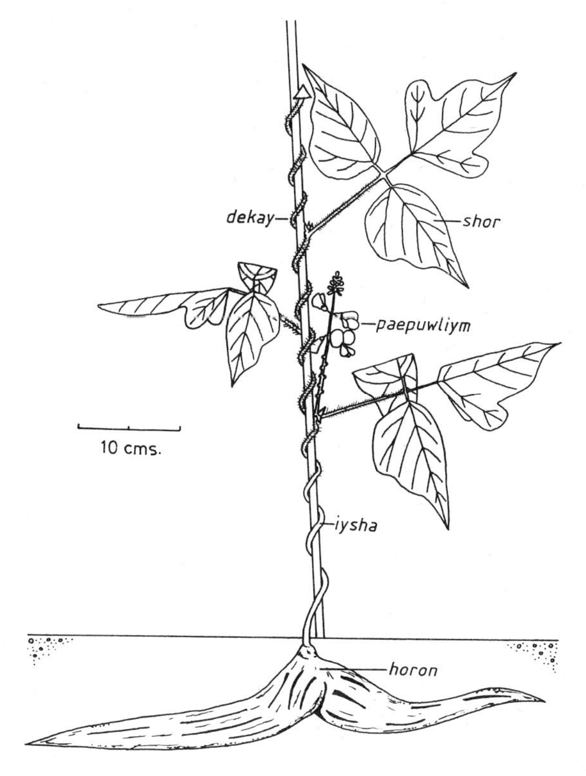

Fig. 7: Kudzu (*Pueraria lobata*)

This climbing plant is a legume, closely related to beans (see Chapter 5), and it is of agricultural interest, and perhaps of some significance, because it fixes nitrogen in the soil. The Wola are unaware of this benefit though, and do not cultivate the kudzu in specific locations to improve the soil for the next crop.

The mature stems or *iysha* of kudzu are somewhat woody in appearance. They grow vigorously and gardeners sometimes train them up poles, while at others they leave them to climb pollarded trees or other natural supports. Some of the roots of the kudzu develop into thick, fleshy, starchy tubers, which the Wola call *horon*. These tubers may grow a metre or more long and up to 10 cm in diameter (although those I saw were considerably smaller). They are pointed at the growing end, and grow out horizontally from the stem, running more or less parallel with the surface of the soil and only a few centimetres below it. They do not grow down deep into the soil. The skin of the tuber is brown and the flesh is whitish. A single vine might grow two, three or more tubers, running from it in different directions and all joined together at their apexes where the stem emerges.

Soft orangish hairs called *dekay* cover the stems and leaf stalks. The compound leaves, or *shor* consist of three leaflets and resemble those of beans. The leaflets vary considerably in shape; they often have waved or lobed margins, the middle apical leaflet commonly having two lobes, while the two asymmetrically oval side ones have a single lobe on the stem side. The flowers, or *paepuwliym*, blossom singly from longish stalks that grow out from where the leaf stalks join the stem. They are fragrant and violet-purple in colour, and in their butterfly-like shape resemble those of beans. The fruits are like bean pods too, being long and narrow with green bean-like seeds.

The Wola distinguish two kudzu cultivars.[21] One they call '*horonmasol*' and the other simply '*horon*'. They are identical except that '*horonmasol*' grows only one clump of tubers and a single stem, *ma sol* meaning literally 'long growing shoots', a reference to the long vine of this cultivar. On the other hand '*horon*' sends out secondary woody stems called *horon iysha*, which creep along the ground and root themselves some distance from the parent plant, in turn developing clumps of tubers called *pora*, that grow their own climbing stems and later send out further secondary terrestrial stems.

Only men plant kudzu. They propagate them vegetatively in the same way as yams, taking the tuberous apex where the roots join, together with 15 cm or so of stem. They plant this cutting, with the piece of vine sticking out of the ground, in both newly-cleared sweet potato gardens and established ones replanted for a second or subsequent time. They plant it at the same time as beans and such crops, following the planting of sweet potato, taro and Highland *pitpit*. Another place where men cultivate kudzu is on the edge of their houseyards, where they climb surrounding trees. It requires little husbandry while growing except the provision of a pole to climb, if there is no tree nearby.[22]

The Wola eat only the tuber, which is tough and fibrous. They cook it by baking in ashes or steaming in an earth oven, and sometimes in an attempt to soften it, they cook it two or three times. Nevertheless, it is always tough to eat and as a result is a marginal crop which figures rarely in the diet. It is considered more of an emergency source of food eaten in famine, not normal times. An indication of the tuber's toughness is that it has given rise to the Wola saying *horon hiywa* (lit: kudzu sago) for strong people, persiflage setting them off against the contrarily soft and jelly-like *hiywa*, the sago from Lake Kutubu.

[21] As the Hageners do too (Strathern 1969). But unlike these people, and the Kakoli (Bowers 1964), the Wola do not associate the kudzu more closely with the sweet potato than they do any other tuber bearing crop (they do not, like these Western Highlanders, use the same prefix for both).

[22] See Strathern (1969: 194—6) for details of Melpa kudzu cultivation practices. When they plant this crop, unlike the Wola, they observe certain taboos and say spells to ensure its healthy growth.

GINGER

Zingiber officinale (family: Zingiberaceae)
Wola name: *shombiy* (No. of cultivars = 3)

This plant was probably domesticated somewhere in Asia. The Wola think that they and their ancestors have always grown it.

This aromatic herb is a perennial which produces characteristically ginger-flavoured pungent rhizomes. It is a slender plant, growing to between 30 and 50 cm high, with a robust knobbly rhizome growing horizontally near the surface of the soil. The rhizomes, called *shombiy* by the Wola are lobed, covered with small scales and have small roots called *pil*. The scaly skin of mature rhizomes is red and their flesh is a brownish yellow; those of young plants are both light orange. The rhizomes are usually irregular in shape, straight ones only growing in very loose friable soils which offer little mechanical resistance.

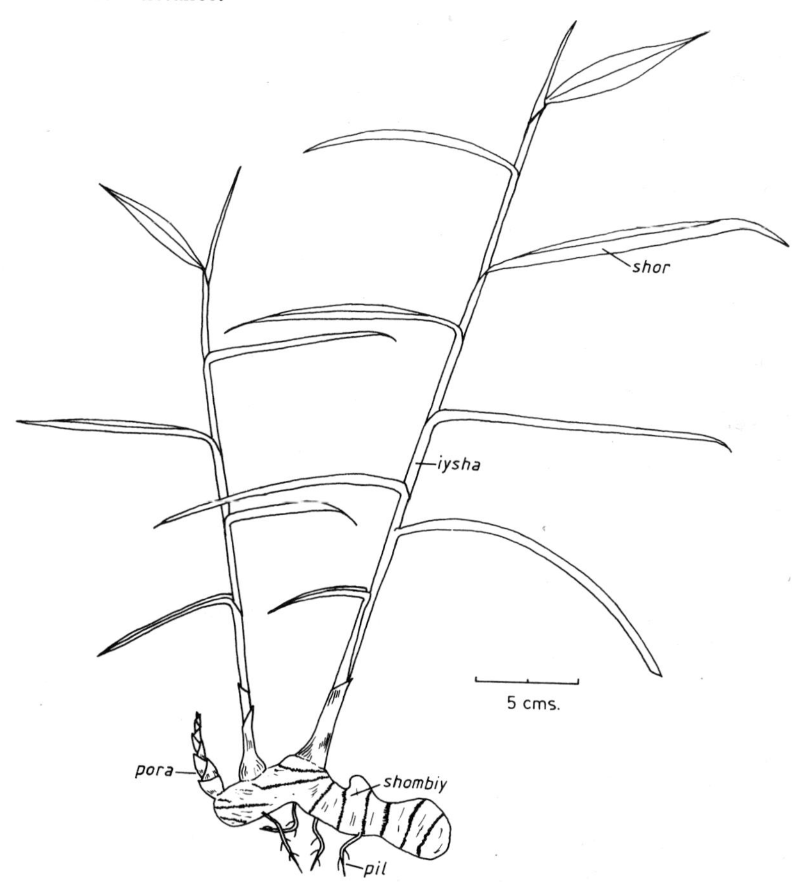

Fig. 8: Ginger (*Zingiber officinale*)

The rhizomes give rise, along their length, to annual leafy shoots or *pora*, which develop into erect stems or *iysha*. These consist of the long overlapping sheaths of leaves. They are green in colour, except near the rhizome, where they are light orange with thin red streaks. The leaves or *shor* are arranged alternately in two opposed rows, sticking out more or less horizontally on either side of the central sheath-stem. They are light green in colour, narrow and oblong in shape, terminate in a point, often with both edges rolled upwards like a gutter. They usually have a pronounced mid-rib.

The Wola say that *shombiy* never flower nor fruit, and even in those parts of the world where this has been observed it is rare, and the production of seeds is even rarer. When they do flower, these plants send up a cone-like spike direct from the root stock which produces pale yellow, short-lived flowers.

The number of ginger clones is apparently limited worldwide; the Wola distinguish three.[23] They base their identification of these on the size of the plants and the strength and hotness of their ginger flavour.[24] The mildest-flavoured rhizome is produced by the '*ogimb*' cultivar, which also has the narrowest leaves and thinnest stem.[25] The largest plant, with the broadest leaves and thickest stem, produces the strongest-tasting rhizomes and is called '*ponjip*' or '*munk*'. The middle cultivar, in size and strength of flavour, is called '*taishombiy*'.

Ginger is planted only by men. They propagate it vegetatively by taking a piece of rootstock from an old plant (a bit from a new one would only rot), which has a bud, although not necessarily one that has started to develop into a shoot. They drop this into a shallow drill hole in the soil. Once it has rooted itself the cutting sends up suckers and the rhizomes grow to maturity. Men plant ginger in the dry soil of newly cleared sweet potato and mixed vegetable gardens, but never on waterlogged sites because it will not tolerate wet conditions. Also they only cultivate it in newly-cleared gardens, it does not do well in older-established ones. Ginger may be planted anywhere in a garden, mixed up with various other crops. It is planted after the major root crops, at the same time as beans and maize.

The Wola say that the hotness of a ginger rhizome varies according to the temperament of the man who plants it. If a bad-tempered man, who is often angry and involved in disputes, plants a rhizome then it is much hotter to eat than when a mild-mannered man plants one.

The rhizomes reach maturity in about nine months. A gardener might pull up the entire plant and eat it, or alternatively he might expose part of the rootstock and break off a piece, earthing up the plant again so that it will

[23] See Fischer (1968: 272—3) for an account of the six cultivars distinguished by the Kukukuku.

[24] The Wola call this its *krai*.

[25] It is interesting in view of the hot stinging taste of ginger that the Wola also call certain wasps *ogimb* (*Polistes comis* and *Ropalidia* sp.).

continue growing. As a perennial plant, the Wola say that they can go on harvesting in this manner for two years or more.

Ginger figures little in the Wola diet. Relatively few plants are grown and their rhizomes never reach any large size. It is eaten raw, but only a little at a time because of its pungent property. People often eat ginger rather like we eat sweets, on its own during the day between meals. At other times they eat it as a relish with something else, giving it a ginger flavour, in which case it is common to wrap a piece up in the other food.

Men often hang ginger leaves round the top of the walls inside their houses to give a sweet ginger smell. It might also act as a curtain reducing the wind blowing in under the eaves, although men denied that the hotness of the plant had any effect, practical or symbolic, in warming their houses by reducing draughts.

IRISH POTATO

Solanum tuberosum (family: Solanaceae)
Wola names: *aspus; kagow; saemow* (No. of cultivars = 2)

This crop, of South American origin, was introduced into the Wola region following contact sometime in the 1960s.

This familiar annual herb produces clumps of tubers which the Wola call *aspus, saemow* or *kagow*. These vary in size and shape from round to oval. The plant's stem is erect and branched with fine hairs. The compound leaves are arranged in pairs along a stout stalk and have small leaflets interjected between them, they are more or less oval in shape. The flowers vary in colour from white through yellow to purple, and they produce a small tomato-like (although inedible) berry.

This crop has many cultivars in temperate regions, but the Wola distinguish only two, on the basis of the colour of the tubers, plant stem and leaf stalks. One cultivar is a 'white' one called '*aspus*' and the other is a 'red' one called '*kaegaliywa*'. The skin of the '*aspus*' tuber is pale yellowish brown and the flesh creamy. The plant stem and leaf stalks are pale green. The skin of the '*kaegaliywa*' tuber is maroonish red and the flesh white. The roots and base of the plant stem are maroon, changing further up where the leaves start, to a patchy maroon and green, which is also the colouring of the leaf stalks. The red skinned cultivar tends to grow larger tubers with tougher skins and so boils more successfully without disintegrating.

Women largely plant Irish potatoes. They take old sprouting tubers and drop them singly into dibbled holes. They plant them in both new and re-worked gardens, and in small mixed vegetable gardens. In the larger gardens, they tend to cultivate them around the edges, usually in mounds or small *paen* beds of broken up soil. They plant them at the same time as ginger, that is following the major root crops.

The tubers are ready after about four months. The Wola bake them in the ashes or roast them over the embers of an open fire, steam them in earth ovens or boil them. They are not popular, though, with the Wola, who find the tubers flavourless, and on those rare occasions when people do eat them they usually cook them with a pile of sweet potatoes. Settlements near to Nipa patrol post tend to grow more Irish potatoes for sale to Europeans at the weekly market.

Section I: Crops described

Chapter 3
GREENS

INDIAN SPINACH

Amaranthus tricolor (family: Amaranthaceae)
Wola name: *komb* (No. of cultivars = 5)

This plant is probably indigenous to Papuasia. The Wola say that it is a traditional crop which they have cultivated since ancestral times.

This herbaceous annual has an erect angular stem or *iysha*, which is often branched. Indeed the Wola induce bushy-branched plants by plucking off their growing tips; this results in bushy branches they call *taenday* (lit: sprouts). The plant's stem is maroon at its base, changing higher up to a patchy pale green and pink. Its leaves are oval and have rounded ends, with a small notch in them where the leaf rib ends.[1] They are green, often tinged with maroon (depending on the cultivar, see Table 8), and sometimes they become variegated red, yellow and very pale green when they are old and about to fall.

Distinctive spikes, bearing flowers and later seeds, characterise amaranths. The small petalless flowers, green and sometimes tinged with maroon, are clustered into them, the one terminating the plant often being large and exceeding 12 cm in length; other smaller compound spikes grow out from where the leaf stalks join the stem. Later the flowers dry out and become light brown when, with their three egg-shaped husks each, they are reminiscent of wheat. Later still, the elliptic fruits burst open to release their seeds. These flowers and seed spikes are diagnostic in distinguishing between different species of amaranth, although only for an expert using a lens.[2]

This is a variable species of plant and a number of cultivars, many of them

[1] This is something to look for when identifying this species.

[2] For further information on the botany of amaranths see Sauer (1950) and Womersley (1978).

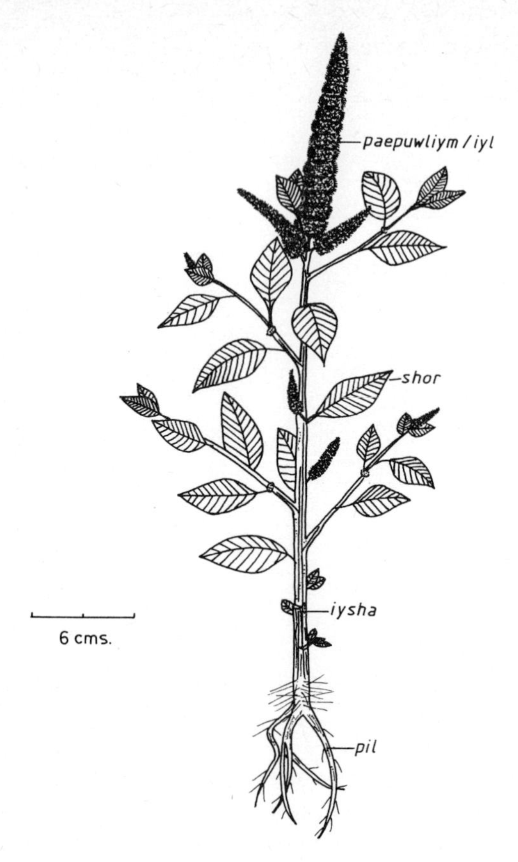

Fig. 9: Indian spinach (*Amaranthus tricolor*)

garden ornamentals, have been described. The Wola distinguish five cultivars of *komb*. The shape and size of the plants and their leaves are the same, and they are distinguished between according to the colour of their leaves, and the form of their seed spikes (see Table 8).

These greens are planted only by men. They propagate them largely from seeds, although they occasionally dig up an entire young plant, which has grown at an unfavourable place (often self-sown outside a garden or in an old one) and transplant it. When a man requires seed, he leaves a few plants until they have shed most of their leaves and the small ears composing the spikes have started to turn yellow and dry out. He then pulls them up and dries them out on the fireguard in the roof of his house. Wrapped up in a leaf parcel to protect them from soot, seeds can be kept here for many months. They are stored undifferentiated by cultivar in a single bundle. When dry though, the seeds of all *komb* cultivars look identical, which means they are planted and grow mixed up together.

To plant *komb* spinach, a man takes a small bundle of dried spikes and

Table 8: Indian spinach cultivars

CULTIVAR NAME	LEAF	RIB	LEAF STALKS	SEED SPIKES
'*Amiykomb*'	Green	Pale green	Pale green near leaf and pink near stem	branching
'*Kaerob*'	Green	Pale green	Pale green near leaf and pink near stem	single
'*Kombab*'	Green with small maroon blotches	Pale green	Pale green near leaf and pink near stem	variable
'*Kombilol*'	Green edges and maroon centre	Pale maroon	Pale maroon	variable
'*Kwai*' or '*Kombnaegimb*'	Green centre with maroon edges	Pale green	Pale green near leaf and pink near stem	variable

walks around his garden twisting it and rubbing the husks together, so splitting them open and releasing the seeds. The Wola call this broadcast sowing *taentaen bay* (lit: shake-shake do). They make no attempt to cover up the seeds or prepare beds for them (beyond generally breaking up the soil, as they will have done in clearing the garden and planting other crops). These greens are the last crop planted, and once their seeds have been scattered people ought not to walk across the garden until they have germinated and started to grow, or else they may trample and ruin them.

Men plant these greens in newly-cleared sweet potato gardens largely; rarely in established ones, unless they have pockets of exceptionally fertile soil. They cultivate them all over a garden, the plants reaching maturity before sweet potato vines cover the surface. They also cultivate these greens in small mixed vegetable gardens, and another popular place for them is the site of recently pulled down houses where the Wola say the refuse of human and porcine occupation (pig faeces, food waste, rotten leafage etc.) makes good compost and gives a very fertile bed for greens.

The plants require a little weeding to protect them from fast growing competitors, but this is not onerous for they are ready to be harvested from within two months. They yield edible leaves for about two months, and after five months they will have seeded and died back. When the plants are young the Wola usually pluck off their leaves and tender stems, leaving them, as mentioned, to develop a bushy growth of side branches. Older plants, sometimes with some seed development, are pulled up entirely, unless the gardener wishes to leave them for seed.

A new garden produces a flush of various early crops, and at this time *komb* figures prominently in the diet of a gardener's family. For most of the time though they will not be harvesting from a new garden; hence, overall, this spinach does not often appear in Wola meals. When it does they eat the leaves and tender stems largely, although when they pull up the entire plant they eat the flower spikes too. They steam this spinach in earth ovens, or in small leaf parcels with hot stones, simmer in bamboo tubes, boil in tin cans, and occasionally bake it wrapped in a leaf parcel, in the ashes of a fire.

LOVE LIES BLEEDING SPINACH

Amaranthus caudatus (family: Amaranthaceae)
Wola name: *mbolin komb* (No. of cultivars = 1)

This plant is of South American origin, and reached the Wola very recently from some other region where it was introduced. It is apparently a rare crop throughout the Highland region: this is only the second time that it has been reported under cultivation there (the other was by N. Bowers in the Upper Kaugel Valley of the Western Highlands).

The Wola name for this plant means literally the 'European's Indian spinach', an indication that they relate it closely to the preceding crop, which it resembles.

The size and shape of this annual herb's stem and leaves are the same as *A. tricolor*, and the colouring of the leaves is the same as the '*kwai*' cultivar. But the leaves have a short sharp point where the rib ends, not a nick. The distinguishing feature for the Wola though is the dense chaffy spikes of these plants, composed of numerous flowers, and later seed ears, which droop instead of standing erect. Another difference, though only visible under a lens, is that the egg-shaped fruit is composed of five and not three husks (these burst open later to release their seeds). The new spikes also tend to be more red than those of *komb*.

The planting, cultivation and consumption of this plant is the same as *komb* and it grows indiscriminantly mixed up with this crop, although it occurs far more rarely.

PRINCE'S FEATHER SPINACH (also called PIGWEED)

Amaranthus cruentus
(*A. hybridis* p.p.) (family: Amaranthaceae)
Wola name: *paluw* (No. of cultivars = 2)

This plant is of Central American origin and was introduced into the Wola region subsequent to European penetration. The Wola recognise that it is similar to their indigenous amaranth spinach, although different enough to demand classing as a separate crop and not as a cultivar of *komb*.

It too, is an annual herb with an erect angular stem, which is often branching. Its leaves are the same size as the other two amaranths although more spear-headed in shape, with a short sharp point at the tip where the rib ends. It is a larger plant than the other two, with an erect spike of chaffy flowers like *A. tricolor.* These stiff, dense, spike-shaped clusters (looked at under a hand lens) have five oblong husks to each egg-shaped fruit.

The Wola distinguish two cultivars, which they call '*uwlabay*' and '*paluw*'. Both plants are identical in size and shape and are distinguished by their colour. The stem, leaves and flower spikes of the '*uwlabay*' cultivar are all green, while those of '*paluw*' are red (the redness sometimes extends to the roots and varies between plants, from pale red with green through to scarlet).

Another point of difference between the two cultivars is their propagation. The green '*uwlabay*' may be planted both from seeds (*taentaen bay*) and vegetatively by breaking a piece from a mature plant's stem and pushing it into the soil (*way koba bway*), whereas the red '*paluw*' can only be planted from seeds (cuttings from it only dry up and die).

Men largely plant this crop, although occasionally women do so too with cuttings. They cultivate it in exactly the same way and in the same places as *komb*, and it matures at the same time. It often grows mixed up with *komb*, and is usually cooked with it too. As the largest of the amaranth plants though, and consequently yielding more edible leaves, it is more popular than the traditional *komb* spinach and figures more in the Wola diet.

CRUCIFER SPINACH

Rorippa sp. (family: Cruciferae)

Wola name: *taguwt* (No. of cultivars = 6)

These greens are of unknown origin, although they possibly originated in the Melanesian region, which is one of the few where they are eaten as a spinach. As far as the Wola are concerned, they and their ancestors have always cultivated this crop.

A low annual herb, rarely exceeding 30 cm in height, this plant has an angular (usually four sided) stem or *iysha*, which is hollow through the centre. Young plants, called *huwniy*, have short, almost non-existent stems, and their leaves grow out from the base in a rosette fashion. But picking them and the soft apical shoots encourages the development of branching stems called *taenday*, such that older plants have several stems growing from their single rootstocks. The Wola call these older plants with long branching stems *taguwt el say* (lit: crucifer spinach old becomes).

The green leaves or *shor* of *taguwt* are distinctly lobed with five to seven or sometimes more leaflet lobes, the terminal one being the largest. The large leaves of young plants resemble those of watercress, a related plant species, but those of older ones are smaller and more ragged looking. These later

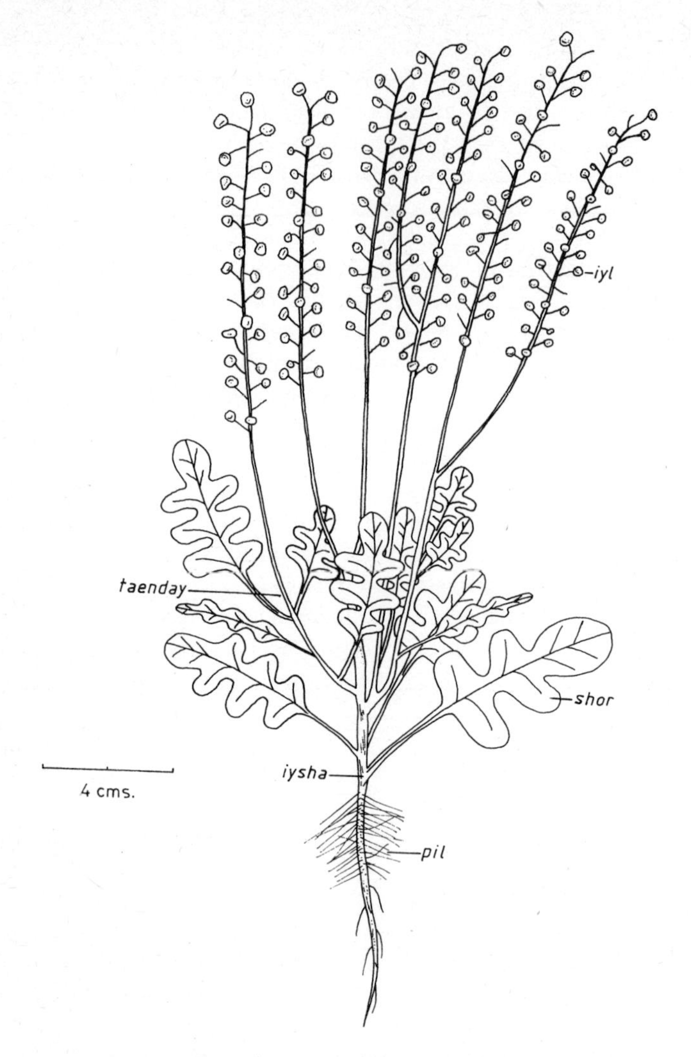

Fig. 10: Crucifer spinach (*Rorippa* sp.)

leaves not only grow out from the base of the plant but also from higher up the stem. The stalks of leaves have a characteristic flange of leaf, sometimes with small aborted lobes, running down either side of them.

The stems of mature plants produce terminal clusters of small yellow flowers, the *paepuwliym*, which develop later into thin-skinned, spherical seed pods, the *iyl*, which are rather like flimsy peas. They are pale green in colour, turning to pale brown when the plant dies. Many of these seed capsules are produced, clustering along the upper half of stems and branches. The roots, the *pil*, are shallow, and consist of either a spreading network of fine rootlets, or a single main root.

The Wola distinguish six cultivars of this spinach,[3] on the basis of the colour of the plants, their size and their lifespans (see Table 9).

Table 9: Crucifer spinach cultivars

CULTIVAR NAME	LEAF	STEM	LEAF STALKS	STEM SIZE	LEAF SIZE	SEEDS	CROPPING PERIOD (months)
'*kongliyp*'	pale yellowish green	pale yellowish green and maroon[1]	pale green	average	average	yes	3
'*laengaben*'	green	green and maroon[1]	pale green	tall	narrow	no	3
'*ogai*'	green	green and maroon[1]	maroon	average	broad	yes	3
'*olbiyaib*'	green	pale green	pale green	thick	average	yes	6
'*taend*'	green	green and maroon[1]	maroon	average	average	yes	3
'*taziy*'	green	pale green	pale green	thin	average	yes	1½

Note
[1] maroon at base of plant, giving way to green at the top.

Men plant these greens. They prepare the seeds in the same way as they do those for the amaranth spinaches, leaving plants from which they wish to gather seeds until they have shed their few remaining leaves and the seed capsules have started to dry out and turn pale brown. They then pull up the entire plants and wrap them in leaf bundles which they hang in their houses to keep dry until required.

This spinach is cultivated largely in newly-cleared gardens. But men do not have to wait until they clear a new garden themselves to plant it, for it is an accepted practice to approach a relative or friend who has cleared one and ask permission to plant a crop there. These greens are restricted mostly to new sweet potato gardens, mixed vegetable gardens and old house site beds, although they sometimes seed themselves in recently reworked gardens and are occasionally planted in particularly fertile areas of them too. Together with amaranth greens *taguwt* is the last crop planted in any garden. It is sown in the same broadcast fashion (*taentaen bay*) all over a garden, except for the '*laengaben*' cultivar, which never flowers or seeds and hence is propagated vegetatively by breaking off pieces of mature stem and pushing them into the soil. (This may possibly be a different species of plant.)

[3] They also distinguish another two varieties of *taguwt*, which they call '*suw taguwt*' (*Solanum americanum*) and '*oluwng*' (*Cardamine* sp.); both are edible but only grow wild.

The plants are ready for cropping after about two months and continue yielding edible leaves for the times shown on Table 9. The Wola do not pull up entire young plants but pluck off a few leaves at a time, harvesting from plants several times over. Older plants they pull up, unless they wish them to seed. They cook these greens in the same way as the amaranths, indeed often mixed in with them. They figure to a slightly less extent in the Wola diet.

WATERCRESS

Nasturtium officinale (family: Cruciferae)

Wola names: *kuwmba; mbolin taguwt* (No. of cultivars = 1)

This plant, of European origin, is a recent post-contact arrival in the Wola region. The lobed leaves resemble those of *taguwt*, a similarity recognised by the Wola, as one of their names shows: *mbolin taguwt* translating literally as 'whiteman's crucifer spinach'.

This perennial aquatic herb grows angular, hollow, trailing stems which either float on the surface of water or root at nodes along their lengths. The stems are hairless and the lobed leaves have a characteristic 'hot' flavour. The plant produces white flowers.

Men plant watercress from cuttings and they say it grows best in running water, which is sometimes found in waterlogged taro gardens; otherwise they plant it in streams outside gardens, where it flourishes and anyone passing may pick it. In water its leaves grow deep green, whereas in soil they are yellowish and of poor quality.

When established the plant grows prolifically, and the picking of its leafy tips induces branching and further spreading. It requires no husbandry; once growing in a stream it may colonise a considerable stretch of water. The Wola do not eat the leaves and stems raw but either steam them in an earth oven or boil them in a bamboo tube or tin. Watercress is not a popular green vegetable and figures little in their diet, usually occurring in meals with other greens with which it is cooked.

ACANTH SPINACH

Rungia klossii (family: Acanthaceae)

Wola name: *shombay; ten shombay* (No. of cultivars = 3)

This is another plant, like *taguwt*, which possibly originated in the Melanesian region, for it is cultivated throughout the Highlands as a spinach. The Wola maintain that they and their ancestors have always grown it as a green vegetable.

It is a perennial herb that grows in clumps when established. Its erect stem (*iysha*) consists of several hollow sections joined together at distinct

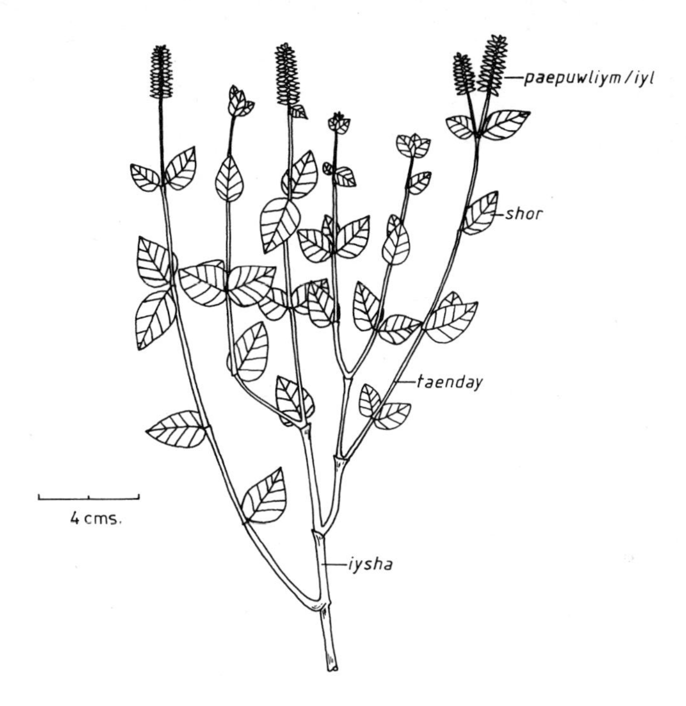

Fig. 11: Acanth spinach (*Rungia klossii*)

nodes, from which both branches and leaves grow. The leaves (*shor*) are simple, oval, thick and somewhat crinkly, and dark green in colour.

The nipping off of the young leaves and growing shoots induces branching of the stem, and mature plants have a bushy appearance. Fine hairs cover the top surface of leaves and young stem. The flowers grow in terminal clusters, each one with attending small leaflets that are pale purple in colour, and which, arranged round the stem in dense bunches look somewhat like the ears of certain grasses.

The Wola recognise three cultivars of acanth spinach which they distinguish by the size and colour of leaves (see Table 10). They consider '*taen*' the best cultivar. The other two they reckon to be the same and cannot in fact distinguish between them when '*saelsael*' is either a young plant without

Table 10: Acanth spinach (*Rungia klossii*) cultivars

CULTIVAR NAME	LEAF RIBS	LEAF
'*mapuwnpogol*'	green	long and narrow
'*saelsael*'	yellow	long and narrow
'*taen*'	green	short and broad

yellow leaf ribs, or when it grows in the shade, which inhibits the development of the yellow ribs. '*Mapuwnpogol*', which literally means 'wallaby thigh' (after the leaf shape perhaps), also grows wild in the forest, whereas the other two only occur in cultivation or semi-wild on recently abandoned garden sites.

Women plant this crop by breaking pieces off a mature plant's stem, 15 to 20 cm long, and inserting three or four of these cuttings into a dibble hole made with the point of a digging stick, firming up the soil around them. They plant three or four cuttings to ensure that at least one takes. The Wola call such vegetative propagation from cuttings *way koba bway* (lit: cutting pull-up plant). When established these cuttings send up shoots from their roots and so develop into bushy clumps. Women plant *shombay* after sweet potato, at the same time as Highland *pitpit*, and they often plant these two crops together in alternating clumps. *Shombay* may be planted anywhere in a garden, intermingled with other crops. A popular place for it is around the edge of sweet potato gardens, and also along the internal boundaries marking the areas cultivated by different women. It prefers a dry and friable soil, which rules out planting in waterlogged taro gardens. Otherwise it may be cultivated in any kind of garden, including established sweet potato ones worked for a second or subsequent time; to replant this acanth spinach in such a garden is called *shombay way orkay*.

This prolific plant requires little husbandry, growing so densely that it inhibits competition from weeds. It takes about six months to reach maturity and then it may be harvested from heavily for twelve months or longer. Women pluck off leaves, young stems and shoots, leaving the plant for periods to regenerate. The Wola eat these raw, or cook them in an earth oven, a bamboo tube or tin, or sometimes wrapped in a leaf parcel with meat and baked in the ashes. *Shombay* is often cooked and eaten mixed together with various other greens. But, as a prolifically growing robust perennial, which occurs in many gardens (including established ones) and stands harvesting over long time periods, it figures more prominently in the Wola diet than any of these other spinaches.

ACANTH MEAT SPINACH

Dicliptera papuana (family: Acanthaceae)
Wola name: *omok* (No. of cultivars = 1)

This plant, like the previous acanth spinach, is possibly also of Melanesian origin. The Wola maintain that they and their ancestors have always cultivated it.

It is an erect perennial herb with a stem composed of sections joined by prominent nodes, from which grow leaves, branches and flowers. Leaves tend to grow in opposed pairs from these nodes, together with clusters of the

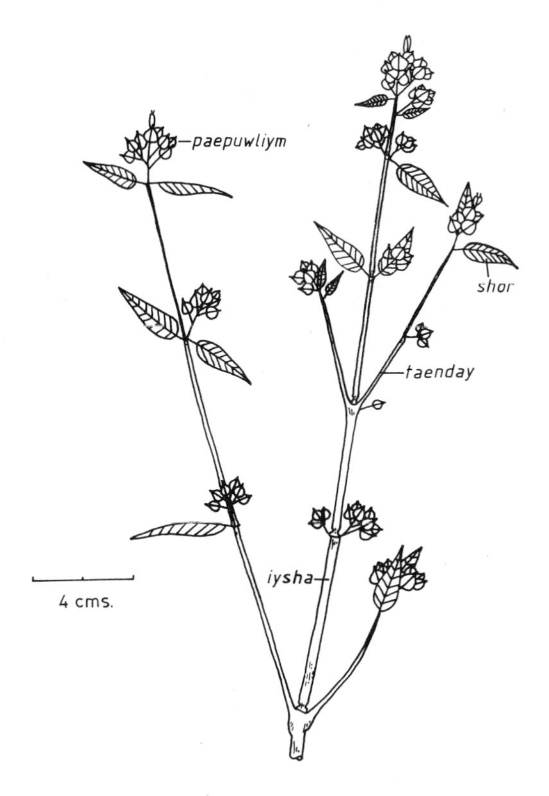

Fig. 12: Acanth meat spinach (*Dicliptera papuana*)

leaflets that accompany the flowers. The leaves, like the stem, are dark green. They are entire and distinctly spear-like in shape, tapering from an oval to a long point. Fine hairs cover the leaves, stem and flower leaflets. Mature plants have branches, which like the leaves, grow in opposed pairs from either side of nodes on the parent stem.

The flowers, or *paepuwliym*, occur in clusters, both at the top of the plant and from nodes up the stem. They are accompanied by characteristic circular bract leaflets that terminate in a small point, and which are more prominent than the flowers themselves. They are blue, while the flowers are whitish, small and short-lived.

The Wola distinguish only one cultivar of this plant, which they call '*kenjom*' to distinguish it from a wild form called simply *omok*. Both plants are identical except that the leaves of the cultivated one are narrower than those of the wild one.

Both men and women plant this crop. It is propagated vegetatively by stem cuttings taken from mature plants, dibble planted in the same way as those of *shombay*. It grows in well-drained and friable soils, which rules out waterlogged taro gardens, otherwise it may be cultivated anywhere. Only a few plants are cultivated in any garden, and favourite spots for them are around obstacles such as tree trunks and rocks, or at the edge, near the fence. It is planted at the same time as the other acanth spinach *shombay*.

It is a tall erect plant which can compete successfully with weeds and grasses (as indeed it has to in the wild state) and requires little husbandry when established. It is some months before there are sufficient leaves worth picking. It is not a commonly eaten vegetable, contributing little to the Wola diet, and is usually only picked to be eaten with meat or screw-pine nuts. Its leaves may be eaten raw with these or cooked with them; a popular culinary preparation is to bake the leaves in the ashes in a parcel with salt and pork or marsupial meat, so that they become impregnated with salt and grease.

JAVANESE DROPWORT

Oenanthe javanica (family: Umbelliferae)
Wola name: *taziy* (No. of cultivars = 2)

This parsley-like plant, which occurs wild throughout the Papuasia region, was possibly domesticated in the New Guinea Highlands where it is now widely cultivated as a green vegetable. According to the Wola, they and their ancestors have always grown and eaten it.

It is an erect plant with a hollow, branching stem that has a furrowed surface. It has compound leaves consisting of two rows of opposed compound leaflets, which in turn consist of small opposed folioles that are roughly diamond-shaped, sometimes lobed, and have finely serrated margins. It is a hairless plant.

The stalk which carries flowers grows from the stem opposed to the stalk of a leaf. These flower stalks terminate in several small stalklets which grow radiating out like umbrella spokes, and each of these divides again in a similar manner, and these fine terminal stalklets carry the small white flowers, which, tightly packed together, produce flat-topped clusters or umbels.

The Wola grow two cultivars of this 'parsley' which they call '*kog taziy*' and '*taziy hundbiy*'. Both plants are the same size and shape, and are distinguished between by their colouring. '*Kog taziy*' is all green, whereas '*taziy hundbiy*', as its Wola name suggests (meaning literally 'Javanese dropwort yellow'), has leaves which turn varying shades of yellow when mature (from yellow through pale yellow to greenish yellow); its stem also has a pale maroon base. The Wola also distinguish a third kind of *taziy* which they call '*iyb taziy*' (lit: water Javanese dropwort). This plant only grows wild in the forest or on abandoned garden sites, although it is edible and collected like the two domesticated cultivars. It is all green, and is distinguished from '*kog taziy*' by its spindly tall stem and sparser leaf growth.

Women plant this crop largely. They propagate it vegetatively from stem cuttings taken off mature plants, two or three of which they firm into dibble

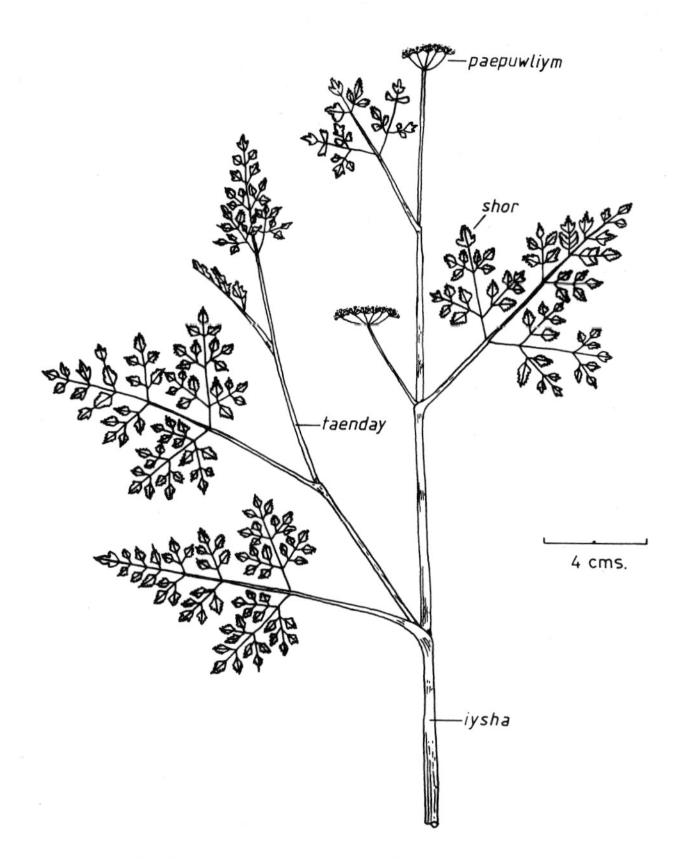

Fig. 13: Javanese dropwort (*Oenanthe javanica*)

holes. This herb grows well in most situations, although as the name of the wild type suggests, it prefers damp places. It may be found growing in any type of garden. In large sweet potato gardens, both newly-cleared and established ones, it is usually planted around the edge near the fence or as a marker along any internal women's boundaries; another favourite spot is around tree stumps and rocks where other crops do not grow well. Women plant *taziy* at the same time as Highland *pitpit*, following the main sweet potato crop.

When established this tall crop requires little husbandry, competing successfully with weeds and grasses. It takes about six months to develop into a robust plant, from which women may then pluck off leaves if they do not pull up the entire thing. If they pick off leaves they may continue to harvest from a plant for a year or more. The leaves, stem and flowers are all edible. They may be eaten raw, or cooked in an earth oven, a bamboo tube or tin, or tied in a leaf parcel and baked in the ashes. The latter is a popular way of cooking it with meat, with which it is commonly eaten, together with salt.

Overall though, this plant contributes little to the Wola diet, figuring only occasionally in meals, often mixed up and cooked with other green vegetables.

CABBAGE

Brassica oleracea var. *capitata* (family: Cruciferae)
Wola name: *cobaj* (No. of cultivars = 5)

This vegetable originated in the Mediterranean, and Europeans have only recently introduced it in the Wola area.

This familiar biennial plant of temperate regions has a short stout stem from which grows a more or less dense 'head' of thick overlapping leaves. These may vary in colour from green to red, may be smooth or wrinkled, and the compact head may be pointed or round in shape. The stem has a stout taproot and the plant is hairless. In the first year of growth the arrested terminal bud develops into a swollen head, and in the second year the plant flowers and produces seeds. The lemon-yellow flowers grow in clusters from branching stems and have long-clawed petals. When pollinated by insects they produce cylindrical pods which taper to short beaks and contain spherical seeds.

There are many cabbage cultivars recognised in temperate climes, but the Wola recognise only five, which they distinguish on the basis of the size, shape and colour of plants (see Table 11).

Although cabbages grow reasonably well in high-altitude tropical regions, they do not fare well in the Wola area, possibly because the soils do not suit them (they prefer rich sandy loams). The region's unsuitability results in the growth of many plants without compact heads, indeed the Wola recognise two cultivars with tall stems which never develop heads. The Wola call such bolted cabbages *cobaj sinjay*, and they say that some plants not only 'grow wrongly' like this but also flower and seed before growing many leaves. Plants that grow properly and develop dense heads of leaves they call *dwem bay*.

Both men and women plant cabbages. They propagate them vegetatively from the small lateral buds which develop on a plant's stem when they pick off its terminal head to eat, except for the '*caebij*' and '*shondoltiy cobaj*' cultivars which develop branching stems instead. The Wola call these lateral buds *taenday* or *doliymba*, and they leave cabbage stems in the ground after harvesting their heads to allow them to develop. When planting, they simply break these buds (which they also eat) off and firm them into dibbled holes.

Cabbages will only grow in well-drained and crumbly soil. This rules out waterlogged taro gardens, but otherwise they can be planted in any garden, even in established sweet potato ones which have been planted a second or subsequent time. They are put in at the same time as maize and beans, which is usually after Highland *pitpit*, although it is sometimes simultaneously. The

Table 11: Cabbage cultivars

CULTIVAR NAME	HEAD DEVELOP-MENT	LEAF STALK	LEAF SHAPE	STEM SIZE	STEM AND STALKS
'boluw cobaj'	Yes	no stalk — all leaf	round	short and stout	pale green
'caebij'	No	patchy flange	oval	tall and spindly	green and maroon
'cobaj'	Yes	flange	oval	short and stout	pale green
'poroma cobaj'	Yes	no flange	round	tall and stout	pale green
'shondoltiy cobaj'	No	patchy flange	round	tall and stout	green and maroon

Wola plant cabbages anywhere in a garden and favourite spots are where garden refuse has been burned. Sometimes women make *paen* beds at these places, and on recently abandoned house sites, which they plant with cabbages and other green vegetables. The only problem with planting these crops close together is that it encourages a small white butterfly,[4] the larvae of which are a pest that eat the leaves not only of cabbage but also of *paluw* and *taguwt* greens. In an attempt to discourage this pest, which the Wola call *ed*, women spread tree fern[5] leaves over these greens, which they say kills the grub.

Other than this, cabbages require little husbandry while growing, although beds of greens and mixed vegetables, like small mixed vegetable gardens near houses, are weeded until the plants are well established. Cabbages take from four to eight months to grow and are harvested ideally after a dense head of leaves has developed. The Wola usually cook cabbages in earth ovens or boil them, chopped up, in bamboo tubes or tins. They figure in the Wola diet to about the same extent as some of the spinaches.

CHINESE CABBAGE

Brassica chinensis (family: Cruciferae)

Wola name: *kwa* (No. of cultivars = 2)

This crop domesticated and grown widely in East Asia, only arrived in the Wola region in post-contact times.

It is a biennial herb cultivated as an annual. Unlike the cabbages described above, it does not grow a compact head on a short stem. It sends up a taller, more spindly stem along the length of which leaves grow loosely overlapping one another, with the lower ones on the outside. The leaves are elongated and

[4] Possibly *Ascia monuste* or *Pieris* sp.

[5] Of the variety the Wola call *henk* (*Cyathea magna*).

oval with coarsely toothed margins. They are thinner than cabbage leaves, more like those of lettuce. They are green, while their stalks and prominent central ribs are pale green, almost white and translucent. It is a hairless, shiny plant. The spindly top of the stem grows above the leaves, and produces branches which bear small clusters of pale yellow flowers. When pollinated these develop into pale green cylindrical pods with pointed ends.

The Wola differentiate between two Chinese cabbage cultivars which they call '*kwa*' and '*paekiyba*'. The size and shape of their leaves, flowers and seeds are the same, and they distinguish between them by leaf colour and texture. The leaves of '*kwa*' are thicker, and dark green with white stalks and central rib; while those of '*paekiyba*' are thin, more lettuce-like, and paler green with a yellowish tinge (the leaves of poor plants of this cultivar turn entirely pale yellow), they also have pale green stalks and ribs.

Only men plant this crop, which they propagate from seeds. To obtain seed, they allow plants to die back, as with other greens, and wait for the seed pods to turn brown and start to dry out, then they pick them and dry the plants thoroughly on the fireguard indoors, where they store them until required. Chinese cabbage, together with the other seed-propagated greens, is one of the last crops to be planted in a garden. Men plant it in newly-cleared gardens largely; it does not grow well in established ones unless they have especially fertile pockets. They often plant it, like the spinaches, in mixed vegetable gardens near houses and on recently-abandoned house sites. And they plant it in the same broadcast way as these other greens (*taentaen bay*) sowing it anywhere in a garden.

It is a quick-growing vegetable, ready before the staples approach maturity. Indeed it is the fastest-growing crop cultivated by the Wola and is usually the first harvested from a newly-planted garden, being ready for picking within one and a half months (it produces seeds and is ready for pulling and drying within only four months). Women usually pick off the larger leaves and leave the stem rooted to continue growing for harvesting from again later, as they do with other greens. When they pluck the growing tip this induces branching and it is usual to see Chinese cabbage with several *taenday*.

The Wola steam the leaves of Chinese cabbage in earth ovens, and boil them in bamboo tubes or tins. This rapidly growing and prolific green vegetable only features significantly in a family's diet, though, when it is harvesting from a newly-cleared garden, which is not regularly.

ONION

Allium cepa var. *aggregatum* (family: Alliaceae)
Wola name: *^{e}nyun* (No. of cultivars = 1)

This crop was probably domesticated in Central Asia in antiquity. It is a new crop for the Wola, though, only reaching them recently since contact.

A biennial plant, usually grown as an annual, it gives off a characteristic onion odour when any part is crushed. It has a dense mat of thin fibrous roots growing like hairs from the bottom of an attenuated and condensed stem which grows largely below the soil. This root system is shallow. The leaves are hollow cylindrical blades which are dark green and terminate in points. The plant produces a succession of alternate leaves, each new one growing up inside the previous sheath and then bursting through it.

When the plant is established and has grown a number of leaves, the leaf bases around and immediately above the condensed subterranean stem become swollen with food reserves and develop a bulb which the Wola call the *iyla*. The bases of the outer leaves remain as thin dry membranes, which serve to protect the fleshy inner ones of the bulb. Onions are long-day plants requiring sunlight for considerable periods to develop bulbs. The Wola region, though, being near the equator scarcely gives this critical amount of daylight. This hinders bulb formation so that onions cultivated here develop only small bulbs, never larger than the size of robust spring onions. Under this short photo-period they will grow leaves indefinitely without ever really producing bulbs.

Once a plant produces a bulb it ceases growing leaves and becomes reproductive. It produces a ball of greenish-white flowerlets on a hollow stalk resembling a leaf. Initially this ball is enclosed in a membranous spathe which splits open later. The fruits are globular capsules containing small irregular black seeds. However, few onion plants flower and seed in the Wola region. In order to do so, onions require a cold spell, the likes of which only occur irregularly here. This, combined with the retarded bulb formation and picking of plants before they commence reproduction, explains why few plants ever flower.

Although there are many onion cultivars recognised in temperate regions, the Wola distinguish only one. They say though that there are two noticeably different types of plant, one with broad and the other with thin leaves, but they do not give them different names.

Women plant onions mainly, although men do so on occasion. They propagate them vegetatively from small lateral bulbs. The variety of onion grown by the Wola multiplies freely, producing many lateral shoots and clusters of small bulbs, which women split up and plant singly by firming into shallow dibble holes. These sets then develop clumps themselves. Onions like a soft damp soil but not one which is too wet, hence they are planted in newly-cleared and established sweet potato gardens, mixed vegetable gardens and on old house sites. They may be planted anywhere in a garden, often in mounds among sweet potato or in beds of mixed greens. They are put in at the same time as beans and cabbage.

Onion plants are rarely ready for cropping before six months, and then the Wola only harvest from them a few bulbs at a time, doing so for a year or more and earthing up the plant each time they tamper with it. They eat the entire plant, leaves and bulbs. They sometimes eat them raw, although they find them *krai* (lit: hot-tasting) like this and prefer them cooked. They cook onions by steaming them in earth ovens, boiling in bamboo tubes or tins, roasting gently over embers or baking briefly in ashes. Although they are prolific and available nearly all the time, the Wola do not like onions much and they only feature marginally in their diet.

SPIDERWORT

Commelina diffusa (family: Commelinaceae)
Wola name: *hombiyhaem; kondow* (No. of cultivars = 1)

This is a widely-distributed tropical weed, the domestication and cultivation of which probably occurred independently in different places, including New Guinea. The Wola say that they and their ancestors have always grown it.

It is a perennial herb with a creeping and straggling stem, composed of sections jointed at prominent nodes. The leaves occur alternately on either side of the stem at these nodes, and are attached, not by stalks, but by sheaths which enclose the stem. The leaves are oval and spear-shaped, taper

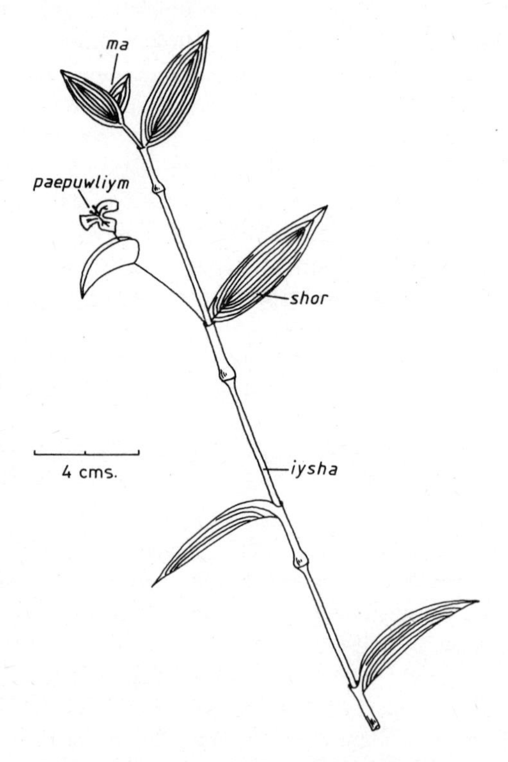

Fig. 14: Spiderwort (*Commelina diffusa*)

to a point and have ribs that follow their ovoid outline. The flowers emerge from boat-like leaves or bracts which grow on the end of slender stalks projecting from the leaf sheaths. They have three petals, are delicate and soon wither. The fruit which develops from them is a capsule having a cavity which splits open.

The Wola recognise two kinds of spiderwort. One is wild, and the other cultivated (although it grows wild too, particularly in abandoned gardens). They do not give these plants separate names, but they can distinguish between them by their size and colour. The cultivated one has a thicker, greenish white stem and larger leaves. The wild one is a smaller plant with a spindly stem blotched pale green and maroon.

Women plant this crop, which they propagate vegetatively from stem cuttings 15 cm or so long, taken from mature plants and simply pushed into the soil. They may plant it in any type of garden, usually around the edge near the fence, where it is left untended. It can compete successfully against weeds with its spreading and swiftly growing grass-like stems, such that it requires no attention and can even go on flourishing after the garden's abandonment. They plant it at the same time as the other greens propagated from cuttings like *shombay*.

This plant may be cropped throughout and after the life of a garden. The Wola eat the new shoots and leaves, plucking off the first 3 cm or so of the growing tip, which they call the *ma*. They usually cook spiderwort in an earth oven or bamboo tube and it is a favourite accompaniment of pork, the grease of which it soaks up during cooking. Overall though it is an insignificant plant which is eaten rarely and is associated more with times of hunger than normal everyday meals.

HIBISCUS SPINACH

Hibiscus manihot

Wola names: *huwshiy; ol shombay; hezarat; hwiyziyhobil* (family: Malvaceae) (No. of cultivars = 5)

A native of China, this plant reached Melanesia sometime in antiquity. The Wola consider it to be a traditional crop which they and their ancestors have always cultivated.

It is a perennial bushy shrub, which when established develops a woody stem covered with bark. This may be stout, and the bush grow to over two metres in height. It has large, simple, single-bladed leaves that are varyingly lobed, with from five to seven lobes, which on deeply cut leaves are slender and pointed. The leaves resemble the palm of a hand with fingers spread open, and their edges are often coarsely toothed. The bush produces large and showy flowers that grow singly on the end of stalks which emerge from the stem where leaf stalks join it. They are hermaphrodite, have five petals

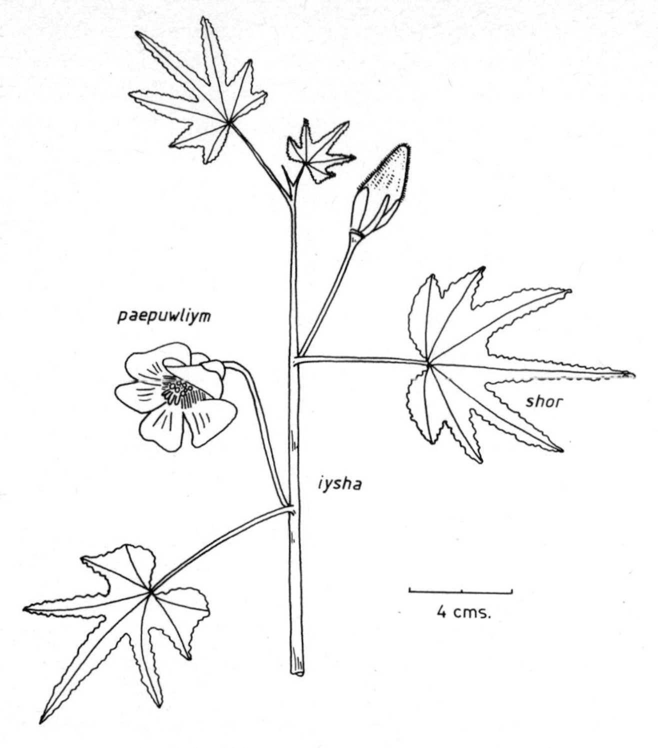

Fig. 15: Hibiscus spinach (*Hibiscus manihot*)

and are bell-shaped, and are yellow with a crimson centre. The fruit is a dry capsule. The altitude and climate of the Wola region inhibit flowering and fruiting, and people only know that hibiscus do this from what they have seen or heard at lower altitudes towards Lake Kutubu.

The Wola distinguish five cultivars, largely on the basis of stem and stalk colour and leaf shape (see Table 12).

Only men can plant hibiscus; in fact one of the names for this is *ol shombay*, literally meaning 'man *shombay*', which contrasts it with the all-female acanth spinach *Rungia klossii*, which is called *ten shombay* meaning 'woman *shombay*'. They propagate it vegetatively by taking a cutting from a mature bush, which must be woody or it will rot. They trim off all the soft new stem and leaves so that the cutting, called *huwshiy way* (lit: hibiscus spinach cutting), looks like a piece of wood about 25 cm long. They simply firm one or two such lengths upright into a dibbled hole.

Men plant hibiscus in newly-cleared gardens only, and not in established ones. They cultivate this bush on the edge of sweet potato gardens, dotted about small mixed vegetable gardens, and on old house site beds. It requires a good depth of well-drained fertile soil, and a good place, the Wola say, is under a casuarina tree, or where one stood recently, because this tree's rotting

Table 12: Hibiscus spinach cultivars

CULTIVAR NAME	COLOUR OF NEW STEM AND STALKS	LEAF SHAPE*	LEAF SERRATIONS	REMARKS
'kwiydol'	green and maroon	deep average lobes (I)	Yes	
'shumbuwhond'	green and maroon	deep slender lobes (I)	No	
'siyshwa'	green	deep very slender lobes (I)	No	Lowest bush (<1 m) and fastest growing
'waenuwpaen'	green and maroon	shallow lobes (G)	Yes	
'washumbuw'	dark maroon (and small green patches)	deep stout lobes (I)	Yes	From Kutubu

* The letters in this column refer to Fig. 4 and indicate the sweet potato leaf shapes which these hibiscus leaves must closely resemble; the leaf drawn is that of *'kwiydol'*.

needles give an excellent soil. A shady position does not, according to the Wola, inhibit this bush's growth, it simply reduces the depth of the leaf lobes, which cut deeper in direct sunlight. When planting a new garden, they put hibiscus cuttings in at the same time as beans and cabbage. While growing the leaves are especially susceptible to insect attack, and are often badly eaten.

As a perennial bush, hibiscus may be harvested from over a period of years. Sometimes one or two bushes near a house mark all that remains of an earlier mixed vegetable garden, and where there are a number of such bushes after all the other crops have been consumed, men on occasion maintain a rough enclosure, giving them a little hibiscus spinach garden. They pick both leaves and soft young shoots, and steam them in earth ovens, boil them in bamboo tubes or tins, and bake them wrapped in leaf parcels with meat in the ashes of fires. When cooked, and particularly when boiled, they are soft and very slimy. Considering that hibiscus leaves are available all year, they are not as popular as might be expected, figuring to only a small extent in the Wola diet.

FIG

Ficus wassa (family: Moraceae)

Wola names: *poiz; tuluwp* (No. of cultivars = 1)

According to the Wola, this tree has grown in their region since time immemorial and has always been a source of food for them. It is probable that its domestication took place in Melanesia long ago.

This tree may grow up to 20 m high, although usually it is only 5–8 m, similar to a temperate fruit tree, which it also resembles in shape, with its rounded and bushy canopy. Its leaves are leathery and rough to touch, sometimes having fine white hairs on stalk and midrib. They are elliptic to egg-shaped with stalk attached at the narrow end, the other tapering to a point. Sometimes they are toothed. The leaves of saplings are considerably larger (up to 26 by 10 cm) than those of mature trees and they are markedly lobed,

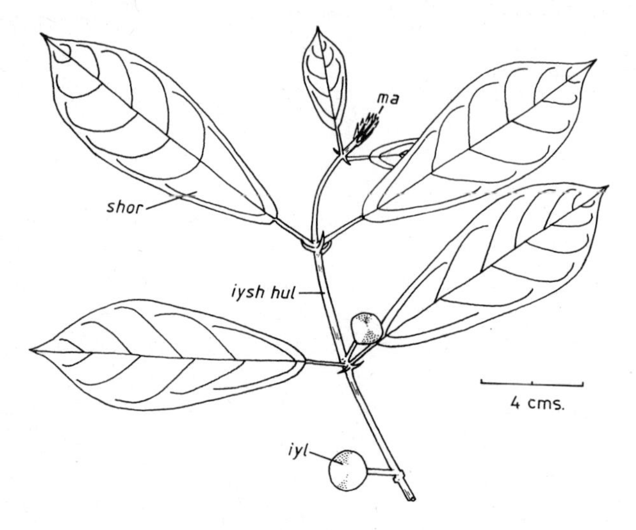

Fig. 16: Fig (*Ficus wassa*)

looking as if insect pests have eaten lumps out of their margins. There are pairs of spear-shaped leaf-like appendages (stipules) where the leaf stalks join the twig, as there are with all figs. The leaves grow either in a spiral round the twig or in more or less unequal pairs alternately at right angles to it.

The characteristic feature of figs is the development of flowers on the inner surface of hollow fleshy receptacles that have an orifice in them closed by small scales; they contain both male and female flowers. On the *poiz* tree these figs grow from where leaf stalks join twigs, from branches and from the old wood of the trunk. They have a somewhat spherical body, are rough to touch like the leaves, and ripen to a white, yellow, pink or purple colour. They grow on short stalks and at their apex, opposite the stalk, they have a small orifice (1 to 1.5 mm wide) closed by several small scales. The hollow inside has no bristles.[6]

The Wola recognise only one cultivar of *poiz*, although they say that at lower altitudes around Lake Kutubu several others occur. They maintain that there is no wild variety of this tree growing in the forest. It does occur

[6] F. wassa is very similar to *F. copiosa*, which is also cultivated in the Highlands of New Guinea, and is distinguished from it by its smaller leaves and figs — see Corner (1967). It requires an expert to distinguish positively between them, and I am grateful to Professor Corner for identifying *poiz*.

semi-wild though in areas of secondary growth, but always as a result of human activity, either having been planted in an old garden or near a one-time house, or occasionally establishing itself as a seedling from a nearby cultivated tree.

Only men plant this tree. They either break a piece of branch off a mature tree and push this cutting into the soil, or they find a small self-sown seedling and transplant it to their garden or house-yard. A man may cultivate *F. wassa* on the edge of any garden, or on the verge of a houseyard. In later life, if he does not occupy the site continuously, it will, as mentioned, stand in an area of regrowth. Saplings, as well as mature trees, yield edible leaves and figs.

This tree grows for years, often beyond the lifetime of the man who plants it. While it stands in his producing garden or near his occupied house, then only members of his household may freely harvest from it, whereas older trees standing on abandoned sites are more like semi-wild resources, which anyone living in the area may exploit, although they remain the property of the planter or his descendants, whose permission ought to be sought before harvesting considerable amounts of leaves and figs. These are not a popular food though (featuring rarely in the diet), hence people's *laissez-faire* attitude to harvesting from mature trees on abandoned sites, which contrasts with their jealous guarding of rights to the nuts of screw-pines. The Wola cook both tender leaves and figs in earth ovens, bamboo tubes or tins, and leaf parcels baked in the ashes. They also eat the figs raw.

HIGHLAND BREADFRUIT

Ficus dammaropsis (family: Moraceae)

Wola name: *shuwat* (No. of cultivars = 1)

According to the Wola this tree, like its fig relative above, is a traditional source of food which they and their ancestors have always exploited. It is probable that its domestication also took place in Melanesia sometime in the distant past.

Highland breadfruit grows up to 18 m high, although it is usually considerably shorter at 5 to 6 m. It has a greyish smooth bark and five or six branches, which together with its large leaves, results in a less bushy canopy than *poiz*. When any part of it is cut or broken it exudes a copious sticky white sap. A characteristic feature of this tree is its large leaves (up to 70 by 60 cm) which have a glossy green top surface and yellowish green underside. They also have a distinctly waved margin which gives them a crinkly appearance. They are egg-shaped with, for their size, short stubby stalks growing from their broad end. The leaves are spirally arranged round twigs, or else grow opposite one another.

This tree also produces characteristic figs which superficially resemble the

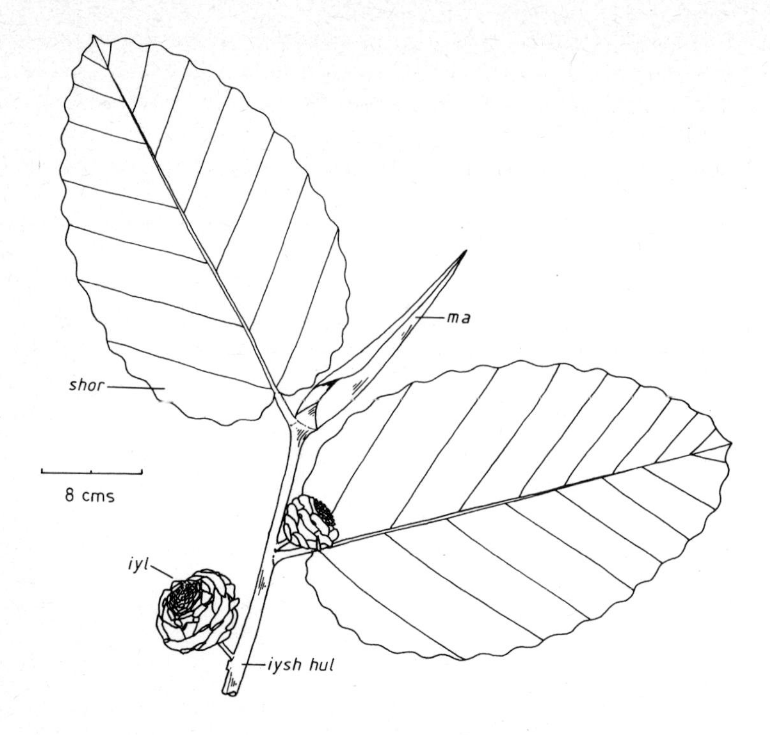

Fig. 17: Highland breadfruit (*Ficus dammaropsis*)

fruits of the breadfruit (another tree in the Moraceae family),[7] hence the popular name of Highland breadfruit. Another common name for this tree is Highland cabbage, again a reference to the fig, which when immature and green resembles a small cabbage. The figs (up to 13 cm wide) grow on robust stalks which emerge from where leaf stalks join the twig. They are somewhat spherical in shape, and covered by broad scales which diminish in size towards the apex where they cover the orifice. When ripe the fig is a reddish colour and the flowers inside are bottle-shaped.

The Wola distinguish only one cultivar, which they can identify from wild *shuwat* trees. Both cultivated and wild trees have the same name and are identical in every way, except that the new leaves and stalks of the former have fine hairs on their underside.

Only men plant these trees, which they propagate solely by transplanting self-sown seedlings. They cultivate them on the edges of gardens and house-yards like *poiz*, and the rights covering their exploitation are the same as those described above too. They also grow and may be harvested from for many years, like the above. The Wola eat only the tender new leaves, of both cultivated and wild *shuwat* trees. They are not a popular source of food

[7] See Massal and Barrau (1956: 18—21).

though, and are associated more with periods of food shortage and famine than with normal everyday meals. They gain in importance to the north-west towards the Huli people with whom they are a more popular crop. On the rare occasions when the Wola eat Highland breadfruit leaves they usually cook them in earth ovens, although they also boil them chopped up in bamboo tubes or tins.[8]

[8] For comparative material relating to the green vegetables discussed in this chapter see Fischer (1968: 271—2), Martin and Ruperte (1975), Massal and Barrau (1956: 34—6), and Powell *et al.* (1975: 28—30).

Section I: Crops described

Chapter 4
SHOOTS AND STEMS

HIGHLAND *PITPIT*

Setaria palmifolia (family: Graminae)
Wola names: *deken; kot; pombiy* (No. of cultivars = 9)

This vigorous herb is found throughout Asia and the Pacific, although rarely is it cultivated outside Melanesia, which suggests it was possibly domesticated here (where today both wild and cultivated forms occur together). According to the Wola anyway, they and their ancestors have always grown Highland *pitpit*.

This robust perennial grass grows in thick clumps. It has a stout stem composed of overlapping leaf sheaths which the Wola call *suwpuw*, and in the centre it has a thickened young edible shoot. Where the leaf sheaths unite there is a fibrous knob called the *iyla*. The leaves or *shor* are broad blades, spear-shaped and tapering to long points. A distinguishing feature of these green leaves is their prominent corrugation. They grow alternately opposite one another, first from one side of the plant and then from the other.

Cultivated forms of this plant do not flower in the Wola region, although they do at lower altitudes towards Lake Kutubu. When it does flower, Highland *pitpit* sends up from the centre of its stem a terminal grass inflorescence which may be 30 cm or more long. This is branched, each branch producing clusters of grass ears, some of which have hairs growing from them.[1]

The Wola distinguish nine *kot* cultivars, largely by the colouring of plants, and also by size and hairiness (see Table 13).[2] They also distinguish one wild form, which is a spindly plant, called '*dikiytagot*'. They consider the largest cultivar, '*tabiy*' to be the best and the second largest, '*nduwm*', to be the next

[1] For further botanical details on *Setaria palmifolia*, see Henty (1969).
[2] See Fischer (1968: 268–9) for comparative data on the Highland *pitpit* cultivars recognised by the Kukukuku.

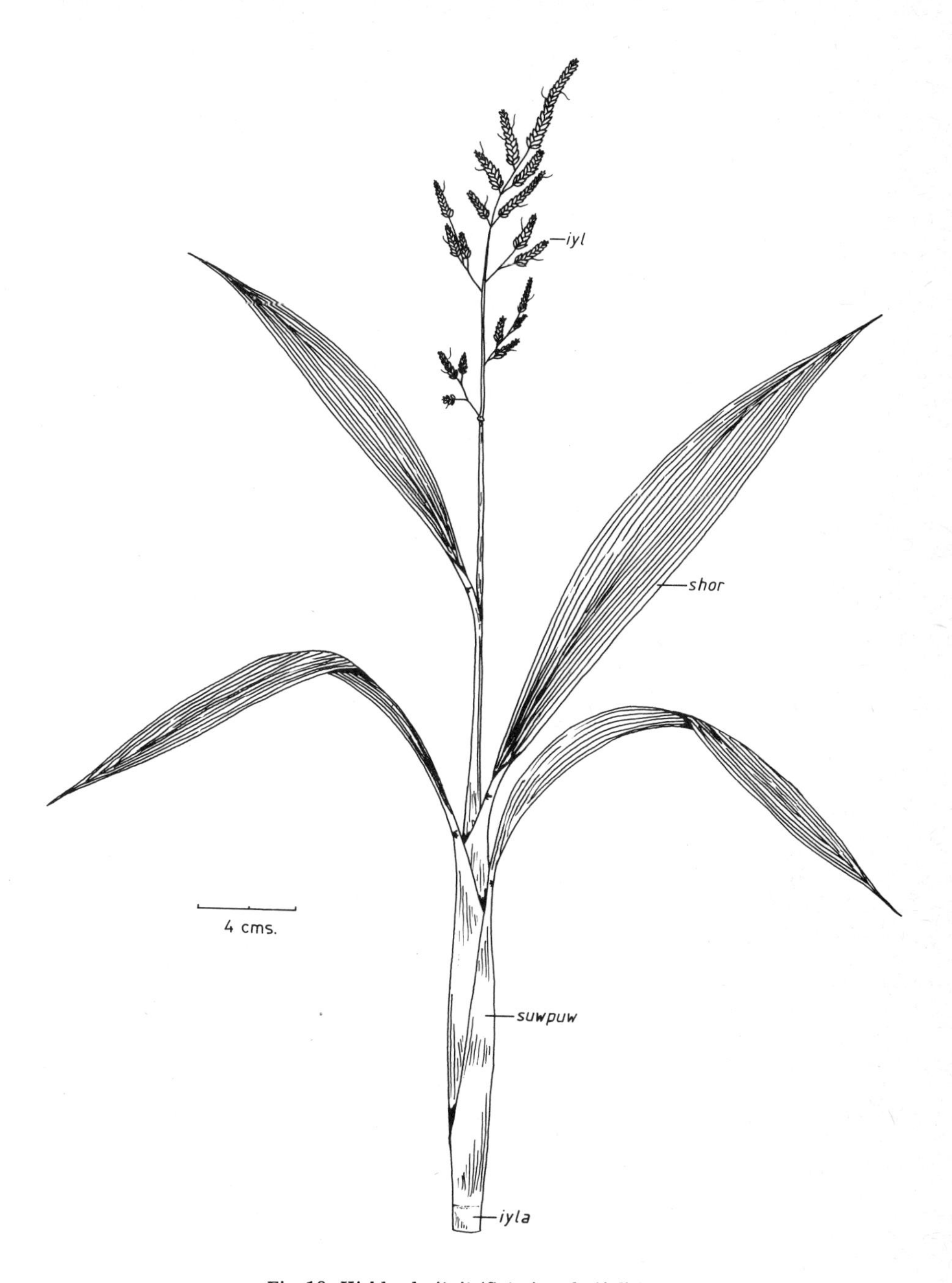

Fig. 18: Highland *pitpit* (*Setaria palmifolia*)

Table 13: Highland *pitpit* cultivars

CULTIVAR NAME	COLOUR OF LEAF SHEATHS				COLOUR OF LEAF		HAIRS	STEM SIZE
	OUTER	INNER	WHERE JOINS LEAF	MARGINS	MARGINS	MIDRIBS		
'*Augiyba*'	maroon	pale green and maroon streaks	maroon	maroon	green	white; pale maroon near sheath	Yes	4th largest
'*G^{e}mb*'	maroon and green streaks	pale green	maroon	green	green	white	Yes	3rd largest
'*Haez*'	very pale green (pale maroon base)	white	very pale	green	green	white	No	middling
'*Mobiyael*'	pale green	white	pale green	green	green	white	Yes	middling
'*Nduwm*'	green	pale green	green	green	green	pale green	Yes	2nd largest
'*Ol*'	green	pale green	green	green	green	pale green	No	tall and thin
'*Tabiy*'	maroon	pale green and maroon streaks	maroon	maroon	green	maroon	No	largest
'*Tenaisiybil*'	maroon and green streaks	pale green	maroon	maroon	maroon	white; pale maroon near sheath	No	middling
'*Tindiy*'	maroon	pale green and maroon streaks	maroon	maroon	green	pale maroon	Some-times	smallest

best. This ranking depends on size alone, all cultivars having the same flavour, except perhaps for '*tabiy*' which is a little more moist than the others (for example, when fed cooked pieces of different unknown cultivars, people were unable to distinguish between them).

Only women plant Highland *pitpit*. They do so by breaking off leafy lateral shoots called *shoba* or *taenday* from mature plants, which they firm singly if large, or in pairs if small, into holes dibbled with the pointed end of their digging sticks. Although it cannot tolerate conditions that are too wet, this plant likes damp situations and thrives in shady ones. It is adaptable and will grow well in a large range of soils. Hence it may be cultivated in any kind of garden except a waterlogged one. In large sweet potato gardens, although they may plant it anywhere, women commonly put it around the edges and along internal boundaries between their areas. They also regularly cultivate this crop in established gardens worked for a second or consecutive time, by either planting new leafy shoots as described (called *kot way orkay*) or banking up the soil around established clumps (called *koluw hum bay*) so giving them a new lease of life. Highland *pitpit* is usually the second major crop they plant, after sweet potato, although they may put in some other lesser crops at the same time.

Once planted, *kot* requires little husbandry. It takes about six months to grow sufficiently for harvesting, from which time shoots may be taken at intervals for two years or longer. To harvest from plants is *kot tiyay*. Sometimes women even return to harvest shoots from gardens some years after their abandonment. These are spindly, such cultivated plants eventually disappearing under secondary regrowth. They never though, according to the Wola, degenerate into wild '*dikiytagot*' plants, which are altogether separate and self-propagating.

The Wola eat the thickened young shoots or stem hearts of Highland *pitpit* as a vegetable. To obtain these they tear off the plant's leaves (*kot pongor payay*) and then peel off its outer leaf sheaths (*suwpuw daebay*), which they commonly feed to pigs. They eat the soft shoots both raw and cooked. Unshucked they steam them in earth ovens or bake them lightly in ashes, and shucked they boil them in bamboo tubes or today tin cans. The Wola eat *kot* frequently and it figures significantly in their diet.

BAMBOO

Nastus elatus (family: Gramineae)

Wola name: *taembok* (No. of cultivars = 1)

Bamboos are widely cultivated throughout the tropics and probably originated somewhere in Asia. According to the Wola theirs is an indigenous crop which they and their ancestors have always grown.

This bamboo is a woody perennial which grows in clumps. Its root system

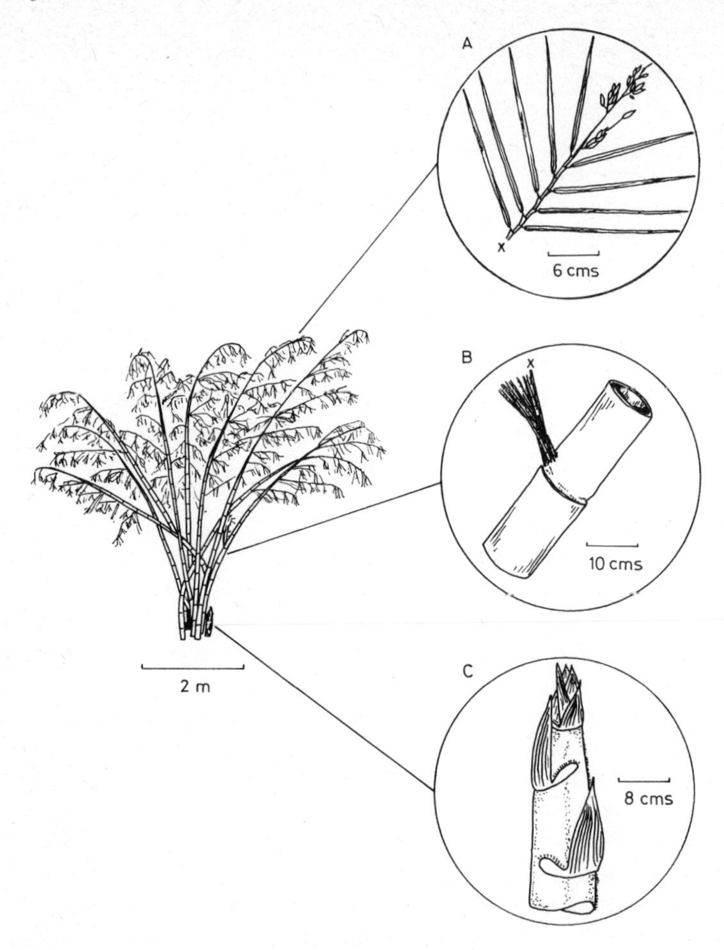

Fig. 19: Bamboo (*Nastus elatus*)

centres on a rhizome which develops laterally below the soil and from the top of which sucker shoots push up at intervals. The plant expands, so giving rise to a clump of canes, as these new rhizomes and stems develop from buds at the base. The young shoots are massive, solid and packed with food. They are bullet-shaped and protected by overlapping rigid maroon sheaths that are modified leaves which drop away as growth proceeds (see circle C on Fig. 19).

The stem comprises several hollow sections joined by nodes having strong cross walls, and consists of a very hard wood with a glossy green outside surface. These hollow stems stand erect and are somewhat curved which, with the plant's hanging leaves, gives a drooping appearance. The nodes at the base of the stem are closer together and they sometimes produce a ring of roots, where higher nodes have a sheath subtending a bud or branch spray. Only mature stems which have reached full height grow, on their upper parts, branches from their lateral buds. These consist of sprays of thin grass-like stems bearing leaves (see circle B on Fig. 19, the X indicating that the stem continues from this break in circle A).

The leaves are long, narrow and lance-shaped with tapering pointed ends. They have short stalks which join sheaths that attach them to their thin branches, along which they grow alternately opposite one another (see circle A). According to the Wola these plants never flower nor fruit. Indeed bamboos seldom flower in any region of the world and when they do, some die. And there is no record of them producing fruits anywhere, although *Nastus* do produce small flowers enclosed in spikelets that resemble the ears of grass seeds. So this bamboo is one of those plants that apparently exists in a steady vegetative state, depending on man for its propagation and movement to new regions, and consequently known only in cultivation.[3]

In the Bambuseae tribe, the most primitive in the Gramineae family, there are many genera, species and cultivars, but the Wola cultivate only one kind of plant in their region.[4] They are aware of a second cultivar, which they call '*tegelab*', occurring at lower and warmer altitudes towards Lake Kutubu. The wood of this cultivar's stems they say is exceedingly hard and thick compared with that of '*taembok*'.

Only men plant bamboo. They may cultivate it in any garden, although they prefer ones near their homesteads. Another popular place is on the edge of houseyards, where again it is readily accessible. This is a significant consideration because the hard wood of bamboo, which gives a razor-sharp edge, is in regular demand for making numerous artefacts. Men propagate bamboo vegetatively by digging up a tuft from the base of a mature plant, of spindly grass-like stems, together with their roots, which they transplant. The stout grass-like stems of this immature clump die back as it sends up new shoots. The first of these new shoots are small, developing into canes no longer than *Miscanthus* grass, but they gradually increase in size until a mature clump of bamboo is established. Men may transplant bamboo at any time, not necessarily when their women-folk are planting a garden with other crops. They locate it on the edge of gardens, never in the centre, because of the shade it casts.

Once established, a bamboo can be harvested from for many years. It can withstand severe gathering, not only of edible shoots but also of mature stems for making such objects as knives, containers, arrows and so on. The Wola dig up new shoots when they are 50 cm or so long. They peel off their layers of protecting sheaths to reveal their creamy coloured tender hearts, which they may eat raw or cooked; either steamed in an earth oven or baked in ashes. Bamboo shoots are something of a delicacy not eaten very often. Although several clumps of bamboo may be seen in any area, they do not

[3] For further information on the botany of bamboo, see Holttum (1967).

[4] See Fischer (1968: 285) for comparative data on the bamboo cultivars distinguished by the Kukukuku and the uses to which they put them.

yield many shoots and when a man sees one on his plant he may carefully watch its development and warn off poachers in his family.

SUGAR CANE

Saccharum officinarum (family: Gramineae)
Wola name: *wol* (No. of cultivars = 12)

According to the Wola, they and their ancestors have always cultivated this crop, and in this case their assertions are probably literally true, as botanists think that sugar cane was first domesticated and cultivated in New Guinea in ancient times from the closely-related cane species *Saccharum robustum*,[5] spreading from here throughout Oceania and on to Asia.

Sugar cane is a large perennial herb of the tropics, cultivated for its thick stems which contain a sugary juice that is its store of carbohydrate held as sucrose. The cane stem or *wol iysha* is solid, unbranched and erect, and may grow up to 6 m high and 6 cm in diameter. It consists of short sections, which the Wola call *el*, separated by distinct nodes or joints, which they call *lor*, the same word as they use for knots in wood. These internodes are shortest at the bottom of the cane. When sharing out sugar cane men usually ensure a fair distribution by counting the number of internodes each person receives (for example if three of them share a length with twelve sections the one cutting it will say '*Koton el mak*' (lit: Kot's sections four) and so on). The tough hard skin of sugar cane is dull glossy and varies in colour from yellow through various shades of green to pink, red and purple. And sometimes it has a black coating which the Wola call *haeruwk* (lit: soot).

It is from the nodes that roots or *pil* develop. At each node there is a band of root initials, which below the soil produce a widely spread network of roots, and above at the bottom of the cane produce aerial roots which act as props. There is also a lateral bud at each node and on established plants some of these develop into tillers below the soil, from which stems grow to give dense clumps of cane. On the joints above ground the roots and buds remain dormant, as the leaves do below ground where they are represented by rudimentary scales.

The leaves or *shor* grow in two opposite rows along the cane, one growing from each node and attached by a tubular sheath that encloses the cane, which the Wola call the *paziy*. The leaves are long and lance-like in shape, tapering to a finely pointed tip and having fine teeth along their margins. They may be up to 1 m long and 10 cm wide, and have a prominent mid-rib. They grow up from the plant and then bend over so that their fine points hang towards the ground. They are green, becoming yellow as the crop

[5] See Warner (1962), who speculates that this occurred when people who chewed succulent young *S. robustum* stems decided to cultivate sweet ones, so preserving them.

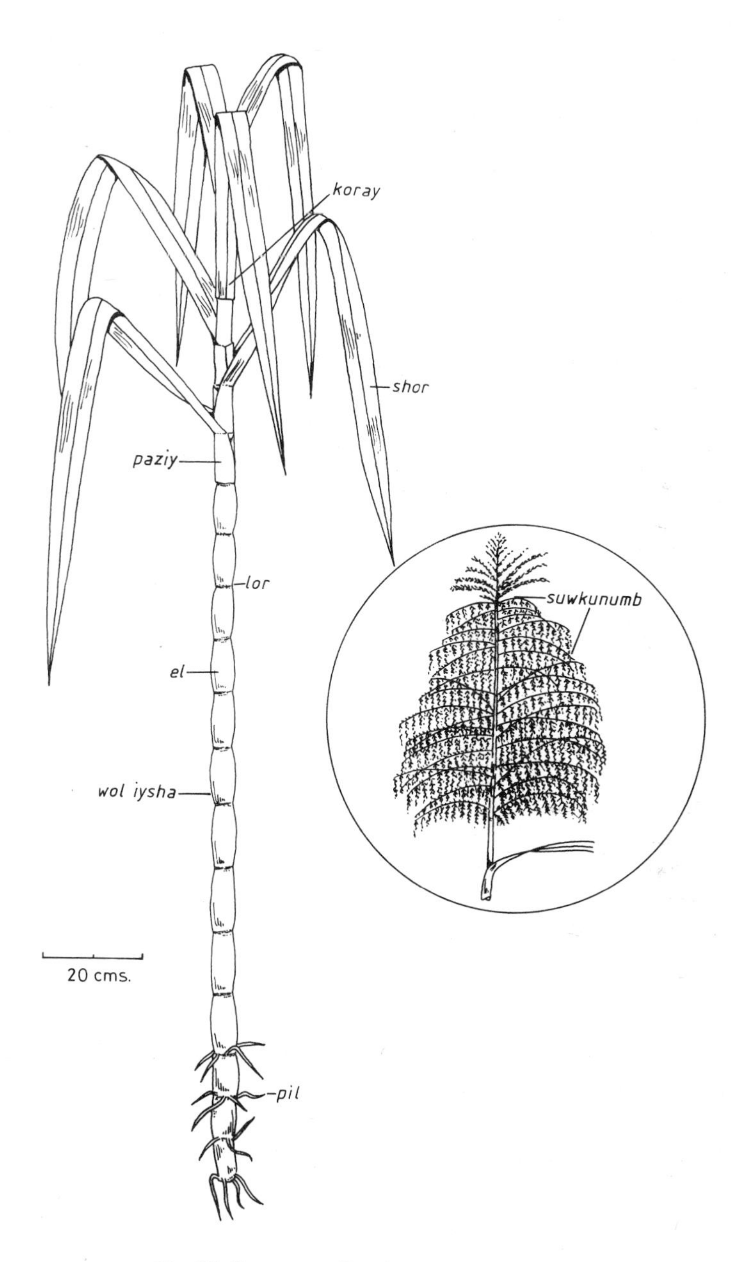

Fig. 20: Sugar cane (*Saccharum officinarum*)

matures, and old ones become brittle and develop dry brown patches before falling to leave a length of naked cane.

Sugar cane produces flowers on a terminal tassle, which is arrow-like in shape. The spikelets making up this tassle occur in pairs (one with a stalk and one without), each enclosed by a sheath containing its florets. They have long silky hairs which give the plume-like tassle an attractive fluffy appearance. Sugar cane is reproductively sensitive to light requiring short days to flower. This is not the reason for the absence of flowering in the Wola region though because plants reproduce nearby at Lake Kutubu where day length is the same. The Wola say that it is higher and cooler where they live, and they are probably correct in pointing to altitude and climate as the factors inhibiting reproduction. The absence of flowering, contrary to expectations, is something of an advantage because flowers consume sugar stored in the cane, so increasing its fibre content, and decreasing its quality.[6]

Throughout the world a vast number of sugar cane cultivars are recognised with a 'bewildering range of local names and astonishing assortment of colours' (Purseglove 1972: 215).[7] The Wola distinguish twelve cultivars, primarily by the colour of the cane and the nature of the leaf sheath (see Table 14). All of them produce the same sized cane, except for '*ar*', which the Wola consider the best cultivar, with taller and larger diameter canes. They say that it grows bigger because it sends up fewer tiller stems, concentrating its energy on a few canes only. The leaves of the various cultivars are also the same size, except for '*homb*' with noticeably wider ones and '*maesep*' with markedly longer ones. The canes all taste the same when chewed too. A point to remember when distinguishing between sugar cane cultivars, as Table 14 indicates, is that the colour of the skin varies depending on whether it is covered up or exposed to direct sunlight. The white powder which covers some leaf sheaths and is diacritical in the identification of some cultivars is called *taeng biy* (lit: ash-like do).[8]

Only men plant sugar cane. They propagate it vegetatively with tops from mature canes, which they cut off together with two or three mature internodes covered with tough skin (the soft-growing top alone would only rot). They trim the leaves off these cuttings near to their sheaths, and then simply push them erectly into the soil leaving their growing tips exposed. The mature nodes send out roots and the cane establishes itself. Men always plant cuttings in pairs, so that, as they say, if one fails to grow the other will compensate. They may cultivate sugar in any garden except a waterlogged one. It is, together with bananas, one of the last crops they plant in a new garden, put in just prior to, or a while after, the sowing of spinach seeds. However, they

[6] For further information on the botany of sugar cane, see Barnes (1964) and Stevenson (1965).

[7] See Grassl (1969).

[8] For comparative ethnographic material, see Fischer (1968: 267) on the sugar cane cultivars identified by the Kukukuku, and Powell *et al.* (1975: 26–7) on those distinguished by the Melpa.

Table 14: Sugar cane cultivars

CULTIVAR NAME	COLOUR OF CANE		LEAF SHEATH				OTHER FEATURES
	COVERED	UNCOVERED	HAIRS	WHITE POWDER	COLOUR	EDGES	
'*Ar*'	pale yellow	brownish yellow	No	Yes	green	green	tallest cultivar with largest diameter cane; grows few tillers
'*Bayow*'	green and yellow streaks	green and yellow streaks (more green)	Yes	Yes	green	brown	reddish pith
'*Boray*'	pale yellow tinged green	pale green tinged pink	No	Yes	green	maroon	
'*Homb*'	greenish yellow	reddish orange	Yes	Yes	green	green	widest leaves; reddish pith
'*Huwlhaeruwk*'	greenish yellow and green streaks	green and pale brown streaks	No	No	yellow and green stripes with maroon flecks	yellow or green	
'*Injiluwmb*'	yellow and green streaks	pale reddish brown	No	No	green	green	
'*Kobal*'	maroon	dark maroon	No	No	green	maroon	
'*Kolbort*'	greenish maroon	dark maroon	Yes	No	maroon and green stripes	maroon or green	
'*Komben*'	green	yellowish brown	Yes	Yes	green	green	
'*Maesep*'	maroon and green stripes	dark maroon	No	Yes	green	green	longest leaves; yellowish pith
'*Saezuwp*'	yellowish green	red	No	No	green	green	reddish pith
'*Tigipshaend*'	pink and maroon stripes	dark maroon	No	Yes	green	maroon	

may plant sugar cane at any time during the life of a garden, not only when it is newly planted; for instance men commonly collect the tops from canes when they are eaten and replant them. They may plant it anywhere in gardens: it is usual to see clumps of cane dotted about all over them. Although canes may be planted in established gardens, they cannot survive in abandoned ones. Sometimes though, together with other slow maturing plants like bananas, they are the only crops remaining in otherwise grass-covered gardens, which will be abandoned once they are eaten.

Once sugar cane establishes itself men support it with long poles, tying the canes together in an erect clump. They refer to this propping up of cane as *wol tongay* (lit: sugar cane tie up). They say that it is necessary to tie up and support cane in this way or else it will hang over and grow poorly. They point out that a sprawling clump of cane would take up more garden than necessary, preventing other crops from growing near it. Also, they say that when tied up cane develops long sections between nodes, untied it has more nodes which means a poorer quality cane for chewing.

It is possible to harvest sugar within a year or so, although it is usual to leave it for two years or more to establish into a sizeable mature clump. Although they may cut and consume cane before, the Wola say that a plant is fully mature when its leaves become smaller and yellowish, a stage which they call *el* (lit: mature, old).[9] The first cultivar from which they harvest is the fastest growing '*bayow*', and the last is '*ar*', which they leave considerably longer than the others to develop its large cane. Men harvest sugar cane piecemeal, cutting one or two canes from a clump at a time and leaving the rest growing.

The Wola think of sugar cane as a source of refreshment rather than as a food, which they may enjoy at any time of day, not only at meal times. It figures significantly in their diet through such snacks. For chewing, they cut the cane into manageable lengths and peel off its tough outside skin with their teeth. They then snap off pieces in their mouths, spitting out the chewed fibre when they have extracted its juice. Pigs sometimes pick up and eat this waste; otherwise, in a house, it dries, and together with peeled skins and other vegetable refuse, makes up the floor covering. Another part of the plant that the Wola eat is the swollen heart of the growing tip, which resembles a large Highland *pitpit* shoot and is prepared in the same way. If they eat this soft heart from the growing tip, which is called the *wol koray*, men cannot, of course, use the top for planting.

[9] They also use this term for mature greens when they seed and for old screw-pines.

Section I: Crops described

Chapter 5
FRUITS

THE PULSES

The Wola group together all their pulse crops into a single family which they call *taeshaeniyl* (lit: bean fruits), and within this category they distinguish recently introduced beans from others by the collective name *towmow taeshaen* (lit: ghost beans). The pulses are the only crops which they bring together into such higher-order named groups.

HYACINTH BEAN

Lablab niger[1] (family: Leguminosae)
Wola name: *sokol* (No. of cultivars = 4)

This crop is of Asian origin, and possibly evolved in India. It reached Melanesia sometime in antiquity. According to the Wola it is one of their traditional crops which they and their ancestors have always cultivated.

This variable perennial herb has a stem which climbs by twining round a support. This is very long, branching higher up, and is round in cross-section with fine hairs. This plant has the typical trifoliate leaf of a bean, that is a compound one composed of three leaflets, one at the top of the leaf stalk and two on the side growing opposite one another. The leaves grow alternately on either side of the stem, on grooved stalks that are somewhat swollen where they join. Their leaflets are a broad egg-shape, the two lateral ones often asymmetrical and lop-sided, and they taper to a short point; small leafy appendages or stipules subtend them.

[1] At previous times called *Lablab purpureus, Lablab vulgaris* and *Dolichos lablab*; an example of how revisions in botanical taxonomy prompt new names which confuse the uninformed.

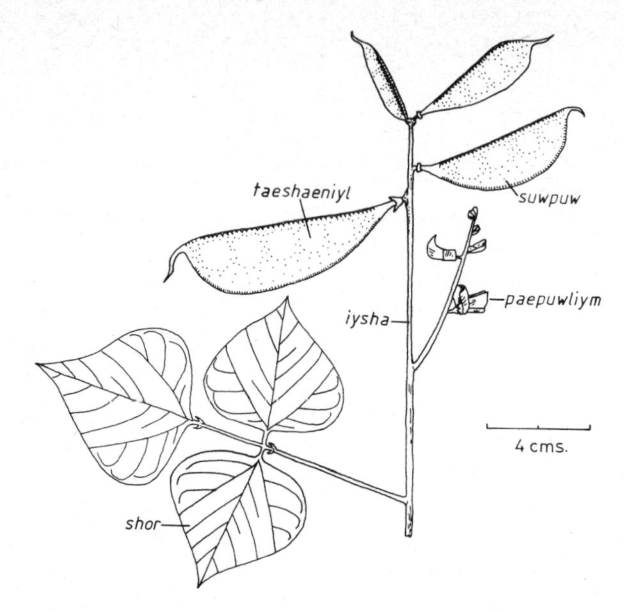

Fig. 21: Hyacinth bean (*Lablab niger*)

Hyacinth beans grow four or five flowers or *paepuwliym* along a single stalk. They are the typical butterfly-like flowers of beans, with two wing and two fused keel petals, and one standard petal fanned out behind. They may be white or purple in colour. They are commonly cross-pollinated by insects, which trigger them with their weight by forcing the keel down and releasing the anthers and stigmas against their abdomens, and they produce stubby, flat, broad pods which terminate in a pronounced beak. The Wola call the pod the *suwpuw*, and the beans inside the *taeshaeniyl*. The pods are green with varying amounts of maroon, and produce up to six flat oval bean seeds which have a prominent raised white scar where they are attached to the pod. The seeds may vary in colour from white to cream through reds to brown and black, and are sometimes mottled; those grown by the Wola, when ready to eat, are very pale green with a maroon tinge along the edge of the scar, and when left on the plant to dry for seed they range from brown to purple.[2]

The Wola distinguish four hyacinth bean cultivars. They say that the stems and leaves of these plants are all the same size and colour, and they distinguish them by the colour and size of their pods and the colour of their flowers and dried seeds (see Table 15).

Both men and women plant these beans. They propagate them from seeds only, which they drop into shallow drill holes. They plant them in pairs to insure against one failing to germinate and grow. To collect seed, they leave

[2] For further botanical information on this and the other pulses discussed here refer to Smartt (1976).

Table 15: Hyacinth bean cultivars (size in cm)

CULTIVAR NAME	POD WIDTH	POD LENGTH	POD COLOUR	FLOWER COLOUR	SEED COLOUR
'*Porolsiziyl*'	2.5	5	green with maroon border	purple	purple
'*Shwimb*'	2.5	5	purple with green spine	purple	purple
'*Sokol*'	2.5	8	green with maroon border	purple	brown
'*Wen*'	2	8	green	white	pale brown

pods on a plant until they and the leaves are dry and brown, and then they pick them and put them on the fireguard indoors to dry out thoroughly and preserve them until required. The Wola plant beans scattered about all over newly-cleared gardens: they flourish in any well drained soil. Although they will tolerate poor soils too, the Wola do not often plant them in established gardens worked for a second or subsequent time, unless they have a pocket of particularly good soil (for example in a fold or at the bottom of a slope). They may also plant them in mixed vegetable gardens near houses, but not often in taro gardens which are usually too damp. They put in beans at the same time as the other drill planted seed crops such as cucumbers, maize and gourds, before they broadcast the seeds of greens.

Once established, hyacinth beans require long poles or pollarded trees up which to climb. Their beans are ready to harvest after three or four months, and they continue, as perennial plants (unlike the other pulses), to yield beans for many months, sometimes for a year or longer until a garden is replanted or abandoned. The Wola say that they bear well during the *ebenjip* season of drier weather, and that when the *torwatorwa* plant[3] of the forest produces its small bean-like seeds *sokol* beans yield their maximum. During the wetter *bulhenjip* season yields fall off, but a plant, if left to the next drier season, will again bear prolifically.

The Wola cook unshucked beans by steaming them in earth ovens or, sometimes, by baking them briefly under hot embers, and shucked ones they simmer gently in bamboo tubes, often together with other food such as greens. Since the arrival of the kidney bean though, the hyacinth bean has declined in popularity because it produces smaller beans. It is not seen as often in gardens as formerly, nor does it figure like it used to in the diet (even though, unlike introduced beans, it produces for long periods of time).

[3] *Desmodium sequax* and *D. repandum.*

COMMON BEAN

Phaseolus vulgaris (climbing) (family: Leguminosae)
Wola name: *taeshaen pebway* (No. of cultivars = 4)

The kidney or common bean was domesticated in south and central America, and is a recent arrival in the Wola region, coming with European contact.

It is a highly polymorphic annual species of plant which varies considerably in its habit, vegetative character and flower colour, and in the size, shape and colour of its pods and seeds. It grows two or three metres high, twining round a support, commonly a pole.

The kidney bean has a large tap root, together with extensive lateral roots in the upper soil. Its stem is slender, more or less square in cross section, twisted, angled and ribbed, and often streaked with purple. Its alternate trifoliate leaves are large and its butterfly-like flowers vary in colour from white through yellow to pink and purple. They are self-pollinated and produce slender pods between 10 and 20 cm long, which are straight or gently curved, terminate in long pointed beaks and hang downwards. They are green, sometimes with splashes of pink or purple, and contain from four to six variably sizes seeds, which are a characteristic kidney shape and vary in colour from white through yellow to red, purple, brown and black; sometimes they are mottled.

The features which clearly distinguish kidney from hyacinth beans are that they are smaller, short-lived and produce long slender pods. There are many hundreds of kidney bean cultivars throughout the world, but the Wola distinguish only four, according to the colour of their flowers and seeds, and the size and colour of their pods (see Table 16).

The cultivation of this crop is exactly the same as that of the hyacinth bean. It yields beans within three months of planting, and two or so months later it dies. The Wola do not pick the beans of the *'pebway pombray'* cultivar

Table 16: Climbing common bean cultivars

CULTIVAR NAME	FLOWER	SEED	POD	POD LENGTH (cm)
'kabhorol'	white	white	green	18
'kigalow'	yellow	white	green	8
'pebway pombray'	purple	purple	green with maroon streaks	14
unnamed*	white tinged purple	mottled white and maroon	green with maroon streaks	12

* The most recent cultivar to arrive, which although recognised as different to the others currently has no name.

until their pods have turned brown and dry; the others they pick when green or turning yellow. When a man has a new garden stocked with beans they will figure prominently for a short while in his family's meals. The remainder of the time though they are eaten rarely, so that overall their contribution to the diet is relatively small.

COMMON BEAN

Phaseolus vulgaris (dwarf) (family: Leguminosae)
Wola name: *taeshaen suwlshaeriy* (No. of cultivars = 4)

The common climbing bean as described above also grows as a dwarf plant, which the Wola classify as a separate crop. These are exactly the same as the climbers, except they grow as low bushy plants between 20 and 60 cm high, which terminate in a flower bearing stalk.

The Wola distinguish four dwarf cultivars by the colour of pods and beans. Three of these arrived recently and, although recognised as different, have not yet received separate names (see Table 17). All the cultivars are the same size and produce white flowers.

Table 17: Dwarf common bean cultivars

CULTIVAR NAME	BEAN	POD
'*naykit*'	purple	green with scarlet border
unnamed	red	green
unnamed	pale brown	green
unnamed	blue	green with maroon streaks

Again, the cultivation of dwarf beans is the same as other pulses, except that they require no pole to climb. They mature earlier than the climbers, though they do not yield beans for as long. The Wola do not pick the '*naykit*' cultivar, like the '*pebway pombray*' climber, until its pods are dry and brown.

PEA

Pisum sativum (family: Leguminosae)
Wola name: *mbin* (No. of cultivars = 1)

This crop was probably domesticated in south-west Asia. It is a recent arrival in the Wola region, coming sometime after European contact. Its classificatory status is a little ambiguous; as an introduced low bush producing pods some people think that it is only a *taeshaen suwlshaeriy* cultivar, while others think that its markedly different structure and leaves mark it off as another crop.

This bushy plant, 30—150 cm tall, is a short-lived, climbing annual with a slender weak stem that requires support. Its compound leaves terminate in a tendril and consist of one to three pairs of opposed egg-shaped leaflets. The flowers are white and winged, and self-fertilised. They produce straightish green pods on short stalks, containing from two to ten round, smooth, green pea seeds.

The Wola cultivate peas in the same way as other beans, but they fare poorly in their region. The soil here is probably too acidic and the sun too intense and hot (see Tables 4 and 5). As a consequence, peas are rarely seen growing, and in the area on the Was river where I conducted fieldwork we could not find them growing in any garden. As people said, they are '*bumhaez*' (lit: lost), that is they grew badly and have now disappeared. This illustrates how the Wola experiment with any new crops, dropping or losing those that do not grow well.

WINGED BEAN

Psophocarpus tetragonolobus (family: Leguminosae)
Wola name: *wolapat* (No. of cultivars = 1)

The winged bean probably originated in tropical Asia, and arrived in New Guinea sometime in antiquity. So far as the Wola are concerned they and their ancestors have always cultivated it.

It is a perennial climbing herb with persistent tuberous roots that send up new stems each year. These grow to about the same height as the stems of common beans, that is 2—3 m, and twine about a support or trail along the ground. They bear alternately on long stalks the characteristic trifoliate leaves of beans, composed of egg-shaped leaflets terminating in short points. Their papilionaceous flowers are blue or white, and develop into characteristic winged pods which are roughly square in cross section with four prominent, jagged wavy flanges running along each corner for their length. The seeds are round and dark blue to brown, and they are embedded in a soft flesh that fills the pod, which the Wola call the *pat henget*.

The winged bean is only grown in a small part of the Wola region, in the lower areas to the south; in the Nembi valley for example towards Poroma patrol post. The higher altitudes composing most of the region are unsuitable because they are too cold; in the Was valley, for instance, where I conducted fieldwork there were no winged beans. As a consequence, the majority of Wola speakers distinguish only one *wolapat* cultivar, although they know that the colour of the beans and the flesh in which they are embedded varies from pale green to dark blue. Those living at lower altitudes in the south, neighbouring the Kewa, who cultivate winged beans, distinguish several

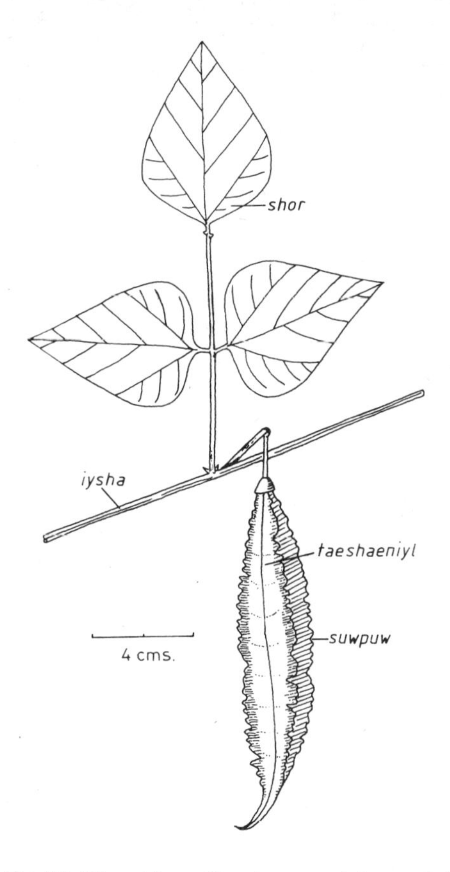

Fig. 22: Winged bean (*Psophocarpus tetragonolobus*)

cultivars by size, shape and colour (as the Hagen people do forty-eight cultivars — see Powell *et al.* 1975: 25).[4]

The winged bean is cultivated in the same manner as other pulse crops. In addition to yielding beans though, its tubers are edible and so is the soft flesh lining the inside of its pods. Indeed when picked young and tender people eat the entire pod, which is so for all the beans, although it is not a popular way to eat them. Considering all Wola, the winged bean is insignificant in their diet, growing as it does in a small and restricted area to the south.

[4] See also Strathern (1976) for further information on winged beans both here and at Pangia.

THE CUCURBITS

BOTTLE GOURD

Lagenaria siceraria (family: Cucurbitaceae)
Wola name: *senemiyl* (No. of cultivars = 3)

This crop probably originated in tropical Africa. It is thought to be one of man's first cultivated plants; among the few that have been common to both the Old and New Worlds since very remote times (Whitaker 1971). It reached New Guinea over two thousand years ago (Powell 1970), and so, as far as the Wola are concerned, has always been cultivated by them and their ancestors.

This annual herb has a long-running stem which may either trail along the ground or climb a support. The Wola grow it as a trailing plant. The stem is robust, has furrows running along it and has soft hairs, particularly on its new parts. It has coiling tendrils growing out from where the leaf stalks join the stem. Its leaves are large and roughly fan-shaped, with wavy edges giving a shallowly serrated look.

The flowers also grow out from where the leaf stalks join the stem. They are showy and short-lived, consisting of five distinct white petals. Male and female flowers grow separately although both occur on the same plant. The female ones grow on a shorter and stouter stem, the swollen part of which covered with fine hairs is the ovary that will develop into the fruit when fertilised by a pollen carrying insect, probably a bee.

The fruits have a dark green speckled skin and vary considerably in size and shape. They can be from 10 cm to over 100 cm long; globular, bottle or club-shaped, with a thick or thin, long or short neck which may be straight, crooked or coiled. They consist of a pale green, moist and soft flesh, in the centre of which are many tan or white, tooth-like seeds. If the fruit is harvested dead ripe, that is left until the plant dies back, it will have a hard, brittle durable rind which is impervious. Emptying out the seeds and rotting flesh gives a useful container, which the Wola use largely to carry drinking water. Fruits left until dead ripe are called *senemway* (lit: gourd plantable, because of the viable seeds they contain), in contrast to those gathered for eating while young and tender which are called *senemiyl* (lit: gourd fruits, the immature seeds of which never germinate).

The Wola distinguish three cultivars of bottle gourd on the basis of the shape and colour of their fruits; their stems, leaves and flowers all being the same. Both the '*huwlpay*' and '*pila*' cultivars have fruits with dark green skins. Those of the former though are club-shaped with long thin necks that are sometimes crooked, whereas those of the latter are globular or bottle-shaped with thicker, generally shorter necks (the fruit shown in the drawing is a '*pila*' one). The third cultivar is called '*bordorwiy*' or '*senemhaez*' and produces fruits which have pale green, almost white undersides where they

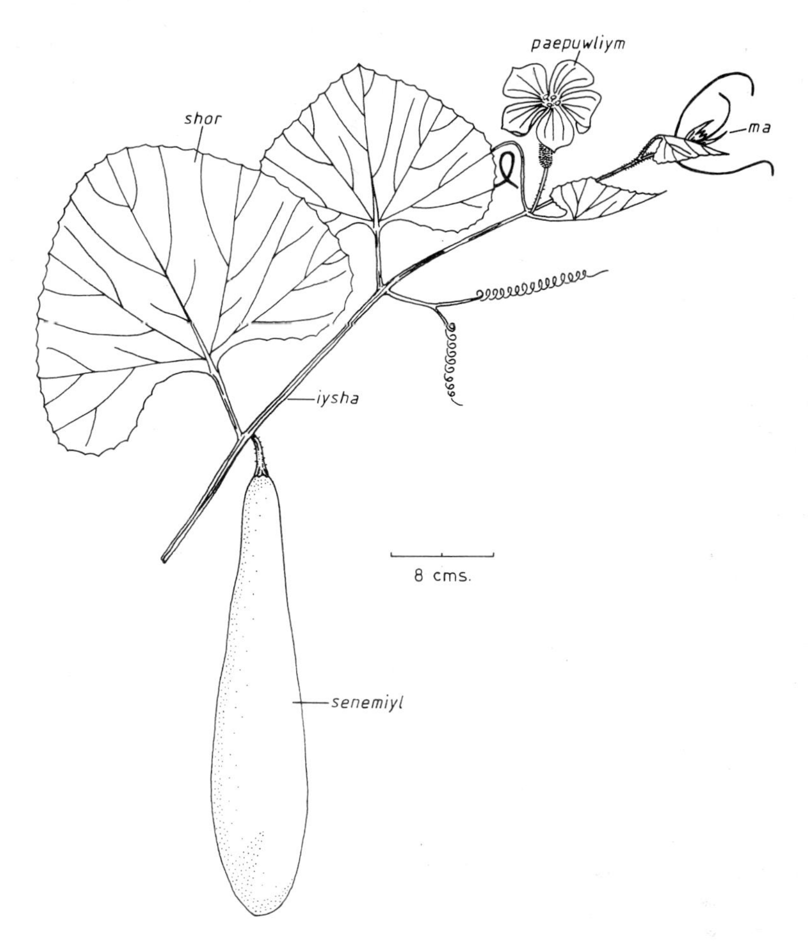

Fig. 23: Bottle gourd (*Lagenaria siceraria*)

rest on the soil (the name '*senemhaez*' means 'the white gourd'); their upper side is dark green, like the others. These fruits may be either club-shaped like '*huwlpay*' or more globular like '*pila*', but these differences are given no cognisance, 'white gourd' plants are not distinguished between and given separate names according to the shape of their fruits.

Both men and women plant gourds. They propagate them from seeds, which they separate from the mushy rotting flesh of dead ripe, inedible gourds and dry either over a fire or in the sun, and store until required on top of a fireguard. They plant seeds in pairs in shallow drill holes, usually made with the fingers; again this is an insurance against one failing to germinate. The Wola cultivate gourds in newly established sweet potato gardens, small mixed vegetable gardens and in small beds located on recently abandoned house sites. They rarely plant them in established sweet potato gardens unless they have a particularly fertile spot because gourds prefer virgin soil. They may plant them anywhere in a garden, and put them in at the same time as beans and maize, before they sow spinach.

Once established, gourds require little attention. Their young fruits are ready for eating about four months after planting, and they continue yielding for another two to three months. Those fruits to be used as containers take about twice this time to develop a hard rind. The Wola cook the young juicy fruits of gourds in earth ovens, bake them in the ashes of a fire and peel off the burnt skin before eating, or boil them occasionally in a tin. Like other crops which, by and large, they plant in newly-cleared gardens only and which yield for short periods, the Wola do not eat gourds often: they are relatively insignificant in their diet (especially as gourd containers are brittle and break easily, so demanding the forgoing of several fruits to produce replacements).

PUMPKIN

Cucurbita maxima (family: Cucurbitaceae)
Wola name: *pompkin* (No. of cultivars = 1)

This crop originated in South America, arriving in the Highlands of Papua New Guinea only recently. Europeans introduced it in the Southern Highlands at Lake Kutubu and the Wola say that it found its way from here into their region sometime in the 1950s, via settlements at lower altitudes. Today they laugh at their early ignorance of the plant which they thought yielded edible young leaves only. They threw away its fruits, which seeded themselves to give dense pumpkin growth around their houses.

The pumpkin is an annual plant with long trailing and branching soft stems, which are hairy and more or less round in cross-section. It is similar in many ways to the bottle gourd. The leaves are large and roughly fan-shaped, although not as broad as those of the bottle gourd and only shallowly lobed. They have finely serrated margins, are dark green, sometimes with characteristic white blotches, and have a covering of fine hairs. They are carried on long stalks which have coiling tendrils growing out from where they join the stem.

The flowers also grow out from where the leaf stalks join the stem. They are large, yellow and showy; short-lived and trumpet-shaped with five lobes. The same plant bears male and female flowers; although fewer of the latter, carried on shorter and stouter stalks, the swollen part of which behind the flower develops into the fruit when fertilised, most likely by a hymenopterous insect. The fruits vary in size, shape, colour and markings. They may be round, elongated or oval, with a narrow neck, which may be straight or crooked. They may have green, yellow or orange skins, often with a striped or flecked pattern in two or more colours running along them. Their flesh is soft, thick and orange-coloured, and in the centre has many white or brown, flat, plump seeds.

The Wola distinguish only one pumpkin cultivar, even though plants produce a variety of differently shaped and coloured fruits. They distinguish between immature fruits like courgettes, which they call *monuw* and sometimes pick for eating, and fully grown mature fruits which they call *pompkin*.

Women plant pumpkins by and large, although men sometimes do so. They propagate them both from seeds and by cuttings. The former they collect from mature fruits, or ones they miss harvesting which turn squashy and rotten, and they dry and store them on a fireguard until required, when they plant them like gourd seeds in shallow finger-made drill holes, usually in pairs as an insurance against one failing. For the latter method of propagation they break off 50 cm or so of stem from an established plant which has a healthy growing tip and simply push this into the soil for 15 cm or so, the cutting sprouting roots from the nodes where leaf stalks join it. The Wola say that there is no difference between plants grown from seeds or cuttings, both grow equally well.

Pumpkins may be cultivated in any newly-cleared garden, except a waterlogged one. They tolerate poor soil and are commonly replanted in long-established gardens too. According to the Wola they grow best in a moist soil, and they look for damp locations in which to cultivate them. They tend to keep pumpkins to the edge of gardens, notably across the bottom of the slope, from where they spread downwards, because they are prolific plants whose trailing stems cover a considerable area when established and crowd other crops out. They also cultivate them where other crops do not grow well, around obstacles like tree stumps and rocks, over and around which their trailing stems will grow. Women sometimes plant pumpkins following Highland *pitpit*, and at others a little later when they put in the other dibble-planted seed-propagated crops like beans and maize.

Once established pumpkins need little attention. They yield small courgettes or *monuw* within three months of planting, and fully grown mature fruits after about five months. In addition to the fruits, the Wola eat the plant's tender new leaves and shoots. They cook these often mixed with other greens, in earth ovens or simmer them in bamboo tubes or tins; the fruits they also cook in these ways and they bake them in ashes too (particularly the young *monuw*). They eat tender young *monuw* whole, whereas they remove the seeds from mature fruits before eating. Sometimes they thread these on to grass stems and gently roast them to eat (this is especially popular with children, the roasted seeds having a nutty flavour). Pumpkin plants go on yielding fruits for several months, usually for the lifetime of a garden. They yield well, and their fruits and leaves figure significantly in the Wola diet. Indeed, in times of sweet potato shortage, which sometimes occur for no apparent reason, they become almost a staple source of food, people relying heavily on them to see them through the lean period.

CUCUMBER

Cucumis sativus (family: Cucurbitaceae)

Wola name: *laek* (No. of cultivars = 2)

This crop was probably domesticated in northern India and found its way to New Guinea sometime in antiquity. According to the Wola it has been with them since time immemorial.

It is an annual herb with a trailing or climbing stem; the Wola cultivate it spreading over the ground. The stem is rough, bristly and somewhat furrowed, and from each side of it the leaves grow alternately. These are considerably smaller than those of the gourd or pumpkin, are roughly triangular in shape and varyingly lobed (like the sweet potato leaf types E and G in Fig. 4). They are scabrous and rough to touch, and are supported on rough bristly stalks.

As with the other cucurbits described, which the cucumber resembles on a smaller scale, coiling tendrils grow out from where leaf stalks join the stem. The flowers do too, the same plant producing both sexes. The male ones open first and predominate; they are carried in clusters of two or three, whereas female ones occur singly. All flowers have five yellow petals fused together and spreading to form a bell shape. After fertilisation, in which bees are again usually the pollinating agents, the swollen inferior ovary at the top of the stubby female flower stalk develops into a fleshy fruit, which the Wola call a *laek*.

Cucumber fruits vary greatly in size, shape and colour throughout the world. They are generally green and may have a smooth or warty skin. Those grown by the Wola are particularly short and stubby (about 8 cm long by 4 cm wide), and by European standards they have many large tough seeds set in their juicy flesh (which develop because the plants demand fertilisation to bear fruits, unlike the virtually seedless cultivars bred elsewhere).

The Wola distinguish two cucumber cultivars, largely on the basis of the colouring of their fruits. The '*laek*' cultivar has fruits with a pale yellow skin, which turn almost orange if left until the plant dies and picked dead ripe. The other cultivar called '*paerep*' has slightly larger leaves, and fruits streaked green and yellow (more yellow at their tip and more green round the stalk) which if left until dead ripe turns a deep yellow.

Women plant cucumbers largely, although men sometimes do so. They propagate them from seeds, which they plant in pairs in shallow dibble holes. To obtain seed they leave the fruits until dead ripe and then dry them entire on the roof guard above a fire. When they require seeds for planting they simply break open the shrivelled and brittle casing of the fruit, one supplying a large number of the lance shaped pip-like seeds.

The Wola cultivate cucumbers largely in new gardens, both sweet potato and small mixed vegetable ones, and on recently abandoned house-site beds. These plants do not grow well in waterlogged gardens, nor in established ones planted over again, unless they have particularly fertile pockets. They may be planted anywhere within a garden, although the Wola say that they do best on the sites of fires, where the vegetable rubbish from the clearance was burnt. Women plant cucumbers at the same time as the other seed-propagated cucurbits.

Once established, cucumbers require little attention. They grow quickly, yielding fruits within three months, and producing them for a further three months or so before dying. They are mostly eaten raw, often for refreshment during the day. Sometimes they are cooked in an earth oven. They are popular eaten with salt. However, like other crops which, by and large, occur solely in newly-cleared gardens, cucumbers are only available for limited periods of time. Hence, overall they make a small contribution to the Wola diet; although when available they are eaten in considerable numbers.

CLIMBING CUCURBIT

Trichosanthes pulleana (family: Cucurbitaceae)
Wola names: *tat; puliyba* (No. of cultivars = 1)

This genus of climbing cucurbit is found in Asia, although this species was probably domesticated in New Guinea, where it also occurs in the wild. The Wola anyway say that they and their ancestors have always cultivated it.

This climbing herb has a furrowed stem which is angular in cross section. It climbs with the aid of spring-like coiled tendrils, which are often split into two or three strands. These tendrils, as with the above cucurbits, grow out from where the leaf stalks join the stem. The leaves vary from a rounded heart shape to somewhat more angular, with shallow lobes. They taper to a point and sometimes have small irregular serrations around their edge.

In keeping with the other cucurbits too, the flowers of this climber grow out from where the leaf stalks join its stem. Both male and female flowers are carried on the same plant, the male ones in small clusters which soon fall and the female ones singly. All the flowers are five lobed and bell-shaped, with petals that have finely dissected top edges. The female ones have stems that are swollen for half their length with the inferior ovary which, when fertilised, develops into this plant's characteristic oblong and pendant fruit.

This is large and has a reddish-orange skin when ripe, which is somewhat warty and slightly furrowed. Young fruits have fine hairs. The flesh inside is orange, turning red at the centre. It is also stringy at the centre where many brown tooth-shaped seeds are embedded.

The Wola distinguish only one type of cultivated climbing cucurbit. They also recognise two wild kinds. One called *deraen tat* (lit: outside climbing-cucurbit), which is similar to the cultivated plant, only more hairy with thicker scabrous leaves that have deeper lobes. And another called *towmow tat* (lit: ghost climbing cucurbit), which is a smaller and hairier plant than

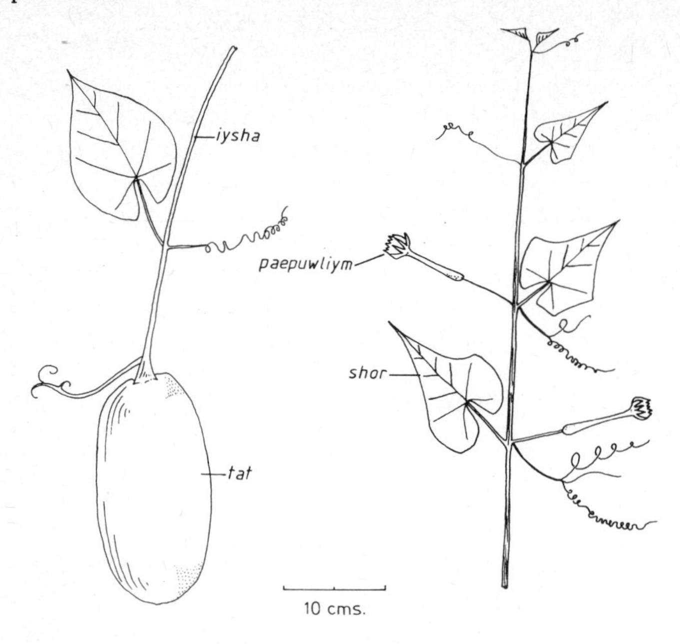

Fig. 24: Climbing cucurbit (*Trichosanthes pulleana*)

the other two with smaller fruits and markedly lobed and toothed leaves (like type G in Fig. 4).

It is men who, by and large, plant *tat*, although women may occasionally do so. They propagate them largely from seeds, which they drop either dried or fresh into shallow dibble holes. They also occasionally cultivate them from self-seeded young plants found wild, which they dig up and transplant. The Wola cultivate climbing cucurbits in all types of garden and on the edge of house yards; they grow relatively well in any situation. They are planted at the base of trees, up which they climb as they grow.

When established the plant requires no attention. It takes some months before it yields any fruit, but it will go on growing and producing for a number of years. The Wola eat the fruit when it is ripe and orange, although they sometimes pick them green and unripe and keep them in their houses to ripen: this is especially so when they find wild ones which someone else might discover and pick if left to ripen on the plant. They never eat them raw but always cooked in earth ovens, baked in the ashes of fires, or today boiled in tins. The climbing cucurbit is not very popular though and is rarely cultivated; in some settlements there are none to be seen at all. They are rarely eaten and contribute little to the Wola diet.

SCREW-PINE (*KARUGA* TYPE)

Pandanus brosimos &
Pandanus julianetti[5] (family: Pandanaceae)
Wola name: *aenk* (No. of cultivars = 45)

This tree with its large ball of nuts was domesticated in New Guinea sometime in antiquity, possibly semi-cultivated in the first instance by nomadic hunter-gatherers. Today the Wola cannot conceive of a time when their ancestors might not have cultivated it.

This tree grows up to 20 m high and may carry a single crown of leaves, or branch several times and carry up to six crowns, rarely more. Its trunk, or

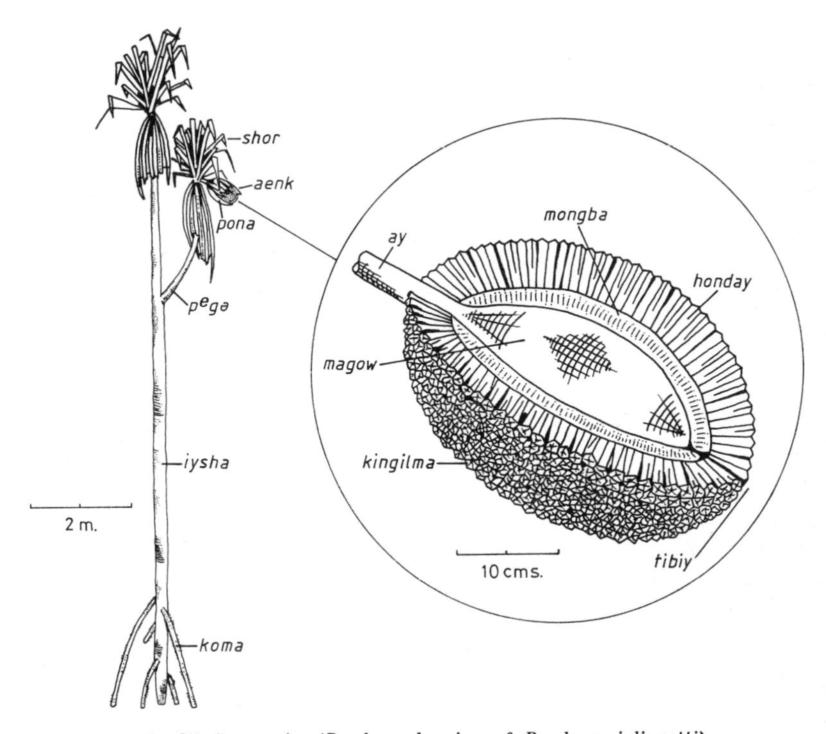

Fig. 25: Screw-pine (*Pandanus brosimos & Pandanus julianetti*)

iysha, as the Wola call it, is grey mottled with white. It is relatively smooth, though it may be warty and have small knobs on it. It bears as rings around it the annual scars where leaves were previously attached. The trunk is devoid of cambium and inside consists of a stringy, fibrous white pith. It stands erect

[5] *Karuga* is the Pidgin term for the screw-pine. These two species are difficult to distinguish. Although no proof is available they can possibly hybridise, and the latter may be derived and domesticated from the former (B. Stone 1974, 1982 and pers. comm.).

and often has growing from around its base aerial stilt roots, which the Wola call *koma*; these are not wide-spreading, although they can be quite long, and are prickly.

The leaves grow from the crown in a spiral (hence the common English name for the tree). They are long and narrow (up to 3.5 m in length), and are linear, almost sword-like in shape with parallel margins tapering off to a point. They are hairless, a little glossy on their upper surface and stiff. They commonly grow up rigidly at a steep angle and then bend over near their ends with their point hanging downwards. They are dark green, with a duller, more bluish-green underside. They have a pronounced midrib running along the bottom of a marked runnel dividing them longitudinally in two. There are sharp spines along the edges of the leaf and the underside of its midrib. The leaves are attached to the stem by semi-circular sheaths which swell out at their base. When they die the leaves turn brown and limp, falling down to hang around the trunk below the crown like a frill; they are then *shor kabiy* (lit: leaves dry).

Screw-pines are unisexual, producing either male or female flowers. It is the latter which develop into clusters of nuts, and hence these are the ones the Wola cultivate. Male trees are wild and found in the forest only. Their flowers are whitish and occur in densely crowded, compact oblong heads. Several heads grow alternately from a single long stout stalk, and are enclosed by leaf-like spathes which are white with green ends. These male flowering organs are two metres or more long and hang down from the tree rather like bananas. The Wola call them *aenk kor* (lit: screw-pine bad) because they consider the trees that bear them 'bad' or deformed for growing no nuts. They are interested in them simply for their white spathe leaves, in which they wrap their valuable pearl shells (they are popular for this because they are both durable and tough, and turn bright red when rubbed with ochre). No one ever indicated to me that they thought male trees played any part in the development of nuts in other trees, that they are anything other than queer screw-pines which grow 'wrongly'.

In contrast, the clusters of flowers on female trees develop into large spherical balls comprising many finger-sized nuts. Each pistil (the reproductive part of the female flower) develops into a single nut, which the Wola call a *honday*, and these cohere together into a ball, with their bases set in a soft yellowish pulp that extends some way up the outside of their shells. The Wola call this pulp the *mongba*, and the swollen fibrous stalk surrounded by it the *magow*. This ball of nuts hangs down below the leaves from a stout stalk, which the Wola call the *ay*, and is encased in cup-like, pale green leaves or bracts, which they call the *aenk pona*. The nut heads, which are all that is visible on a whole cluster, are angular and are raised up in a pyramid-like shape. They are a pale bluish-green colour. The Wola call them *kingilma* or *el* (lit: eye, after their resemblance to many eyes). When they trim off the 'eyes' to expose the nuts underneath, they reveal brown bristles which they call the *iriy* (lit: hair). Each nut has a hard brown shell called the *suwpuw*, which when cracked open reveals the pale creamy-coloured soft flesh or endosperm of the nut.

The Wola distinguish at least forty-five *aenk* cultivars.[6] They discriminate between them largely according to variations in their nut clusters, although they sometimes refer also to leaves and trunks (they cannot distinguish between cultivars by these latter features alone though, as they sometimes can by the nut cluster). Table 8 lists ten of the cultivars recognised in the Was valley and gives their distinguishing features (those I saw with nuts; to describe the others without seeing them proved too difficult and inaccurate). Although incomplete, it shows what the Wola look for when identifying cultivars.[7]

The identification of *aenk* cultivars proved difficult, as the preceding comments indicate. One reason was that some are particularly rare in occurrence, growing in restricted areas, and their identification is known to only a few people. The '*hap*' and '*peliya*' cultivars for instance are found only at the settlement of Haelaelinja in the Was valley; the people there tell the following anecdote to explain their origin. 'Long ago, a man was out hunting and his dog found these *aenk*. Excited it barked and barked and barked. Coming to see what was the matter, its master found the trees. He felled them and cut off their crowns, which he took home and planted in his garden. In this way the "*hap*" and "*peliya*" cultivars came to Haelaelinja only.'

Something else making the documentation of cultivars difficult was men constantly disagreeing over the identification of screw-pines; a problem with other crops too, raising important questions, taken up in Chapter 8, about the nature of this classification.

The way Wola classify wild *aenk* screw-pines reflects interestingly on this issue. They call wild trees by any one of the following four synonyms: '*shorluwk*', '*saez*', '*kutuwp*', or '*poroliy*' (these refer to female nut-bearing trees only, wild male ones are '*kor*'). Any screw-pine that grows in the forest of its own accord, to which no man can claim ownership, is a '*shorluwk*' and any member of the community on whose territory it stands may harvest the nuts it bears. These wild trees vary in size and shape; three examples measured had round nut clusters 65 by 23 cm long, thin clusters 55 by 15 cm, and very long clusters 65 by 23 cm. They were all called '*shorluwk*' though, and not classified further into different types. Many wild trees are identical to cultivated ones, but they do not take their cultivar names, they are '*shorluwk*'

[6] There are probably many more, because some cultivars are peculiar to very restricted and small areas. The forty-five recorded here occur in the Was river valley where fieldwork was conducted; in another community the list would be somewhat different, with cultivars unique to that place.

[7] See Powell *et al.* (1975: 31) for a list of cultivars recognised by the Melpa. The other cultivars named by the Wola, omitted from the table, are: '*baerel*', '*dobiyael*', '*dor*', '*emonk*', '*hap*', '*honal*', '*hones*', '*humbuwm*', '*kaba*', '*kagat*', '*kambiyp*', '*kat*', '*kongop*', '*korhombom*', '*laek*', '*lebaga*', '*maeka*', '*maela*', '*mbul*', '*morguwm*', '*nenjay*', '*nolorwaembuw*', '*obaib*', '*ombohonday*', '*piliyhongor*', '*posjuwk*', '*sayzel*', '*shond*', '*shuwimb*', '*taziy*', '*tiyt*', '*toi*', '*tombelpayliya*', '*tombpayliya*' and '*tomok*'.

Table 18: Some screw-pine (*Pandanus brosimos* and *Pandanus julianetti*) cultivars (dimensions in cm)

	NUTS									
CULTIVAR NAME	SYNCARP SHAPE	SYNCARP LENGTH	SYNCARP WIDTH	SHELL	*MONGBA* PITH LAYER	LEAF COLOUR	LEAF SHAPE	TRUNK	RIPENING TIME	OTHER FEATURES
'*bort*'	round	30	30	hard	very thin	green with white powder	short and narrow	knobbly	second	nut 'eyes' small
'*dob*'	round—oblong	30	25	hard	middling	green with white powder	shortest cv.	few knobs	second	
'*hael*'	pear shape	38	30	softish	thin	green — tips turn yellow as age	average	knobbly	second	very long stem
'*mabiyp*'	oblong	35	25	fibrous — difficult to crack	middling	green	crinkled edges	knobbly	mid-season	several aborted crowns
'*maeraeng*'	elongated—oblong	33	20	hard	thick	green — tips turn yellow as age	narrow	smooth	last	
'*pebet*'	round—oblong	35	30	very hard	thin	yellow and green	short and broad	last		
'*peliya*'	squat—round	30	35	hard	thin	yellow and green	short and broad	mid-season		
'*tabuwn*'	round—oblong	30	25	hardish	very thin	green	long and narrow	knobbly	mid-season	*mongba* disintegrates when nuts fall
'*taeshaen*'	round—oblong	30	25	soft	middling	green	average	smooth	first	eaten raw only
'*womb*'	round—oblong	30	25	very hard and brittle	thick	green	long and narrow	smooth	last	*mongba* turns red when cooked

only. The Wola think that rodents are responsible for propagating many wild trees, and they say that those identical to domesticated cultivars may originate in nuts from these screw-pines, which rats 'stole' and buried in the forest. Conversely, '*shorluwk*' trees may be brought into cultivation. Again though, they do not take the names of the cultivars with which they may be identical, they continue to be called '*shorluwk*', even though it is acknowledged that they might have initially originated from cultivated trees.[8] This affirmed exchange between cultivated and wild stocks, acknowledged by calling any screw-pine that passes into, or comes from, the wild state a '*shorluwk*', and the variation between wild trees, illustrates the variable nature of the classification system.

Only men plant screw-pines, sometimes saying a spell as they do so. They propagate them vegetatively from the crowns of old trees, which bear nuts no longer or poor clusters only. These old trees, which are usually tall and spindly, the Wola call *el*, while young ones are *pak* and mature nut-bearers are *huwniy*.[9] The Wola say that only the crowns of *el* trees grow, those of younger ones having wood that is too soft, which would rot without rooting. Some trees also grow small aborted crowns called *shombor* from the side of their trunks (especially the '*mabiyp*' cultivar), but these are sterile and not used for planting either. For planting, men fell old trees and cut off their crowns, which they stand up with soil piled around them, plus perhaps some pieces of wood to prop them up until they root. Another way in which they occasionally propagate *aenk* is to dig up self-seeded nuts and transplant them. But this is not common practice; a seedling might be a useless male tree, whereas crowns from female trees are sure to produce balls of nuts. The selection of the latter over many generations has predictably resulted in the stock of female nut-bearing trees far exceeding that of the wild male ones.

Screw-pines grow well in any soil and almost any location. Consequently they may be planted in any garden, usually around the edge, rarely in the centre where they will cast a shade. Mature trees, already growing on a site though, before its clearance and still producing nuts, are left wherever they stand. Only *el* trees are felled and their crowns planted along the garden boundary. Men also plant screw-pines around house yards and in clearings in the forest. Sometimes they plant several in one place and so establish a small grove of trees.

Aenk invariably outlive gardens, stands of them dotted about the cane and tree regrowth (marking old sites) and strings of them showing old fence courses. If a man clears a fallow garden site, or a virgin area, on which mature trees stand, these remain the property of those who planted them. Over the

[8] It is not clear what happens when the crown of such a cultivated '*shorluwk*' is planted again, but in all probability the original planter has died and the origin of the tree forgotten, in which case it will take the name of the cultivar with which it is identical.

[9] They also apply these terms of age to certain other crops.

years such shifting land use can confuse ownership rights, which men exacerbate when, as they commonly do, they ask others for permission to plant trees on the edge of their gardens. All screw-pines belong to the planters and their descendants, whether initially planted in a garden, around a house yard or in a forest. Their long life, though, compared with other crops, combined with the shifting use of the land on which they stand and the popularity of their nuts, predictably gives rise to frequent disputes over ownership, often when the planter dies and his relatives argue over who inherited which trees (an issue which they commonly leave until they produce large clusters of nuts over which to fight!).

While growing, a screw-pine requires no attention, but when it produces clusters of nuts these demand protection from giant rats which will otherwise climb up and eat them. The rodents concerned are *Mallomys rothschildi*, *Hyomys goliath*, *Uromys anak* and *Anisomys imitator*, which the Wola group together under the name *ogom*; to some of them they give the specific name *aenknokor* (lit: screw-pine nut eater). To protect a fruiting tree, a man lashes an obstacle round its trunk at about head height. There are two kinds of obstacle (see Fig. 26). One called *aenk kab liy* consists of a rough stick platform with vegetation piled on it, and the other called *showbaenk liy* consists of the stout, cup-shaped leaf sheaths of a wild pandan, which the Wola call *aendashor* (*Pandanus antaresenesis*). If an *aenk* stands in the forest, or an area of secondary regrowth, its owner, in addition to lashing on a barrier, will clear away all the undergrowth and vegetation growing around it, which, leaving them no cover to approach, deters rats. This activity also serves to show others that the owner of the tree is aware that it is nutting, so staking his unequivocal ownership over it and deterring any would-be thief, who might otherwise be tempted if he saw no obstacle and thought the nut was unnoticed. When a large number of trees come on to fruit at the same time, men spend a considerable amount of time erecting barriers and checking on maturing nuts.

The time screw-pines take to reach maturity and yield nuts varies considerably, but on average it is something like eight years from planting. When mature they go on producing nuts at intervals for several years, each cluster taking three months or so to ripen. Sometimes large numbers of trees produce nuts at the same time, although there is no apparent calendrical regularity to the occurrence of these seasons (for example there was one in January 1977, another in November 1977, and another in March and April 1978). The Wola say that they cannot predict them either according to their climatic seasons. Nut flushes occur in both the drier *ebenjip* season and wetter *bulenjip* one, although less often in the latter. Also, the number of nuts which ripen in any season varies: sometimes there are bumper yields and at others there are meagre ones. Nuts coming in *bulenjip* tend to be poorer, containing less edible flesh. When a flush of nuts occurs, not all *aenk* ripen at once; some cultivars are early and others late (see Table 18); also trees growing lower

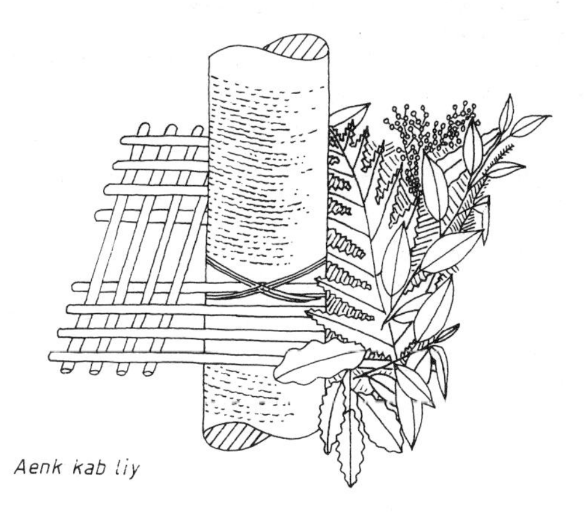

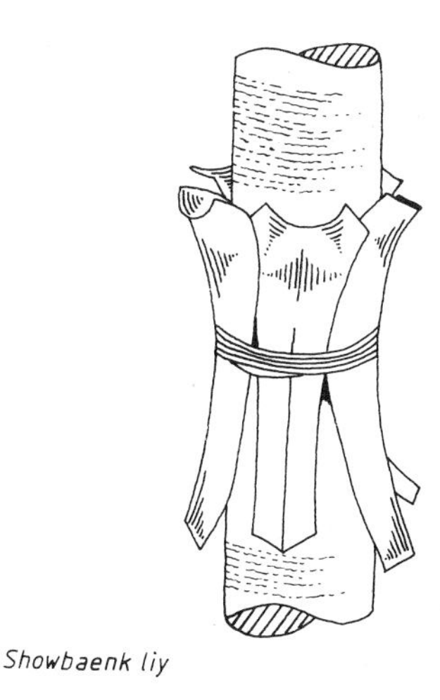

Fig. 26: Obstacles to prevent rats eating ripening screw-pine nuts

down near rivers ripen a month or so sooner than those higher up on the mountain crests. So the crop is staggered through a season. In addition to these seasons, there are occasional trees which nut outside them, which the Wola call *aenk hombuwni nj say* (lit: screw-pines all not ripen).

To harvest a ball of nuts a man puts his feet in a rough vine loop called a *pap*, and with his hands clasped around the tree's straight trunk and the loop pressed against it, climbs up in a frog-like fashion. When he reaches the crown he cuts through the stalk of the nut cluster with an axe and allows it to fall to the ground. If they do not cut down a nut cluster, then as it ages it will release its individual nuts to fall to the ground, and when this happens anyone who finds them can collect them (if rats do not beat them to it).

The nut yields of screw-pines vary. Some produce many nuts containing flesh, while others have only a few. The former, tightly packed in the cluster, are long and fang-like in shape, the latter, with room to expand, are bulbous and oblong. The Wola call those with few kernels *momborlom*. These nut clusters, although less desirable, are not entirely useless because they still have the soft, edible *mongba* pith. The nuts themselves are rich in oil and have a pleasant almost coconut-like taste.[10]

The Wola eat screw-pine nuts both raw and cooked. They prepare a ball of them for eating by trimming off the 'eyes' first, to expose the bristles on the ends of the nuts, and then cut them into different-sized segments. The culinary preparation of these nuts is more varied than for any other crop. The Wola cook them in earth ovens, roast them over the flames of fires or bake them buried in ashes. They sometimes cook the *mongba* pith on its own in a bamboo tube. One way they roast individual nuts is tied on a stick. They wind a length of vine spirally down the stick so that it pinches the nuts' terminal bristles, and then turn this 'spit' bedecked with kernels over a fire to roast them lightly; they call this *aenk kwiysh ombagay* (lit: pandan-nut, head-dress-support, cook), an illusion to their colourful enamelled bird of paradise plume head-dress, to which they attach parrot feathers in the same way.

Another practice is to bury nuts which have hard shells in soft, water-logged soil for a month or so; this is *aenk iyba wiy* (lit: Pandan nut water is). The nuts develop a strong ripe flavour, and when dug up may be eaten raw or cooked (sometimes the *mongba* pith rots too much and is not eaten). When there is a bumper season and more nuts than can be eaten, they are stored in a similar way, only buried in dry soil; a practice called *aenk honday suwl bay* (lit: pandan individual nuts soil do). Buried nuts keep for about a year; they may start to sprout but remain edible. Another way the Wola store nuts is to dry them, which they call *aenk honday leb bay* (lit: pandan individual nuts dry do). They leave their shells on and put them on a fire-guard to dry; they may dry them as a whole cluster or separated. When dry, people either leave the nuts in their shells and eat when required, or they remove the dry leathery kernels and parcel them up in the cup-like leaves that surround a ball of nuts, the *aenk pona*, tied up with slivers of rattan cane. These parcels are kept hanging on the wall, the dried nuts keeping for up to two years. When screw-pines are out of season such dried nuts are a delicacy and men sometimes exchange parcels of them for a pig or a pearl shell or a K10 banknote (when the parcel is about the size of a two-gallon bucket). They may be eaten dried, or reconstituted by cooking with pork in an earth oven, when they swell up and become soft once more.

[10] Indeed those with knowledge of coastal regions liken their *Pandanus* nuts to the coconuts of the *nembis* men.

When in season and plentiful, screw-pine nuts feature almost as a staple in the Wola diet. They are relished and consumed in great numbers. When out of season, though, they are rarely eaten, except for the occasional tree fruiting or the consumption of dried kernels. It is noteworthy that screw-pines supply not only food but also raw materials used in making certain artefacts, such as leaves for raincapes and durable bark sheets for house walls.

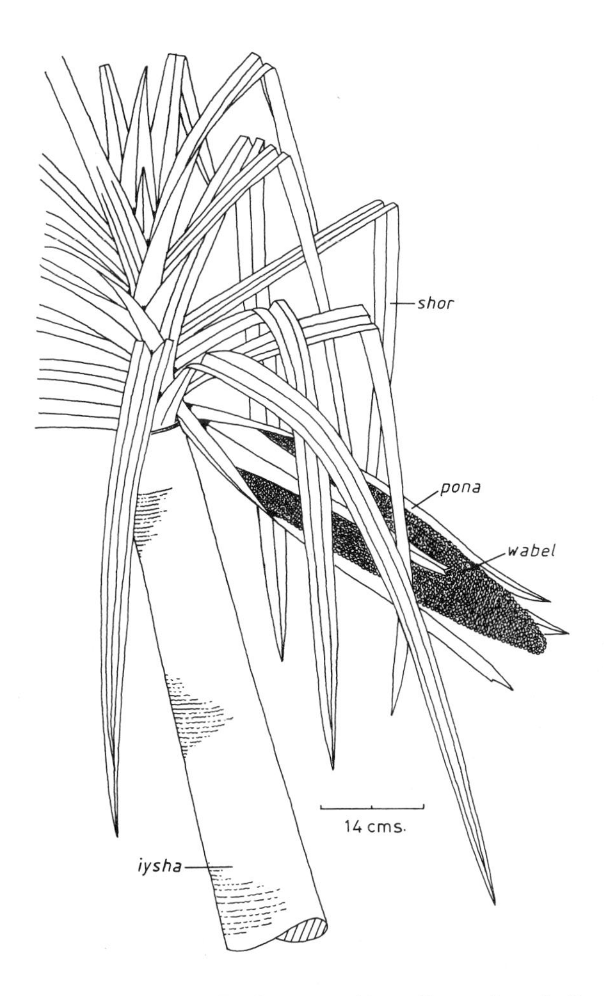

Fig. 27: Screw-pine (*Pandanus conoideus*, discussed overleaf)

SCREW-PINE (*MARITA*[11] TYPE)

Pandanus conoideus (family: Pandanaceae)
Wola names: *wabel; howat* (No. of cultivars = 4)

This tree with its elongated syncarp of nuts was brought under cultivation in New Guinea long ago, like the related *karuga* type discussed above. According to the Wola their ancestors have cultivated it since the beginning of time in the lower altitude zones of their region; its altitudinal limit of 1600 m restricting it to a few places only (such as lower down the Was and Nembi river valleys).[12]

This screw-pine is considerably smaller than the *karuga*, growing to between 3 and 5 m high. The trunk, although similar, is smaller in diameter and armed with upturned prickles. It stands on a few prickly stilt roots called *koma*, is often branched, and carries several crowns. The leaves are the same shape as those of the *karuga* — sword-like, barbed and with a pronounced channel — but shorter and narrower; also, when mature, they are often noticeably damaged by insects.

In addition to its smallness it is the nut cluster of this screw-pine that distinguishes it from *Pandanus brosimos* and *Pandanus julianetti*. It is long, tapering to a blunt rounded point, and almost cylindrical with a somewhat triangular cross section. It is red when ripe, and its individual nuts are considerably smaller than those of the *karuga* (20 mm by 4 mm, compared to 60 mm by 12 mm).

The Wola distinguish four *wabel* cultivars, on the basis of the size and colour of leaves and nut clusters (see Table 19). They consider the '*kwaen*' cultivar the best because it grows the largest cluster of nuts.[13]

Only men plant these screw-pines. They propagate them from crown cuttings off old trees and cultivate them both in and out of gardens, often establishing small groves, especially around house-yards. These screw-pines are hardy plants which require little attention; they flourish in most situations at low enough altitudes. The fruit demands plenty of sunshine to ripen, and once set men commonly cut back the tree's canopy to expose it to the sun and assist its maturation. The tree takes about four years to mature and bear fruit.

The Wola always cook *wabel* fruit before eating, usually in an earth oven. Following this they take the softened syncarp and place it on a scoop-shaped sheet of bark,[14] add water to it and squelch up to remove the red pulp

[11] *Marita* is the Pidgin name of this tree.

[12] Some men in higher altitude settlements recounted how they had tried to cultivate this tree but that it always dried up and died.

[13] Some Wola are aware that at low-altitude Huli settlements towards Lake Kutubu (where some of them have relatives) there are more cultivars, but they are unable to describe their distinguishing features or identify them. The names of five of these for instance are: '*borogay*', '*kinjuwdiy*', '*koluwa*', '*shortgai*' and '*yabayda*'. See Fischer (1968: 274—7) on the *marita* cultivars identified by the Kukukuku.

[14] This bark bowl is called *wabel iysh humbiy* (lit: *Pandanus conoideus* tree bark) and may be made from the bark of any of the five following trees: *waen* (*Trema orientalis*), *wenet* (*Engelhardia rigida*), *shwimb* (*Elaeocarpus dolidrostylus*), *timbol* (*Homalanthus* sp.), or *mul* (*Glochidion* sp.). These are suitable because their bark peels off easily in sheets and, more importantly, does not impart a bitter taste to food prepared on them.

Table 19: Screw-pine (*Pandanus conoideus*) cultivars

CULTIVAR NAME	LEAF			NUT CLUSTER	
	COLOUR	LENGTH	WIDTH	SIZE	IMMATURE COLOUR
'alaenda'	green — yellowing with age	shortest	narrow	smallest	red
'kaenomiy'	yellow and green streaks	shortest	narrowest	smallest	green
'kwaen'	yellowish green	longest	widest	largest	red
'tobor'	green	middling	wide	middling	red

surrounding the seeds. They eat the resulting thick red oily juice, discarding the small nuts from the centre. They sometimes mix the juice with greens; otherwise they drink it with spoons fashioned from screw-pine leaves; when available they add salt. The majority of Wola live at altitudes too high for this screw-pine, and rarely have the opportunity to savour the oily juice of its fruit. For them, it is something of a delicacy, and those with relatives living at lower altitudes occasionally make special trips of a day or more to collect some of these highly prized fruits and carry them home to share with delighted kin.

OTHER FRUITS

PASSION FRUIT

Passiflora edulis var. *edulis* (family: Passifloraceae)
Wola name: *ya iyl* (No. of cultivars = 1)

This plant is a native of southern Brazil. A European agricultural officer introduced it into the Wola region, and it reached the Was valley sometime in the mid-1960s.

A vigorous woody, perennial climber, a passion fruit plant may grow up to 15 m high. Its new stem is green, hairless and somewhat grooved. The leaf stalks are also hairless and grooved, and where they join the stem, spirally coiled tendrils grow out. The leaves are trident-like, consisting of three distinct, elongated egg-shaped lobes, terminating in fine points and having fine serrations around their edges.

Passion fruit flowers are fragrant, showy and large (up to 10 cm in diameter). They occur singly and grow out from where the leaf stalks join the stem. They have five thin white petals from which radiate wavy threadlike purple filaments, and in the centre of which are the flower's yellow reproductive organs. It was this colourful flower that prompted early

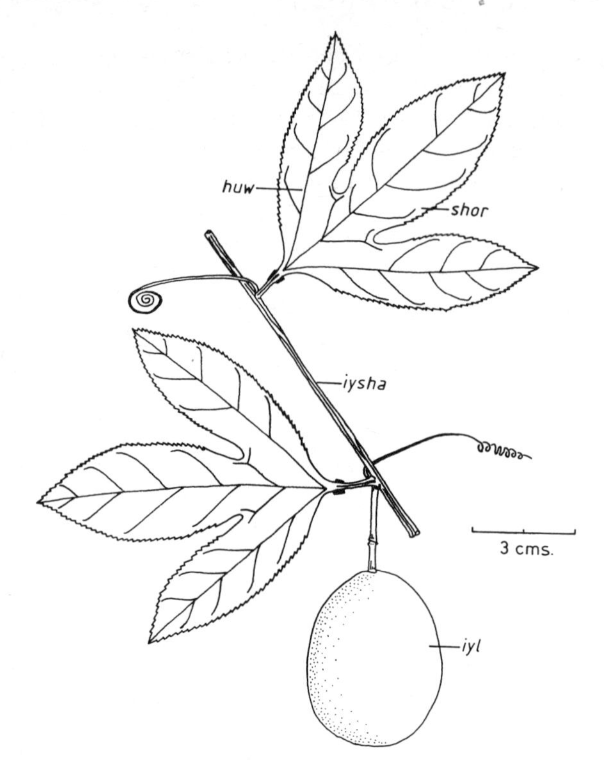

Fig. 28: Passion fruit (*Passiflora edulis*)

missionaries in South America to coin the name passion fruit, from a fanciful representation they saw in it of the crucifixion. The flowers open at dawn and close by midday; they have a sticky pollen and secrete nectar (bees are their principal agents of pollination).

Fertilised flowers develop into oval berries, the size of which depends on the number of pollen grains placed on the stigma. They have a hairless, leathery skin, which is purple when ripe, and contain many small black seeds embedded in a yellowish juicy pulp which has a tart flavour. The fruits are produced only on new growth.

The Wola distinguish only one cultivar of passion fruit, although they include under the name *ya iyl* (lit: vine fruit) the recently arrived Peruvian tree tomato *Cyphomandra betaceae* (family: Solanaceae) which bears a similar-sized fruit, though more elongated and reddish-yellow. This is a small (3–6 m high) tree which is short-lived and has large egg-shaped leaves about 28 cm long. It is exceedingly rare in the Wola region (I have only ever seen one, in the Nembi valley), and currently no more than a novelty.

Both men and women plant passion fruit. They propagate it from seeds,[15]

[15] Although the Wola do not do so, it may also be propagated from cuttings of mature woody growth.

which they plant at the base of trees, for the plant to climb as it grows. A popular place for these climbers is on the edge of houseyards, although they are also planted in gardens. The Wola say that the plants propagate themselves too, either directly by dropping ripe fruits which seed themselves, or indirectly from fruits eaten by humans and sown when they defecate. This fairly prolific plant often occurs uncared for on abandoned garden and house sites where it continues to grow and climb secondary regrowth.

Once established, these climbers require no attention, and they yield fruit for several years, although this falls off after about six years. The Wola do not usually wait for the fruits to drop to the ground to collect, but climb the host tree and pick them. This is especially so with children who like to eat them unripe and green. Only some adults like them, others find them too sour. Those who do eat them do so for refreshment during the day between meals; they do not eat many though, as a surfeit results in a sore mouth. They are eaten raw by splitting open the skin and licking out the flesh and seeds. These fruits make only a marginal contribution to the diet of the Wola.

TOMATO

Lycopersicon esculentum (family: Solanaceae)
Wola name: *tomasow* (No. of cultivars = 1)

This plant is of Central and South American origin. It is a recent introduction to the Wola area, reaching the Was valley sometime in the mid-1960s.

The tomato is a branching annual plant with either a solid erect stem or a thinner trailing one. It is covered with fine hairs and has a characteristic odour. The compound leaves grow spirally around the stem and consist of leaflets arranged more or less opposite one another along the stalk. The flowers grow in clusters from short stalks, and consist of six yellow petals that are shed two or three days after opening. Tomatoes are usually inbreeding, shedding pollen on their own reproductive organs. These develop into the familiar smooth skinned red, or sometimes yellowish spherical berries, containing many flat light brown seeds set in a juicy soft flesh. These have a characteristic star-shaped stalk, resulting from the enlargement of the base of the flower (the calyx).

Men and women plant tomatoes. They propagate them from seed, simply by squeezing a ripe fruit on to the soil and then lightly covering the pips. They cultivate them in newly-cleared gardens, and established ones too, if they have fertile spots. The most popular places for them though are small mixed vegetable gardens near houses and recently-abandoned house sites. Tomato plants prefer a less acidic soil than naturally occurs in the Wola region, and they grow best, as gardeners well know, where there were fires during burning off, the ash temporarily increasing the alkalinity of the soil. They may be planted anywhere in gardens, preferably in sunny positions, and are put in either at the same time as other seed-propagated crops, such as beans or maize, or later when spinach is sown.

The fruits mature in something like four months. The Wola eat them raw, frequently for refreshment during the day between meals, but they are not popular, making only a marginal contribution to the diet (indeed they are coarser and less flavourful than those grown in temperate regions). They do not always grow well either. Those who do grow them regularly do so largely for sale to European government officials and missionaries at the weekly market in Nipa.

BANANA

Musa hort. var. (family: Musaceae)
Wola names: *diyr; ebel* (No. of cultivars = 10)

The banana is one of man's oldest cultivated plants: its fruit was among his first crops. It evolved and was domesticated independently in two regions: Malesia[16] and India. The plants of both areas then met sometime in antiquity and developed into the hybrids that predominate today. The Wola assert that their ancestors have always cultivated bananas, although they are entirely unaware, as they are with other crops like sugar cane, of the time-scale involved.

This large perennial has a shallow underground stem or corm from which short rhizomes grow; these send up shoots or suckers near the base of the parent plant. The root system is not extensive; for their large size, banana plants are poorly anchored. Their stems consist of overlapping leaf bases or *paziy* tightly rolled around each other to make a rigid bundle. New leaves grow up through the centre of this pseudostem. They are tightly furled as they emerge, opening out into large oblong blades with pronounced supporting midrib. When newly emerged they stand erect, but as they age they fall to the horizontal and later they hang down, die and fall off. A plant carries ten or more leaves at a time, with new ones appearing every fortnight or so to replace old ones. They have a darker green top surface than underside, with a robust pale green midrib. They are glossy and smooth and commonly tattered and torn by the wind into *dekel* (lit: cuts).

After a plant has produced something like fifty leaves, the apical growing point of the subterranean corm becomes reproductive and sends up a flower stalk instead of leaves. When the neck of a banana plant swells, as the inflorescence makes its way up to the centre of the pseudostem, the Wola say that it is *paeraem puw* (lit: *paeraem* go), and when the flower axis sprouts out between the leaves at the crown they say that it is *diyr beray* (lit: banana sit). When exserted from the pseudostem the robust stalk bends over and hangs downwards. It carries a compound spike of flowers, arranged in several

[16] For a definition of this area see Womersley (1972: 684).

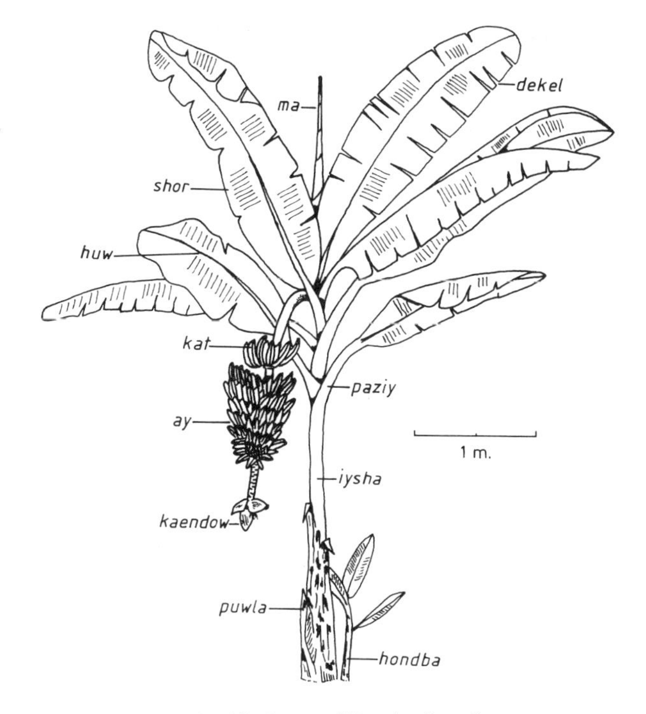

Fig. 29: Banana (*Musa* hort var.)

groups, each enclosed in a maroonish, oval, leaf-like bract. They are arranged spirally and overlap so that the inflorescence, which the Wola call the *kaendow*, is compact and conical. The bracts covering the flowers lift and roll back to expose them as they develop, and eventually fall off to leave a scar on the robust stalk. They do so from the top downwards. When the bracts of the *kaendow* open the Wola say that the plant is *haegaes liy* (lit: *haegaes* hit).

The flowers at the base of the stalk are female, those in the middle are neuter, and those at the top are male. The female ones are like miniature bananas with fine-stranded white tassles on their tips. The male ones though simply fall off, they do not produce viable pollen on cultivated plants. The fruits develop parthenocarpically, that is without pollination, from the ovaries of the female flowers. Banana fruits are seedless elongated berries, slightly curved and more or less round in cross-section. Their soft edible pulp consists of starch partially converted to sugars during ripening. They may have a greenish, yellowish or reddish skin, and are arranged in fan-like groups or hands that the Wola call *kat*, which develop from the flowers enclosed by the same bract. The Wola call the entire stalk of fruits the *ay*, and they refer to a plant carrying a ripening fully developed stem of them as *diyr goiyano* (lit: banana poured out).[17]

[17] For further details on the botany of bananas see Simmonds (1959).

Throughout the world there are many banana cultivars, the identification of which demand expert knowledge because they depend on genetic factors and not morphological ones alone.[18] There is a major division of bananas into two series: the Australimusa series in which the fruiting branch grows erect and the sap is red, these occur mainly in the Malesian region although the Wola do not cultivate them; and the Eumusa series which are plants like those described above. Until the work of Simmonds (1959), banana plants were divided into a number of species and sub-species, which he rejected as inappropriate for sterile and vegetatively-propagated hybrids whose differences result from mutations. He proposed that cultivated bananas are either *Musa acuminata* (which carry the A genome[19] and originated in the Malesian region), or hybrids of varying composition between this and *Musa balbisiana* (which carries the B genome and originated in India). It is probable that several Wola cultivars are *M. acuminata* (which occur frequently in New Guinea), while others are hybrids, but for simplicity I refer to them all as *Musa* hort. var.[20]

The Wola distinguish ten banana cultivars. They do so primarily by variations in the size, shape and colour of fruits and plants plus certain other differences in the way they grow (see Table 20).[21] Their '*banana*' cultivar is probably the commercial 'Dwarf Cavendish', which arrived in the Was valley in 1975 from the Nipa area where European agricultural officers had introduced it. The other six introduced cultivars listed in Table 20 came from lower-altitude settlements near Lake Kutubu (where Europeans may also have introduced them), men bringing back suckers given to them by relatives there.[22]

Men plant bananas. They propagate them vegetatively from small suckers a metre or so high, which they call *diyr hondba* (lit: banana offspring).[23] They dig these up from the parent plant and firm into shallow holes 30 cm or so deep, excavated with sharpened sticks. To plant a sucker is *diyr way bway* (lit: banana plant), and to plant one in an established garden which has been

[18] For help with the banana cultivars cultivated by the Wola I am grateful to *Musa* specialist N. Banks of the Applied Biology Department in Cambridge.

[19] The genome is the haploid set of chromosomes contained in an egg or sperm.

[20] For those who feel confident enough to distinguish between different bananas according to their genetic constitution, Simmonds (1959: 45—51) gives a key; see also Bourke (1975) for a simplified version of it.

[21] For comparative ethnographic data see Fischer (1968: 270) on the twenty-eight banana cultivars identified by the Kukukuku, and Powell *et al.* (1975: 23—4) for a list of the fifty-five distinguished by the Melpa.

[22] Indeed four of these plants ('*kiyabay*', '*paengay*', '*tongorma*' and '*waydow*') were brought to Haelaelinja by one man (Mayka Kot), and two of them ('*paengay*' and '*waydow*') were represented there in 1978 by only one plant each.

At these lower-altitude settlements there are more banana cultivars, some of which the Wola know by name (such as '*hgorlow*' and '*portpogow*'). Few could identify them though; indeed many people are not sure of the introduced cultivars listed in Table 20: they seem to know them only in the sense that so and so has one planted at such and such a place. Men with relatives at lower altitudes are better able to identify them.

[23] The word *hondba* also refers to the offspring of animals, although not humans.

Table 20: Banana cultivars

CULTIVAR NAME	PLANT SIZE	ORIGIN	FRUIT						LEAVES				TRUNK	OTHER
			SKIN COLOUR	FLESH COLOUR	SIZE & SHAPE	CROSS SECTION	WRAPPED UP TO RIPEN	EATEN	SIZE & COLOUR	MARKEDLY TATTERED	STALKS	NEW UNFURLED		
'Banana'	smallest	introduced	yellow	white	2nd largest, curved	round	yes	raw & cooked	shortest, green	no	pale green	pale green	green	
'Hond'	2nd largest	indigenous	yellow – with black blotches	pale yellow; pinkish	shortest & squat	round	sometimes	raw & cooked	longest, green	yes	pale green with black blotches	pale green	green with black blotches	
'Kiyabay'	5th largest	introduced	yellow	pinkish	thin & curved, thick skinned	round	sometimes	raw & cooked	short, green	no	pale green with black blotches	pale green	green	yellowish mid rib
'Paengay'	3rd largest	introduced	green – with black blotches	white	short & squat	round	yes	raw & cooked	longest, green	no	pale green with black blotches	pale green	green with black blotches	when fruits all leaves fall off.
'Ponjip'	4th largest	indigenous	yellow	pinkish	variably sized, straight	round	yes	raw & cooked	long, green	no	pale green with black blotches	pale green	green with black blotches	fruit ripens quickly
'Simbaem'	5th largest	introduced	pale yellow	pinkish	2nd largest, curved	round	yes	cooked	short, green	no	pale green with black blotches	pale green	green with black blotches	leaves have small brown spots
'Tongorma'	5th largest	introduced	green – with yellow patches	white	average, curved	penta-gonal	no	cooked	long, green	no	pale green with white powder	pale green	green	
'Tuwk'	3rd largest	introduced	green	pale yellow	average, curved	round	yes-but not at lower altitudes	raw & cooked	short, yellowish with white powder on underside	no	pale green	pale green	green	slow growing; old leaves remain erect
'Tuwmael'	largest	indigenous	yellow	pinkish	largest & straight	hexa-gonal	yes	raw & cooked	long, green with pinkish underside	yes	yellowish	pinkish	yellowish, stout	
'Waydow'	3rd largest	introduced	green	white	average, curved	round	no	cooked	longest, green	no	pale green with black blotches	pinkish	green with black blotches	

reworked is *diyr way orkay* (lit: banana replant). Bananas may be planted almost anywhere and at any time, so long as there is a reasonable depth of soil and adequate drainage. They are put in when all other crops have been planted, sometimes when a garden is several months old. They are usually put along the edge at the bottom of the slope, where the soil is deepest and their large leaves will cast the least shade over other crops. Later, they often occur as the only crops on otherwise abandoned and grassed-over sites because of the long time they take to mature. Rarely are they cultivated in small mixed vegetable gardens because of the shade they would cast over them. A popular place for them is in and around house-yards, where small groves sometimes develop as suckers proliferate.

Around house-yards is a favourable location because it deters fruit bats and birds from eating the fruit.[24] Plants sited away from homesteads are protected from predation by tying up their ripening fruit stems in a parcel with leaves.[25] This is called *diyr tongay* (lit: bananas tie up). According to the Wola it also encourages the development of large fruits.[26] It is done when

[24] The bare-backed fruit bat (*Dobsonia moluccensis magna*) which the Wola call *tagem*, and various parrots.

[25] This is so for all cultivars except '*tongorma*' and '*waydow*'.

[26] This applies especially when they tie up '*hond*', though its fruits take considerably longer to ripen.

the *kaendow* containing the remaining flowers falls off: this is when ripening starts.[27] A bound-up fruit is called a *jimb* (lit: join), so for example a parcelled-up *'simbaem'* cultivar is a *'simbaem' jimb*. It is also common practice, once a plant has 'poured out its fruit stem', to support this with a forked stick called a *diyr maip* jammed under the stalk to take some of the weight.

All banana plants bear fruit only once, and they are commonly felled after this. The time they take to produce varies, but few do so in less than two years and some take considerably longer. Ripe bananas are called *bor dokor* (lit: ripe burnt). They are a long time growing in the Wola region because at this altitude, with its consequent cool climate, they are approaching the limits of their range and tolerance. They do not feature regularly in people's diet because of the time they take to grow. When a man has a plant ready for harvesting, he and his family may consume large numbers of bananas in a short period, but this happens infrequently. These fruits are regarded as something of a luxury, and are often given as gifts to others.

The fruit of some cultivars may be picked either unripe and cooked, or ripe and eaten raw, while that from others is always cooked (see Table 20).[28] The Wola cook bananas in earth ovens, baked in ashes, roasted over embers, and boiled. When they cook them in their skins or *suwpuw* they always cut off the black spot, the *maendow*, where the flower was attached, or else they will be bitter; the stalk or *punduw* by which it was attached to the stem is left. The soft moist heart at the centre of the pseudostem of the *'hond'* and *'ponjip'* cultivars is also edible, as is that from the wild *kat* plant[29] (so long as they have not fruited; after this they are too bitter to eat). The Wola call this tender heart the *imbil*, and eat it raw, especially with a little salt if available.[30]

MAIZE

Zea mays (family: Gramineae)

Wola name: *kwaliyl* (No. of cultivars = 2)

This large annual grass originated in Central America. It arrived in the Wola region after European contact, the newest cultivar coming in 1973 as relief during the pan-Highland drought of that year.

The maize plant has a single unbranched stem, which is solid and robust, tapering off at its end. Roots grow out from nodes on the stem, both below soil and above, as supporting props. The leaves grow from the nodes, alternately and opposite each other, attached by sheaths which clasp the stem.

[27] Except for *'hond'* and *'simbaem'* which they tie up with the *kaendow* still attached to the stalk.

[28] The use of 'banana' for sweet, ripe fruit eaten raw and 'plantain' for starchy, unripe fruit eaten cooked is not applicable, nor botanically consistent.

[29] This enormous plant grows in the forest, producing small inedible banana-like fruits.

[30] If bitter when first cut, they leave them a day and they become edible.

They are parallel for most of their length, tapering to a point, have a slightly hairy upper surface, prominent midrib and edges which are often noticeably wavy. They grow out from the stem a little above the horizontal, with their ends drooping down.

The organs of both sexes occur separately on the same plant. The stem terminates in an erect or drooping, branched male tassel composed of many small spikes or ears which produce large amounts of pollen. The female cobs develop from buds at some of the nodes where leaves join the stem. These have short, stubby stalks, and are wrapped in concave leaves. They have long silk-like styles, receptive to pollen along their length, hanging out from their top end, like pinkish or creamy-coloured manes. After fertilisation these silks wither and broad wedge-shaped fruits develop, either haphazardly or in rows around the cob. These fruits are various shades of yellow when mature, going darker with age; they consist of a starchy flesh which may be either hard and tough, or soft and floury.

The Wola distinguish between two cultivars by the size of their cobs. The '*kwaliyl*' cultivar, the first to arrive, has the smaller cobs. The other is called '*Hagen kwaliyl*'. Some informants also pointed to a third unnamed group of plants which produce reddish seeds. These cultivar distinctions, though, will probably prove short-lived, plants cross-pollinating and hybridising to give a highly variable population that will blur them.

Both men and women plant maize. They propagate it from seed. They dibble a shallow hole and drop into it two grains or *iyl* picked off the cob (again they plant in pairs to insure against one failing to germinate). A cob intended for seed is left on the plant until it dies and turns brown, when it is picked, dried and stored on the roof-guard over a fire. This crop requires a well-drained soil. The Wola cultivate it in newly-cleared gardens, both small mixed vegetable and extensive sweet potato ones. If the soil in the latter is reasonably good they may plant it again when they rework it. They plant maize dotted about all over gardens because, although it is a tall plant, the shade it casts is only temporary. It soon yields a crop, and is then pulled up. Maize is put in at the same time as beans and other such seed-propagated crops.

Maize plants take about five months to produce mature cobs, and they go on yielding for a month or so; after about seven months their cobs are hard and dry, suitable for seeds only. The Wola steam maize cobs in earth ovens, bake them in ashes and roast them over flames. They are popular with most people and figure noticeably today in their diet. Because of the simultaneous ripening of many plants, it is common for those with new gardens to have more cobs than their families can eat comfortably and for them to give the surplus to relatives and friends.

Section I: Crops described

Chapter 6
INEDIBLES

TOBACCO

Nicotiana tobacum (family: Solanaceae)
Wola names: *miyt; sok* (No. of cultivars = 6)

This narcotic plant originated in South America. It found its way to the Wola long before contact; they consider it a traditional crop which they and their ancestors have cultivated for as long as can be remembered.

This coarse, herbaceous, short-lived perennial has a stout and erect stem, which is covered with fine hairs and tends to be woody at its base. It has a robust tap root with extensive lateral development. Young plants look like rosettes initially. Mature ones remain unbranched, unless their growing tips are nipped off, inducing branches to grow from buds between the leaf stalks and stem. Their leaves are large and roughly egg shaped, tapering to a point. They are usually arranged spirally up the stem, and often have no stalks or stalks with leaf wings.

The flowers or *paepuwliym* grow in groups at the end of stems. They are long and trumpet-shaped, slightly irregular and hairy, and have white or pink petals. They are usually self-pollinating, although insects effect some cross-pollination, and produce an egg-shaped capsule or *iyl* which contains thousands of minute, dark brown, elliptical seeds.[1]

Tobacco plants vary greatly, in their morphology and their tolerance of different ecological conditions. Agronomists have made attempts to differentiate them into botanical varieties, but current practice is to regard them as cultivars exhibiting wide variation. The Wola distinguish six cultivars of tobacco, primarily by leaf size and shape (see Table 21).

Men plant tobacco largely, women doing so only rarely. They propagate

[1] See Akehurst (1968) for further details on the botany of tobacco.

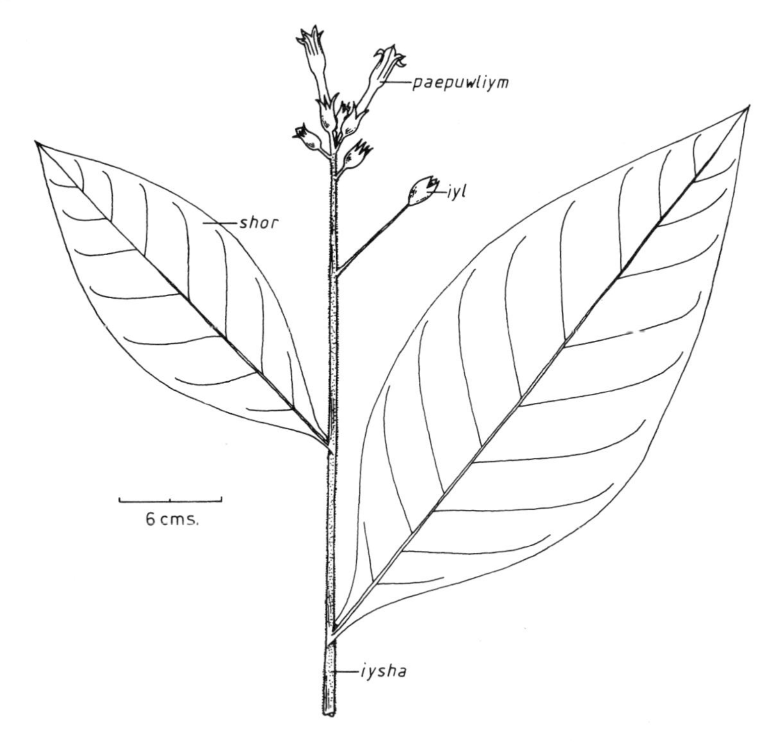

Fig. 30: Tobacco (*Nicotiana tobacum*)

Table 21: Tobacco cultivars (dimensions in cm)

CULTIVAR NAME	PRE/POST CONTACT	LEAF LENGTH	LEAF WIDTH	WINGED LEAF STALK	LEAF SHAPE	OTHER FEATURES
'*haentaymiyt*'	Pre	50	16	yes	long and narrow	Sows itself and grows wild, (especially at rockfaces — '*haentaymiyt*'= 'tobacco of the rock faces')
'*huwlmiyt*'	Pre	45	20	no	oval	
'*shwimb*'	Pre	38	15	yes	oval	
'*sokorsorwiy*'	Pre	50	16	yes	long and narrow	Largest plant with thick stem; branches without nipping; never flowers
'*tobaej*'	Post	60	35	yes	oval	Thick stem
'*tolwaegiya*'	Post	30	22	no	almost round	Short-lived plant; never branches; yellow flowers and large round seed capsule

it both from seed (*taentaen bay*) and from cuttings (*way bway*), depending on the cultivar. They propagate the '*sokorsorwiy*' cultivar by woody stem cuttings only, '*tobaej*' from both cuttings and seeds, and the other four cultivars from seeds only. To obtain seed, they leave fruit capsules on plants until they turn brown and dry, when they pick and desiccate them for storing indoors over a fire until required.

A popular place for men to plant tobacco is in narrow beds along the walls under the eaves of their houses. They break up the soil, often throw down some domestic refuse to fertilise it, and enclose the bed with a low cane fence. They sometimes plant tobacco too on recently-abandoned house sites, which are rich with domestic refuse, and around the edge of mixed vegetable gardens situated near houses. The Wola say that they plant tobacco near their houses so that it is close at hand when they need it. Also, it is the crop most often pilfered by others, and having it in or near house-yards, where there are often people about, protects it from thieves. These locations are also sheltered to some extent, which is important for the tobacco plant as it is weak and susceptible to injury. They are generally well drained too, and have lighter soils which, according to the Wola, the tobacco plant requires.[2]

Men watch over tobacco beds carefully as their plants are growing, and in spare moments weed them. It is three or four months before a plant has mature leaves suitable for smoking. When ready they turn a greenish-yellow colour. Men then either pluck them off a few at a time as required or pull up the entire plant. If they have sufficient tobacco for the time being they hang newly-picked green leaves in loose bundles around the walls of their houses to dry and cure them. But if in a hurry to smoke the leaves, they put them on a fireguard to cure quickly; indeed, men desperate for a smoke will take green leaves and place them on hot embers to dry for smoking immediately. The curing process ought to be one of gradual starvation, in which the leaf loses water slowly and turns a yellow-brown as its chlorophyll disappears and other changes take place in its chemical composition, so that it develops in flavour and odour. The Wola admit that the unhurried drying of leaves is best, that it helps to keep cured tobacco in small parcels for a while before smoking, but they say with a shrug of their shoulders that few can afford to wait that long. Anyway, if they did and relatives heard of the tobacco supply, they would be round to scrounge it: social circumstances encourage hurried curing and consumption.[3]

If a man allows a plant to continue growing, it may last for years; large tobacco bushes for instance sometimes mark the wall outline of abandoned

[2] For comparative ethnographic data on tobacco cultivation in New Guinea see Riesenfeld (1952).

[3] The improvidence of men with tobacco never ceased to surprise me. Addicted to this narcotic and craving a smoke, they often found themselves with no tobacco, and would pester one another for some (and the anthropologist with his stock of twist tobacco); wheedling relatives and friends would pounce on someone with more than a few pieces in his bag.

rotted houses. It is young plants, at the *huwniy* stage of growth, which yield large leaves that are flavourful and prized. Older mature plants, at the *el* stage, yield considerably smaller leaves which, if the branch if flowering, have, the Wola say, a poor taste and are *krai na wiy* (lit: hot-taste not is). When plants reach the mature *el* stage men have difficulty distinguishing between some cultivars because their small leaves all look the same (this is so for '*haentay*', '*shwimb*' and '*tobaej*'). These old plants are invariably bushy, someone having plucked off the end of their stem long ago for its young leaves, so inducing branching.

The majority of Wola men smoke and a few young women too. They prepare a smoke by first placing a piece of tobacco leaf on hot embers to dry thoroughly. Then they crush this on to another green-dried, flavourless leaf[4] which they wrap around the tobacco to make a triangular shaped wad. They tear the top corner off this and push into the small hole in a pipe, of which there are various types.[5] The Wola refer to smoking as eating tobacco: *miyt nay* (lit: tobacco eat).

PALM LILY

Cordyline fruticosa (family: Liliaceae)
Wola name: *aegop* (No. of cultivars = 25)

This plant occurs in cultivation or in semi-spontaneous growth throughout Oceania. It probably originated in the Indo-Malayan region, spreading to the islands of the Pacific sometime in antiquity.[6] The Wola consider it a traditional plant which they and their ancestors have always cultivated.

The palm lily is a tree-like perennial shrub, which may grow up to four metres or so high. It has an erect robust stem, which is usually slender (except for old ones resembling tree trunks) and marked by horseshoe-shaped scars where fallen leaf stalks were attached. It has a tuberous rootstock, tapering to a point and with many roots growing out from it. This plant is an evergreen, with leaves crowded at the ends of its branches in terminal heads. The leaves are elongated and spear-shaped, tapering to a point. They vary considerably in size between cultivars and also in colour, from yellow through various greens to pink, red, maroon and dark maroon verging on black; sometimes they are variegated. They have a relatively pronounced midrib, and a stalk that is deeply grooved, expanding at its end to become almost sheath-like in the way it grips the stem. The palm lily's small flower is white or reddish;

[4] They use a number of leaves for tobacco wrappers, such as *shonon* (*Acalypha* sp.), *hezaembul* (*Rubus moluccanus*), *haen*, *kol* (*Phyllanthus* sp.), and *ya shor* (*Palmeria* sp.).

[5] See Haddon (1947) for a museological account of New Guinea tobacco pipes, including those the Wola smoke.

[6] See Barrau (1965: 289—90) on the antiquity of cordyline in the Pacific and for speculation on it indicating past changes in the economic flora of the region.

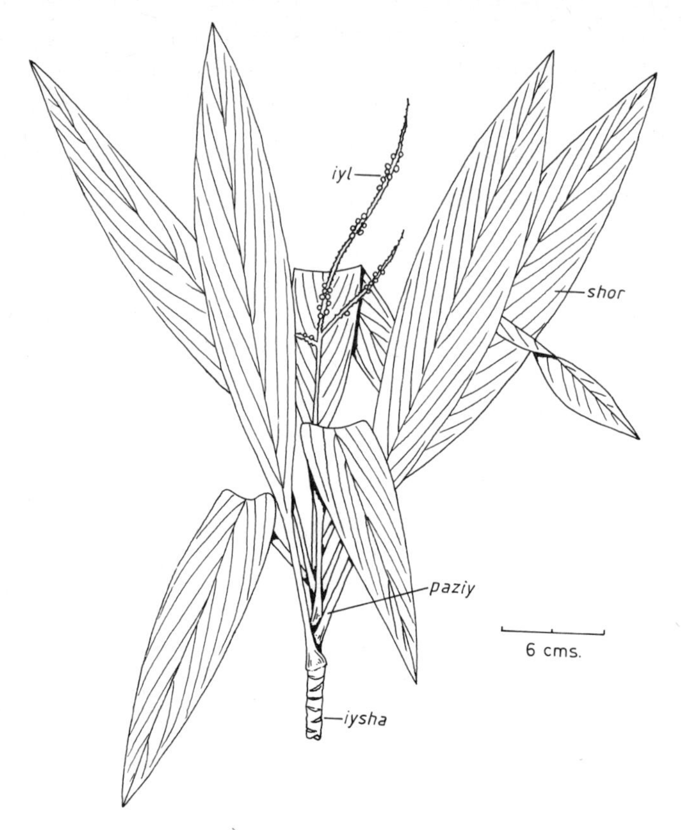

Fig. 31: Palm lily (*Cordyline fruticosa*)

although the Wola say that it never flowers. Its fruits grow clustered around branching stalks which stand erect from the centre of the leaf crown; they are small, round red berries containing three seeds.

The palm lily is a highly variable plant: a large number of cultivars have been raised throughout the world for ornamental purposes. The Wola distinguish twenty-five *aegop* cultivars on the basis of the size, shape and colour of plants' leaves (see Table 22).[7] Men wear many of these as everyday attire, notably the green ones, keeping a few, principally the red and yellow cultivars, for special occasions such as dances and ceremonial exchanges.

The leaves are worn by men, and only they cultivate this plant. They propagate it from cuttings broken off the stems of mature plants, which they simply push into the soil. These must have some woody growth at their base,

[7] This compares with the Chimbu who, according to Aufenanger (1961: 393—4), distinguish twenty-three cultivars using the same criteria. See also Fischer (1968: 286) and Wagner (1972: 124—7) on the thirteen cultivars identified by both the Kukukuku, and Daribi, and Panoff (1972) on the fifty-two distinguished by the Maenge. Barrau (1962: 107) also gives a brief summary of cultivars occurring in the Pacific.

Table 22: Palm lily cultivats

CULTIVAR NAME	LEAF COLOUR: YOUNG		LEAF COLOUR: MATURE		DIMENSIONS (cm) LEAF		STALK	LEAF SHAPE	OTHER
	STALK	LEAF	STALK	LEAF	LENGTH	WIDTH	LENGTH		
'Ago'	maroon	pale green; faint maroon underside	maroon	yellow with red ends	42	8.5	13	broadish	best aegop, worn for exchanges and dances
'Borbray'	maroon	pale green; faint maroon underside	maroon	yellow	40	3	13		
'Dedwael'	maroon	maroon	maroon	maroon with yellow mid rib	38	4	14	narrowish, sharp point	Kutubu palm lily, arrived recently
'Delhaezaegow'	green	green with red border	green	dark red with green mid rib	28	5	12	short, average width	worn at sa, saybel, and wartime dances
'Diy'	green	green tinged maroon	green with maroon edges	green	18	4.5	9	shortest leaf	
'Dom'	green	pale green	green	greenish yellow	29	6	10	short, average width	
'Dom'	green	green	green	green	50	15	20	broadest leaf, long	
'Domkiygil'	green	green	green	green & yellow chevrons	50	15	20	broadest leaf, long	
'Gwaenj'	green	greenish yellow with yellow tip	green	dark yellow	38	7	15	average spear-like	grows into a tree
'Haezaegow'	green	green with red border	green	dark red with green mid rib	44	5	12	longish	worn at sa, saybel and wartime dances
'Haezaegowmawol'	green	green with red border	green	dark red with green mid rib	28	3	9	short & narrow, sharp point	worn at sa, saybel and wartime dances
'Haezaeray'	green	green	green	green with yellow tip	35	8	16	short & wide	
'Kaegat'	green	green with pale maroon edge	green with maroon edges	green with wide maroon border	26	7	10	short & wide	
'Kal'	maroon	maroon	dark maroon	dark maroon	24	5.5	12	short & stubby	worn at dances & exchanges
'Kuwl'	maroon	pale green; faint maroon underside	maroon	yellow, sometimes with red border	45	3	13	long & narrow	
'Maelhaezaegow'	green	green with red border	green	dark red with green mid rib	41	5	16	average spear-like	worn at sa, saybel and wartime dances
'Ndorbayshor'	green	green	green	green & yellow patches	43	3.5	14	long & narrow	
'Pebraysha'	green	pale green with maroon edge	maroon	pinkish maroon	38	2.5	9	long & narrow, closed	from Kutubu, grows tight heads of leaves
'Pidwaeb'	green	green	green	green with yellow ends	35	7	16	wide leaf	worn at reparation exchanges
'Piyt'	maroon	green with maroon mid rib & veins	dark purple	green with dark purple mid rib & veins	24	6	15	short & wide	
'Pobaend'	green	green	green	green	35	9	12	wide leaf	grows into a tree
'Pong'	green	green	green	green	40	6	11	average leaf	commonest cultivar
'Sak'	green	green with maroon edges	green	maroon, green mid rib & yellow tip	32	6	15	average leaf	from Kutubu
'Sepshor'	green	green	green	green	47	7.5	10	large & broad	grows into tree; flat stalks
'Sepwesaembow'	green	green	green	green	45	9	19	large, broad & stalk long	
'Showmayaegop'	green	pale green	green	pale green with maroon edges	39	2.5	13	long & narrow	leaf grows closed
'Soizhumbiyhaem'	green	green	green	green	48	4	15	long, narrow & pointed	
'Wombiyael'	green	green	green	green	60	12	20	longest leaf, broad	grows into a tree
'Yaebabuw'	green	pale green	green	greenish yellow	35	4	12	average leaf	from south Nembi valley

soft new stem alone is no use because it would rot, not root. If the leaves they have worn for a few days are not too ragged and wilted to root, men sometimes take the cuttings and plant them.

The palm lily may be cultivated almost anywhere. In large new gardens men commonly plant it along the internal boundaries between the areas cultivated by different women. Together with Highland *pitpit* it serves as an unequivocal boundary marker, so reducing the likelihood of disputes between women over who has the right to harvest from which areas. A common place for *aegop* in all gardens is along fences, around the edge of the garden; the course of many abandoned garden fences remains traceable as a line of these plants. Another popular place for them is around the verge of house-yards, where they are readily available to men when they need a change of leaves, and where they serve also as ornamentals. The rarer red and yellow plants — those worn on special occasions — men often plant here because they consider them more ornamental. They also put them here to deter would-be thieves. They tend to plant them well inside gardens too for this reason, and not around the edges where they would be accessible to filching by passers-by. It is the common everyday green plants they put along the fences; others can take their leaves without causing annoyance. They occur everywhere, not just in gardens and around house-yards. These ubiquitous green-leaved plants grow prolifically in the secondary re-growth of abandoned gardens, and in the forest along paths too where men in the past have planted a few stems they have been wearing. These comments indicate that this hardy plant will grow in almost any situation. *Cordyline* may be planted at any time, although in newly-cultivated gardens it is usually put in at the same time as bananas and sugar cane, that is after the greens have been sown.

Wola men wear bunches of palm lily leaves on their rear as a buttock covering, their name for the plant, *aegop*, meaning tail.[8] They tuck the leaves, stem first, under their wide bark belts; to wear them is *aegop saemay* (lit: palm lily wear).[9] Before putting them on they beat the leaves against a tree trunk, or a similar hard object, to fray the ends and make them less stiff, and they also bend the stems back to render them pliable. A bunch of leaves lasts three or four days before needing replacement. To keep the leaves crisp and fresh for this time, men remove them at night when they sleep, leaving them outside on the roof. They sometimes wear the colourful cultivars to festive occasions tucked into the top of their bark belts and standing up in a spray of leaves (especially the short cultivar '*kal*'); this manner of wearing them is called *liy korbay*. Occasionally they also crinkle these leaves by folding them concertina-fashion, which is called *huwnduwn menay*. They usually smear them with oil too for these occasions to make them shine.[10]

When a plant produces berries, its leaves diminish in size and men will not wear them, although boys might. A plant with berries is called *el* and one without seeds, producing large leaves, is called *huwniy*. The small leaf phase though during fruiting is only transitory. A plant passes alternately from producing berries to growing new stem and large leaves, and back again. Men do not wear the yellow and red cultivars for special occasions until their leaves are both large and mature, or 'ripe' as the Wola says (*aegop bor day*; lit: palm lily ripe burn), that is markedly yellow or brightly variegated. The large green everyday cultivars on the other hand, they wear at any time, although the shade of green, the Wola say, varies according to the situation of the plant — those growing in shady places are darker than those standing in the sun. The Wola also use the leaves of the larger green cultivars, as they do banana leaves, to line the pits of

[8] The tails of all animals, such as pigs, dogs and marsupials are called *aegop*.

[9] If a man finds himself with no palm lily leaves ready to hand, for example if he is somewhere deep in the forest, then he may wear fronds from other trees or tree ferns.

[10] This oil comes from Lake Kutubu, where people tap it from the tree *Campnosperma brevipetiolata* (see Sillitoe 1979d).

earth ovens; and the yellow and red plants, as mentioned, they value as ornamentals.[11]

SEDGE

Eleocharis cf. dubia[12]

Wola names: *dorow; hurinj; karorgow; tenuwshabort* (family: Cyperaceae) (No. of cultivars = 1)

This is another crop which the Wola consider to be traditional and cultivated by their forebears since time immemorial. It is a plant of considerable antiquity in the New Guinea Highlands.[13]

This sedge is a perennial herb which grows in waterlogged places. It has long hollow and hairless stems which are round in cross-section, resembling those of onions. They are dark green above water and pale green to white below. They terminate in a blunt rounded point, have faint lines running round them at intervals like rings or nodes, and have a creeping, rhizome-like rootstock from which many stems grow to form a clump. When a stem

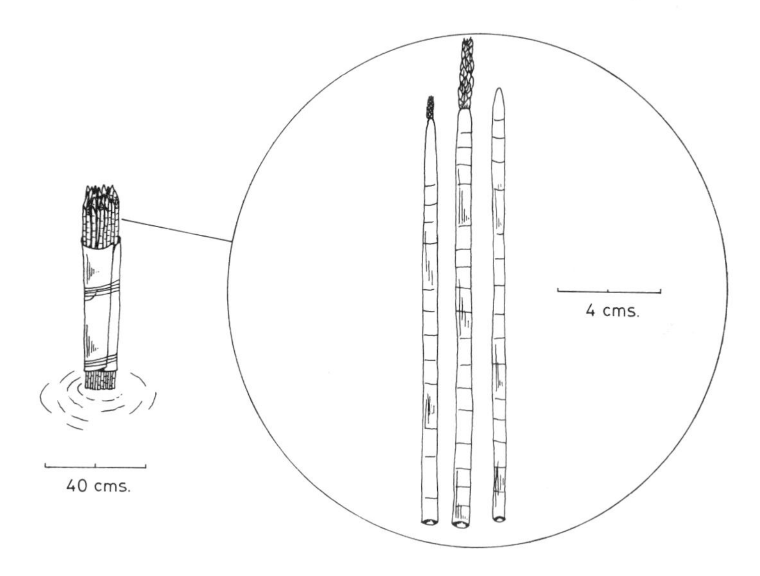

Fig. 32: Sedge (*Eleocharis cf. dubia*)

[11] The Wola do not eat the palm lily's tuberous root, although it is eaten in some parts of Oceania in times of famine (see Massal and Barrau 1956: 39). Nor does this plant have the ritual significance for them that it does for others in Oceania (see for instance Rappaport 1968 on the Maring, Panoff 1972 on the Maenge, and Leenhardt 1946).

[12] Possibly conspecific with *E. sphacelata*, which is Waddell's (1972: 228) identification of the sedge cultivated by the Enga.

[13] Fossil seeds of *Eleocharis* sp. have been found on archaeological sites in the W. Highlands (see Powell *et al.* 1975: Table 1).

flowers it produces a solitary terminal spikelet consisting of small overlapping leaflets with bristles, resembling the ear of some grasses. The flowers are hermaphroditic.

Only women plant sedge. They propagate it vegetatively by digging up a root clump, trimming the stems off it, and then breaking it up into pieces, each with some shoots or buds. They plant these pieces by pushing them into the soft soil of waterlogged places, an act called *hurinj way bway* (lit: sedge plant). As a plant of wet places, sedge does not occur in many gardens, except for waterlogged taro ones where patches of it frequently occur. It is also common to see a few clumps of sedge cultivated in many permanently waterlogged spots, both in the forest and near to settlements. If these spots adjoin paths located outside gardens, then those who plant there usually put a rough fence of underbrush around them to prevent pigs and people from trampling across them.

When it is growing, women tie up clumps of sedge in pandan leaf sleeves to prevent the stems breaking and to train them to grow straight and long.[14] When the stems are long enough, women collect them as required by cutting off with a bamboo knife below water, near the rootstock, which they leave in the ground to shoot and grow more stems. Women use sedge stems to make their grass skirts. First they scrape and flatten them, then they dry them in the sun, and finally they tie them in bundles on a length of string to make a skirt.

COLEUS DYE PLANT

Plectranthus scutellarioides (family: Labiatae)
Wola name: *komnol* (No. of cultivars = 1)

This is another plant the Wola consider to be traditional, which their ancestors have cultivated for all time. It is a plant of considerable antiquity in the Highlands of Papua New Guinea.[15]

It is an erect perennial herb with dark maroon stem and leaves. The stem is hollow and more or less four-sided in cross-section, with grooves running along it. The leaves grow in opposed pairs, and where their stalks join the stem there is a node-like protuberance in addition to small leaflets (stipules). The leaves are simple and egg-shaped, with toothed margins. They are thick, finely hairy and felt-like to touch. According to the Wola this plant rarely flowers — a possible reason for this is that they keep picking off their leaves and growing tips, so inhibiting flowering. When it does produce flowers, these grow in clusters, as whorls around branching terminal stalks. They are small, light blue and fragile.

[14] See the photograph in Womersley (1972: 910).

[15] Fossil seeds of *Coleus* sp. have been found on archaeological sites in the Western Highlands (see Powell *et al.* 1975: Table 1).

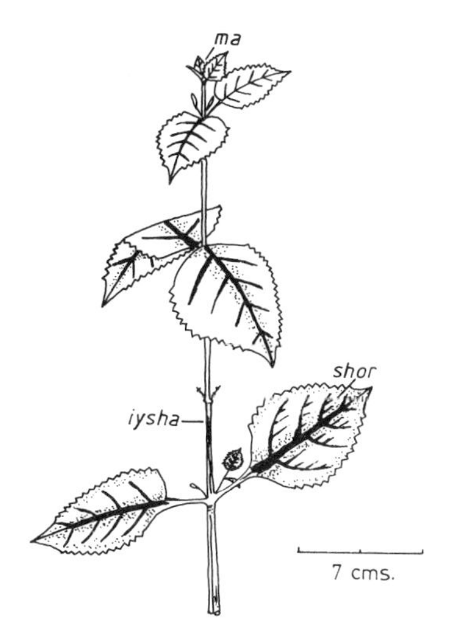

Fig. 33: Dye plant (*Plectranthus scutellarioides*)

Only women plant this herb. They propagate it vegetatively by taking stem cuttings from mature plants, which they push, two or three together, into the soil. They may plant it anywhere in a garden, although usually on the edge. This plant survives the abandonment of gardens and may be found several years later thriving amidst the secondary regrowth; anyone who finds and needs such wild plants may harvest them: they are not the property of the former gardener. Women cultivate only a few plants in some gardens because they are not in heavy or regular demand. They may plant them any time during the life of a garden, not only when it is newly cleared. Sedge plants thrive in almost any location except a waterlogged one.

Women use the leaves of this plant to dye string prior to netting bags and men's aprons. They simply pluck a few leaves, leaving the plant growing, which they bake briefly in the ashes of a fire to soften and make their sap squeeze out easily. To dye a length of string they pull it through a wad of these leaves. The colour of the dye varies from maroon to purple depending, the Wola say, on how thoroughly they heat the leaves — the purple dye coming from the most thoroughly heated ones.[16]

PAPER MULBERRY

Broussonetia papyrifera (family: Moraceae)
Wola name: *korael* (No. of cultivars = 1)

The Wola consider this tree a native of their region, and it is undoubtedly of some antiquity in the Highlands. It is in the same botanical family as the

[16] For comparative ethnographic data, see Fischer (1968: 287) on the Kukukuku, who distinguish eight cultivars of this plant.

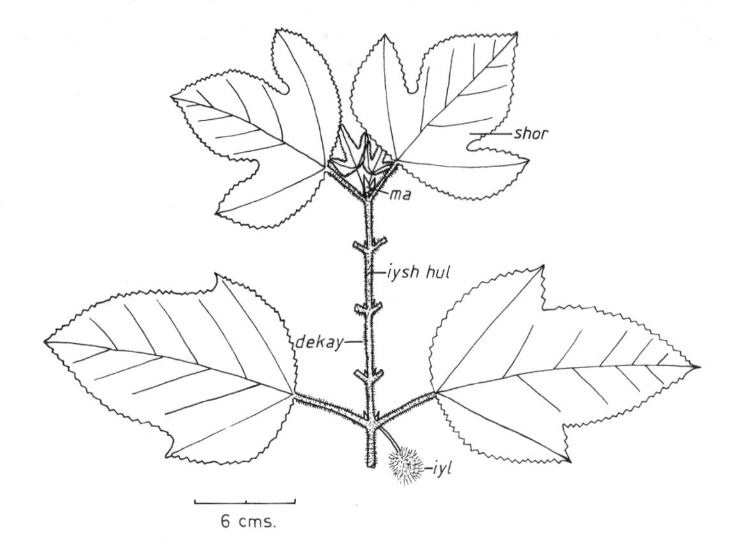

Fig. 34: Paper mulberry (*Broussonetia papyrifera*)

breadfruit (*Artocarpus altilis*) which originated in Indonesia, and perhaps in New Guinea too.[17]

The paper mulberry ranges in size from a shrub to a small tree, ten or so metres high. It has a grey bark and produces a white latex when cut. Its leaves grow in opposed pairs along bark covered twigs, called *iysh hul* (lit: tree bones), which for their last 20 cm or so are covered with hairs or *dekay*. The leaf stalks are also hairy, with small appendages (stipules) where they join the stem. The leaves themselves vary in shape from markedly lobed when new, to negligibly so, more or less egg-shaped, on older parts of the twig. They have toothed margins, and are rough and scabrous to touch. According to the Wola this tree never flowers or fruits. Elsewhere when it does so, it produces pendulous and cylindrical male catkins, and female flowers in round heads. The fruits are orange-red and coalesce into a spherical mass. The trees are unisexual.

Men plant these trees. They propagate them from cuttings pushed into the soil. They are hardy trees and can be cultivated easily in almost any situation. They may be planted on the edge of gardens, where they will continue to grow following their abandonment, or, more popularly, on the edge of house-yards. According to the Wola they require humans to propagate them, they do not seed themselves in the wild. Yet they are not commonly planted and in some settlements there are none at all. Although this tree has a relatively small trunk, the Wola cultivate it for its soft cambium layer, from which they beat bark cloth. They call bark cloth *korael* after this tree (even when they

[17] According to Massal and Barrau (1956: 19).

beat it from the fibre of another one). Men use bark cloth largely to make hats, called *korael tenj* (lit: paper mulberry hat).[18]

SHE-OAK

Casuarina oligodon (family: Casuarinaceae)
Wola name: *naep* (No. of cultivars = 1)

This, of all trees, is the one most valued by the Wola. It has an aesthetic appeal for them: they like to see its drooping fronds swaying in the breeze, and they say that their ancestors have always cultivated it for its beauty. The tree has been present in the Highlands since ancient times.[19]

A she-oak can grow to a considerable height and have a large girth. It has a greyish-green bark streaked with white, and its wood is red. Its branches are slender and drooping, and from a distance, with their long needle leaves hanging downwards, they resemble the feathers of the cassowary bird (the generic name of this tree originating from this resemblance). The leaves are long slender needles composed of many short sections, each about one centimetre long, joined together at prominent nodes. The twigs also have nodes, from which grow scale-like stout hairs. The flowers are unisexual. The male ones occur in terminal cylindrical spikes, while the female ones are arranged along the twig, developing later into globular cone fruits consisting of enlarged thickened, woody bracts.

Men plant she-oaks. They do this by transplanting self-sown seedlings when they find them. They watch closely as their transplants take root, often enclosing them with sticks to mark their whereabouts, to prevent people or pigs from trampling on them, and as a mark of ownership to deter anyone else finding the young plant from digging it up and transplanting it elsewhere. Popular places to transplant seedlings are around house yards and on the margins of ceremonial grounds or *howma*.[20] They are also located around the edges of gardens. When men clear a garden on a site where she-oaks stand they leave them, perhaps pollarding their lower branches. They are loath to fell these trees until they either reach a considerable age or die.

When they do fell them, they value their hard timber as firewood, saying that it is one of the best, burning with a pleasing aroma and giving considerable heat. They sometimes use it in making certain artefacts too, its hardness suiting it to a number of purposes — for example, making axe handles and fence stakes. The Wola cultivate she-oaks for ornamental purposes, though,

[18] The Wola do not eat the leaves of this tree, unlike the Hageners (Powell *et al.* 1975: 32).

[19] Evidence of it has been found in archaeological excavations in the Western Highlands (Powell *et al.* 1975: 47); and Flenley (1967: 309) found a dramatic rise in the number of trees in the Wabag region about three hundred years ago (Flenley quoted by Waddell 1972: 193—4).

[20] For an account of *howma* see Sillitoe (1979a: 48).

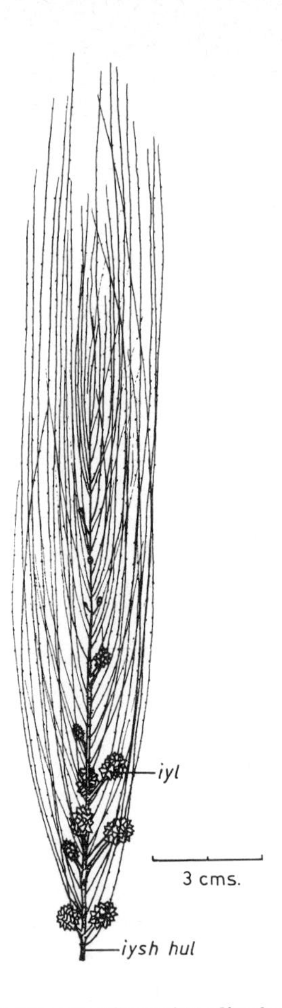

Fig. 35: She-oak (*Casuarina oligodon*)

not for firewood or usable timber. They also say that these trees improve soil fertility, which they attribute to the bed of rotting needles that accumulates below them. There is some scientific evidence to support this claim. The foliage of she-oaks has been shown to have a reasonably high nitrogen content and their roots bacterial nodules, which in some other *Casuarina* species have been demonstrated to fix atmospheric nitrogen in the same way as certain legumes (Mowry 1933; Bond 1957). This vindicates the Wola practice of planting she-oaks in gardens to improve soil fertility. These continue to grow following abandonment, areas of secondary regrowth having them dotted about, helping, the Wola point out, to restore the fertility of the area quickly so that they can garden it again. These, together with screw-pines and figs, are the only trees they plant and control in regrowth. All the others which come up occur spontaneously.

Section I: Crops described

CHAPTER 7
OTHER CROPS

There are a few crops not dealt with in the foregoing account because they occur either rarely under cultivation in the Wola region or not at all, known to the people only as crops grown by those living in neighbouring lower-altitude regions. Few of them are found in the Was valley, where I lived and conducted this research.

The following marginal crops occur as novelties in a few locations, largely in the Nembi valley.[1]

Carrots (*Daucus carota*), in Wola called *kaer*^{e}*s*. These characteristic tubers, consisting of swollen orange tap-root topped by wispy foliage, find their way to the market in Nipa on rare occasions. The groundnut (*Arachis hypogaea*), which the Wola call *piynus*. This annual herb, which grows its nuts under the soil, arrived recently. In 1976 an agricultural worker distributed a small number of seeds in the Was valley, though the few who experimented with them had poor yields. Another crop introduced by agricultural workers was the pyrethrum flower (*Chrysanthemum cinerariaefolium*). The white daisy-like flowers of this tufted perennial herb contain insecticidal substances,[2] but they did not catch on as a source of income and only a few wild plants remain.[3]

The commercial crop on which government administrators have pinned their hopes to lead the Wola — as it has done more developed Highland

[1] The situation in the Lai and Mendi valleys is somewhat different. Some people here grow considerable amounts of these crops, notably for sale in Mendi. Although Wola speakers inhabit these valleys, this study does not deal with these more commercial crops in any detail because, although I have visited these areas, I have little knowledge of them (by and large, for locally consumed crops, the situation is the same as in the Was valley).

[2] As the Wola discovered, spreading them on their sleeping mats and platforms to kill fleas and lice.

[3] People said that the returns on their effort picking them made it ridiculous to waste their time doing so.

regions — into the cash economy of Papua New Guines is coffee (*Coffea* spp.). In the mid-1970s the Department of Primary Industry, in conjunction with the Local Government Councils, planted extensive coffee blocks in their region near Poroma and Nipa, in an attempt to demonstrate to them how profitable it is and induce them to do likewise in their gardens. Following this up in 1977 agricultural workers established one or two small nurseries of coffee plants in the gardens of interested men in the Was valley. It remains to be seen how these will develop.

The choko (*Sechium edulae*), another cucurbit, was seen once in the Nembi valley. This is a climbing, robust vine with tendrils and broad oval, faintly lobed leaves. It produces large, fleshy fruits which are whitish to pale green and have a furrowed and warty surface. The pea and the tree tomato, two crops mentioned earlier, also fall into the marginal crop category.

The following crops flourish at lower altitudes in the region of lake Kutubu and are known to the Wola by hearsay or by visits to the area.

The sago palm (*Metroxylon* spp.): the extracted starch reserves of this swamp tree constitute the staple food of people in these lower areas. The Wola call sago *hiywa*, and on the rare occasions when someone returns with some of it from Kutubu, their relatives consider it a treat (although these who spend any time at lower altitudes soon tire of its bland taste and rubbery texture). Other introduced tropical crops include the evergreen citrus trees of lemons (*Citrus limon*) and oranges (*Citrus sinensis*); the perennial terrestrial plant of the pineapple (*Ananas comosus*) with its narrow spiny leaves and large segmented fruit; the pawpaw tree (*Carica papaya*) with its clusters of melon-like fruits and, from the New World, the short-lived cassava shrub (*Manihot esculenta*) with its edible roots. The inquisitive Wola have tried cultivating all these crops,[4] though with little success, except at lower altitudes in the south where a few occasionally yield small and inferior quality fruits, which sometimes find their way to the weekly markets held at patrol posts.

[4] These illustrate again their willingness to experiment with new crops, however improbable.

[a] The Was valley looking north towards Porsera (in the foreground *Miscanthus* cane hides a garden).

[b] Weeding a *paen* bed containing sugar cane, cabbage, taro and Highland *pitpit* (in the background are sweet potato mounds).

Plate 1

[a] A homestead surrounded by screw-pines (men's house on the left and women's on the right).

[b] Pulling up sweet potato vines from mature garden for transplanting (note sugar cane growing on edge).

Plate 2

[a] Taro, showing corm (note *pora* on right, between parent plant and juvenile banana).

[b] Cutting the tops off taro tubers for replanting.

Plate 3

[a] **Tannia, showing parent corm.**

[b] **A clump of Highland *pitpit*.**

Plate 4

[a] Recently planted sugar cane interspersed with taro (note juvenile screw-pines on the right and immature banana centre rear).

[b] Bottle gourd (cultivar: *'pila'*).

Plate 5

[a] A screw-pine.

[b] Lashing *aenk kab liy* rat deterrent to screw-pine (note man's buttock-covering of palm lily leaves).

Plate 6

Plate 7 →

[b] Palm lily (cultivar: *'kal'*).

[a] Bananas, together with screw-pines, on edge of house yard (fruiting cultivar: *'hond'*).

Sitting in waterlogged taro garden with, in the foreground, sedge growing in screw-pine leaf sleeve.

Plate 8

Plate 9 →

[a] Juvenile she-oaks.

[b] Women's crops: planting sweet potato vines *suwl* method (note woman's sedge skirt).

Men's crops: broad-casting Crucifer spinach seeds (note bundle of dried seed capsules held under arm).

A small mixed vegetable garden adjacent to a house containing taro, sugar cane and various greens.

← Plate 10

Plate 11

[a] Waterlogged taro garden ready for harvesting.

[b] Established sweet potato garden (over ten years old) with new mounds on left, *waeniy* mature ones on the right and *taengbiyp* grass regrowth above.

[a] A sweet potato garden recently replanted containing maize, sugar cane and Highland *pitpit* (separated by *Miscanthus* cane regrowth from established sweet potato garden in *puw* state — note screw-pines).

[b] Laying fire to heat stones for earth oven to cook the clusters of screw-pine nuts in foreground.

[a] Putting taro into palm lily leaf-lined earth oven pit.

[b] Wrapping cabbage leaves round a hot stone prior to parcelling up in *Miscanthus* leaf *ombuwgay* bundle to cook.

Plate 14

Section II: Crops discussed

Chapter 8
THINKING ABOUT CROPS: THEIR CONCRETE CLASSIFICATION

The Wola conceive of their crops in a way that is both familiar to a European and yet, at the same time, different too. The previous chapters convey this feeling clearly.

This impression, of familiarity tinged with strangeness, is commonly conveyed in accounts of how other people classify the natural phenomena occurring in their regions, be they — to mention two thorough series of studies — Mayan Indians classifying plants and animals (see Berlin, Breedlove and Raven 1974; Hunn 1977) or New Guinea Highlanders classifying frogs, marsupials and birds (see Bulmer and Tyler 1968; Bulmer and Menzies 1972; Majnep and Bulmer 1977).

This familiar yet strange impression is explained generally by reference to taxonomic levels (for example see Ellen 1975: 221, and Hunn 1977: 61, who cites Black 1967). It is argued that at the higher and more inclusive levels, where people talk of groups equivalent to 'birds', 'reptiles', 'climbing plants' and so on, there is scope for cultural elaboration and innovation. When grouping various organisms together at these levels, the absence of specific, objective differences allows room for invention. Varying criteria can be used to put flora and fauna into categories and differing classifications result. A good illustration is Bulmer's (1967) discussion of why the Karam do not consider the flightless cassowary a bird.

Conversely at the lower and more exclusive levels of taxonomy, where people talk of the 'black bird', 'grass snake', 'kidney bean' and so on, identifications are founded on constant and objective discontinuities which allow for relatively little variation between cultures (see Bulmer 1970). Be they Eskimos, Londoners or Melanesians, people base their discrimination, if any, between two botanically or zoologically distinct species on the same observed differences between them. The result is a similar ordering at this level, as shown for example by Berlin's, Breedlove's and Raven's (1974) detailed documentation of Mayan Indian plant classification, which parallels in a striking fashion that of botanical science (see also Hunn's 1977 account of their zoology).

The upshot is strangeness at the upper levels and familiarity at the lower. The way in which the Wola order their crops both supports and challenges this accepted explanation of the familiarity and strangeness of exotic classifications. While their crop taxonomy fits the above argument at the upper and middle levels, the strangeness encountered suggests a classificatory system of a notably different order to that of biological science. The confusion encountered at the lower levels reinforces this impression. These observations raise the difficult and tendentious issues about the assumptions that underpin discussions of other peoples' classifications (see Diamond 1979). These appear more different than the above argument suggests, and accepting its explanation of the simultaneous familiarity and strangeness as conclusive may lead to accounts set in a too western-centric, and so distorting, frame. In other words, to what extent are the Wola, and by extension other people socialised in entirely alien cultural traditions, doing something analagous to classifying in Western thinking when they order natural, or any other phenomena, into classes?

THE CROPS

At the upper, all-inclusive taxonomic level the Wola do not, to my knowledge, have an all-embracing term for the plants discussed here and grouped together under the heading of crops. Possibly the overwhelming contribution which these plants make to their largely vegetarian diet conditions them to think of each crop as something too important in itself to be subsumed under a collective higher level name. The occurrence of all-embracing upper-level terms equivalent to the collective word 'crops' for other groups of organisms supports this speculative observation (for instance, the Wola group together all large marsupials and call them *sab*, and all woody trees and call them *iysh*).

The existence of these equivalent collective names for other groups of organisms indicates that the absence of an all-embracing word equivalent to 'crops' does not imply that the Wola cannot conceive of these plants as a group; they know that they go together in some sense, that they have in common the facts that they are cultivated and eaten. If pressed to give a name to all these plants, they will use phrases referring to these shared characteristics; for example that they are *nokmay sem* (lit: edible family — which includes everything that is edible from vegetables to pork, and wild fungi to fish) or that they are *em bort bway sem* (lit: garden at plant family — which is everything grown in a garden).

The significant point is that the lack of a collective term indicates that the Wola do not normally think of all their crops as constituting a single higher-level group. The significance of this terminological hiatus comes over clearly in everyday conversations when people refer to these plants. They talk of this or that crop in their gardens and not all their crops; for instance they say 'my sugar cane is maturing slowly' or 'my spinach is almost ready to pick', and not 'everything is growing well in my garden'.

There is a similar marked absence of any named mid-level groups between all crops collectively and individual crop plants, such as taro and bananas.[1] The Wola for instance, do not lump together all cucurbits or pandans or tuber crops and give them collective names. Again, however, they are aware of the similarities between these crops, that all cucurbits for example produce bulbous fleshy fruits, but they see no reason to group them explicitly. There is only one related group of crops which they categorise sometimes in this way if encouraged to do so, and this covers the pulses. They call any pod-like fruit containing seeds a *taeshaeniyl* and they may refer to any such plant as a *taeshaen*, although they do not usually do so.[2]

The apparent absence of overall names for groups of organisms is quite common throughout the world (Berlin, Breedlove and Raven 1974: 26, 30), and relates to the conundrum associated with Sapir and Whorf (Kay 1970), namely: how does cognitive ordering relate to language? While people can think of things and conceive of relations between them without necessarily giving them names, the problem is: how can outsiders gain access to this unverbalised thought? When approaching this complex problem every effort should be made not to substitute familiar but inappropriate categories to organise this strange, unexpressed thought (albeit acknowledging the use of different criteria to categorise organisms). The absence of any linguistic marker for a category suggests a qualitatively different conception of it than would be so if a label existed for it. So people like the Wola, who do not talk collectively of crops, or other related organisms, differ in a notable regard from those who do, and to talk of 'covert categories' (Berlin, Breedlove and Raven 1968), a concept that has recently gained currency, and to assume an equation with the taxonomic logic of natural historians on the basis of these categories, could be to commit a grave error of interpretation.[3]

The foregoing account of Wola crops is guilty of such distortion. Its arrangement into chapters dealing with tubers, fruits and so on shows this, as does the systematic ordering of each account according to the origin, botany and cultivation of the plants. This organisation of the material is the author's work, although every effort has been made to describe the crops as

[1] According to Strathern (1969: 189), the Hageners have no such mid-level terms either, nor an all-inclusive word equivalent to 'crops'.

[2] Indeed I am not sure to what extent this mid-level category emerged as a result of my persistent enquiries, as something to satisfy me, and to what extent it is a Wola category. However this may be, it shows clearly that they can conceive of such mid-level groups.

[3] This is not to deny the usefulness of this notion for the ordering of others' thinking by outsiders as they perceive it (the previous chapters use it implicitly), but it invariably distorts the insiders' views.

the Wola see and think of them. The imposition of this order was necessary to produce an intelligible account, which might serve as a simplified flora to help others identify the plants described. The Wola do not talk about their crops in this way: they have no reason to do so (which is not to say that they could not do so — they could follow the layout here and appreciate the logic behind it). Therefore a considerable part of the familiarity conveyed by this book actually comes from an outsider ordering the data; in a Wola presentation it would appear stranger. Indeed, the absence of any overt higher-level grouping of crops, coupled with the non-use of specific criteria systematically to recognise them, would sully the accuracy of any attempt to express some of their thoughts in words. The illustrations of the plants are more apt in a sense than descriptions, for it is these, as mentioned in Chapter 1, seen in their entirety as familiar wholes, which count; in fact the reader will achieve a greater sympathy for the unexpressed ideas of the Wola grappled with here by looking at these in the light of the seemingly disconnected comments made about them by informants

The most important taxonomic level for the Wola, though, is the one on which this study rests, namely their classification of crop plants. At this level they give plants names which parallel exactly those of agronomy at the species level. Here they are ordering plants according to objective morphological and other discontinuities, and they are in unanimous agreement over the names that go with particular species (I have never heard anyone confuse plants and names at this level, mixing up for instance yams with sweet potato or taro). The exact correspondence between these names and the labels of scientific botany supports the contention outlined earlier, about familiarity coming from lower taxonomic levels where such parallelism will occur.

Lower down the hierarchy, aberrations become apparent which support the contrary argument, advanced on the absence of terms for any higher taxonomic levels, that the Wola order their crops in a way that differs significantly from the apparently similar logic of biological classifications. At the lowest level of grouping, the Wola divide their crops into cultivars. They base their identification of these on differences in colour, shape and size, micromorphological variations between plants and differences in habitat and growth. They give each cultivar a name.[4]

The names of crop plants, the Wola say, are *semonda imbiy* (lit: family-big names), and those of cultivars are *semg^{e}nk imbiy* (lit: family-small names). These are the same terms as they use for their two levels of local grouping (see Sillitoe 1979: 33), which indicates that they conceive of them as related categories organised hierarchically with one growing out from the other in the same way as local groups represented on genealogies (for instance the crop of

[4] See Panoff (1969: 22—6) on the cultivars distinguished by the Maenge of New Britain; also Fischer (1968) and Powell *et al.* (1975), referred to repeatedly in the foregoing accounts, on the cultivars identified by the Kukukuku and Melpa of the New Guinea Highlands.

sweet potato divides into sixty-four cultivars spreading out like tree branches from the collective or trunk term *hokay*). That the Wola can think and speak of nesting hierarchies, and do so at the lower levels of their crop taxonomy, supports the conclusion reached about the absence of all-embracing names, that the higher-level ordering of crops in this fashion is not important to them.

CULTIVAR CONFUSIONS

Even more telling is the surprising lack of agreement among the Wola over the naming of plants at the cultivar level, which contradicts the argument that at the lower taxonomic levels people are dealing with objective natural discontinuities which impart a concensus to their identifications and make classifications cross-culturally comparable. I first became aware of this absence of agreement over the naming of cultivars when trying to document the hibiscus greens which the Wola call *huwshiy* (*Hibiscus manihot*). Nine men[5] nearly came to blows in an argument which started when I asked them to give me the cultivar names of specimens taken from different hibiscus bushes. They never came to any agreement (although there are only five cultivars of this plant altogether, not many to agree over) and eventually I recorded the majority opinion. Several times later I found myself resorting to this again when describing the cultivars of other crops, and this aroused my curiosity and decided me to conduct the following identification tests in an attempt to understand further Wola thoughts on the ordering of their plants.

The first test concerned sweet potato cultivars.[6] The Wola distinguish between these, as described earlier, by looking firstly at the size, shape and colour of the leaf, which is often enough for identification, and then if necessary, they look at the skin and flesh colour of the tubers. Another point of difference between cultivars is the presence or absence of hairs on the vine stems. In this test twenty-six men and twenty-six women[7] were presented with the same twenty-one sweet potato vines to which they were

[5] The men concerned were Maenget Pundiya, Maenget Kem, Huwlael Kot, Maenget Tensgay, Hoboga Nemb, Wenja Neleb, Maenget Saendaep, Mayka Kuwliy and Wenja Olnay.

[6] See Heider (1969: 80—2) for details of a similar sweet potato identification test that he conducted with the Dani of West Irian.

[7] The following people co-operated in the identification tests, for which I thank them. The men: Huwlael Em, Huwlael Lem, Mayka Gogoda, Mayka Mayja, Wenja Neleb, Hoboga Lem, Mayka Pes, Wenja Olnay, Mayka Kot, Mayka Sorliy, Huwlael Ton, Huwlael Kot, Mayka Muwlib, Huwlael Naenainj, Mayka Duwaeb, Wenja Waenil, Ind Kelow, Puwgael Piyp, Mayka Sal, Ind Ar, Huwlael Shorluwk, Maenget Hiyp, Ind Duw, Ind Kobiyaeb, Maenget Kem and Maenget Saendaep. And the women: Wenja Mabel, Huwlael Pabiym, Puwgael G^{e}nk, Kolomb Towmow, Hoboga Bortnonk, Hoboga Ak, Huwlael Horshiyow, Hoboga Dorwaegol, Puwgael Pul, Hoboga Huwn, Haenda Wen, Huwlael Din, Haenda Kwaen, Haenda Naelomnonk, Puwgael Kolkabiy, Mayka Hendep, Maenget Ibnaewaem, Kolomb Kaelenj, Hoboga Pabiym, Maenget Hundbiy, Piywa Kombaem, Huwlael Aeguw, Ind Pabiy, Huwlael Ponpin, Piywa Kiybaem and Wenja Nonk.

Table 23: The identification of sweet potato cultivars

WOMEN \ MEN	'Balay'	'Bizuw'	'Bordorwiy'	'Daebayda'	'Hagen'	'HULMA'	'Hungabuw'	'Iybob'	'Kauwa'	'KERET'	'KINJUWP'	'KIRAEL' (1)	'KIRAEL' (2)	'KIYGA' (1)	'KIYGA' (2)	'Kiygahaez'	'Kolwep'	'Komiya'	'KONOMA'	'Kuwl'	'Kuwliy'	'MA'	'Maegai'	'Mapopo'	'MAYOW'	'Mbaduw'	'MENDKAUWA'	'Miytbiyp'	'Mormuwn'	'MUKBA'	'Mundiyaem'	'Muwtuwpiy'	'Oba'	'OLSHOR'	'Oma'	'PAENG' (1)	'PAENG' (2)	'Shorhiy'	'Shorwat'	'Sigiliy'	'SIMBIL'	'SOKOL'	'Tat'	'Taengeltwem'	'Taengiyael'	'Toriya'	'Tuwgiy'	'Uwgaiekat'	'Waipgenk'	'WENMUWN' (1)	'WENMUWN' (2)	'Yaebkil'	NO NAME (1)	NO NAME (2)	Do not know
'Balay'						1																																																	
'Bizuw'																																																		1					
'Bordorwiy'												2																																										2	
'Daebayda'												1																																						1			1		
'Denshor'																																					1																1		
'Hagen'										1					1																			1																					
'HULMA'*						6/3						1	3	3				1		1			1				5		1	1				1		1																			
'Hungabuw'																																													1										5
'Iybob'						1					1																			1						8	4																		
'Kauwa'																						1								1																				2					
'KERET'*										7/6				10																																									
																																																					1			2
'KINJUWP'						2					8/8	2	1	1								2		1	2		1			1				1							2				1					1		1			3
'KIRAEL' (1)						1					1	3/7		1	1						1	1		1	1		1		1	1						1									1					1					10
'KIRAEL' (2)						4			1		3		8/8							1		1			1		2									1														1					7
'KIYGA' (1)*						2				14		2	2	24/25					1			1								3				2																				3	1
'KIYGA' (2)									2						2/4			1				1		1			1								3						1								1	2					12
'Kiygahaez'																																		1																					
'Kolwep'										1																																													
'Komiya'						2						1																																						1			1		
'KONOMA'																			21/26						3				1									1			1									1	1		3	1	2
'Kuwl'																																				1																			
'Kuwliy'																																																							
'MA'				1		2		1			1	3	2			1		2				9/4		3	1		1			1													1		1										7
'Maegai'						1																1			1									1		1																	1	1	
'Mapopo'											1	2		1								4			2					3				6			1													1	1		4		
'MAYOW'*						1					1											2			10/23		3				1		1	2			3					1													3
'Mbaduw'																																										1													
'MENDKAUWA'*						5																					18/26										1												1	1				5	6
'Miytbiyp'												1																																											
'Mormuwn'						1				1																																													
'MUKBA'											1	1	2	1		1								1	5					4/5					1		1										1							1	10
'Mundiyaem'																																																							
'Muwtuwpiy'																																																					1		
'Oba'																																																						1	
'OLSHOR'						2								1										1										9/24				1						1											2
'Oma'															1																																								
'PAENG' (1)*							5																							1		1				14/17																			2
'PAENG' (2)				3			3					1				1							1	1	7												6/4	1		1			1		1		1								4
'Shorhiy'																																					1																		
'Shorwat'											1																																							2					
'Sigiliy'						1					1																										2													1					
'SIMBIL'*											1				1							1								1											21/26									1	2		1	1	1
'SOKOL'*										5																																24/26													1
'Tat'												1																																											
'Taengeltwem'																																																							
'Taengiyael'													1																																										
'Toriya'																						1														1	1																		
'Tuwgiy'											1	1	2									1														1																			
'Uwgaiekat'															1																																						1	1	
'Waipgenk'						1																								3																					1				
'WENMUWN' (1)*											1	1			1								1				1											1	3		1									5/15		1	2	1	7
'WENMUWN' (2)*																																									2										19/23				2
'Yaebkil'											3																																										2		
NO NAME (1)		1												1								1			1					1									3		1									5	1		10/12		
NO NAME (2)									1	1				9										2	3		1							3													1							9/5	
Do not know						4				3	3	8	4		12							2			2					6							1													4					

Note: the names of the twenty-one cultivars presented for identification appear underlined; the other cultivar names are 'incorrect' ones given in the identification of these. The cultivars with asterisks were represented by vines only, and the other as vines together with tubers (to ascertain the part played by tubers in identifications).

asked to give cultivar names. The matrix presented in Table 23 records the results.[8] The figures contained in boxes along the diagonal line represent the number of 'correct' identifications (the diagonal line dividing the replies of men in the upper half from those of women).

The matrix reveals a startling lack of agreement over the identifications; in no case did those in the sample agree one hundred per cent over the identification of any cultivar. Taking the replies of men and women as a whole, 49 per cent (534 out of 1092) of their identifications agreed with the majority opinion (taken to identify the specimen). There is a noticeable difference between men and women, with the former scoring only 43 per cent (237 identifications out of 546) to the latter's 54 per cent (297 out of 546). This is understandable and might have been expected to be higher, because it is only women who cultivate sweet potato.

Looking at the individual scores there is no indication that age influences knowledge. Young unmarried persons scored as many 'correct' identifications as mature and old people (of the three top scorers with sixteen 'correct' identifications each, one was an unmarried youth and two were married women in their early thirties, and of the two persons who scored fifteen, one was a young married man and the other an unmarried teenage girl). Indeed, contrary to expectations, the older people did poorest (two old men scoring the lowest with only three each 'correct', one of them making the excuse that his eyes were poor). Similarly, there is no correspondence between a man's status and his ability to identify sweet potato (the man of highest repute in the area for example only managed seven 'correct' identifications).

There is a noticeable trend for those with little idea (especially the men, although some women too), to repeat the name of a single cultivar a number of times in the absence of any firm knowledge (for example the man of renown mentioned above gave the cultivar '*mendkauwa*' six times and another man gave the cultivar '*mapopo*' seven times). This is a reflection of the Highlanders' habit of giving an answer, any answer, to a question (something which sometimes makes fieldwork difficult) — although as the matrix shows, a fair number did say, when presented with a plant which was unfamiliar to them, that they did not know what it was.

Taro plants also exhibit considerable variation and many edible clones are known. The Wola distinguish several cultivars, as described earlier, largely on the colour of tuber flesh, plant stem and leaf ribs. Similar results to the above were obtained in a test with taro, involving the same fifty-two persons, in which they were given nine plants (with tubers) and asked to give them their cultivar names. The second matrix, presented in Table 24 records the results,[9]

[8] When I selected the specimens for this test I thought that a Wola helper and I had collected a wider range of cultivars than turned out to be the case. I though that '*kirael*' (1) was '*bordorwiy*'; *paeng* (1) was '*hungabuw*'; '*kiyga*' (2) was '*oma*'; and '*w^{e}nmuwn*' (1) was '*shorwat*'. I changed these subsequently 'incorrect' identifications to correspond to the majority opinion after the test.

[9] Again, the cultivars given for identification appear underlined, the others are 'incorrect' identifications of these.

showing the same disagreement over naming as found with sweet potato; again there was not one hundred per cent agreement over the identification of any one cultivar.

Taking the replies of men and women, together, 53 per cent (248 out of 468) of the identifications agreed with the majority opinion. And the difference between men and women is the reverse of the sweet potato with men 'correctly' identifying 57 per cent (134 out of 234 identifications) and women 49 per cent (114 out of 234 identifications). The reason for this reversal is that although both sexes plant taro, it is men who take more interest in its cultivation and are concerned with both the rites and spells to ensure its healthy growth and the exchange of tubers which marks its harvest.

Table 24: The identification of taro cultivars

WOMEN \ MEN	'Bobay'	'Daelma'	'Hael'	'Hebiyat'	'Holor'	'Huwshwab'	'Ishma'	'Ishor' (?)	'Iybaib'	'Kaerob' (?)	'Kendkend'	'Keret'	'Kogba'	'KOMBOL'	'Kwiylow'	'Lab'	'MEZ'	'MOND HUNDBIY'	'Mond kaeriyl'	'MOND KONGOL'	'MOND POMBRAY'	'Muwt'	'Naenk'	'Ndobok'	'NEMHAEZ'	'Paela'	'Paym'	'Pindiypindiy' (?)	'Shorhobor'	'SIGAB'	'SIZIL'	'Taegat'	'TAESHAEN'	'Tolop'	'Tongay'	Do not know
'Bobay'																				3																
'Daelma'																				1											1					
'Hael'														1																	2		2			
'Hebiyat'																									3											
'Holor'																																				
'Huwshwab'																	1			2																
'Ishma'																				1											1					
'Ishor' (?)																																				
'Iybaib'																																				
'Kaerob' (?)																															1					
'Kendkend'																																	1			
'Keret'																									1											
'Kogba'																		1													1					
'KOMBOL'			2				1							17/13		4															5		4			1
'Kwiylow'																																				
'Lab'														1				1			2												1			
'MEZ'																	25/23	1																		
'MOND HUNDBIY'																	3	13/11		3	4	1			1	1								4		
'Mond kaeriyl'																															1					
'MOND KONGOL'					2		1	1	1			1		1				6	2	5/2	5			1	1						1				1	6
'MOND POMBRAY'																		2		4	21/20															1
'Muwt'																		1																		
'Naenk'																																				
'Ndobok'																																				
'NEMHAEZ'				1																1					15/18								1			3
'Paela'														1				3		1																
'Paym'																																				
'Pindiypindiy' (?)																				1																
'Shorhobor'																															1		1			
'SIGAB'			1											1	1	1														24/14	1		1			1
'SIZIL'			2						2			2		1			2		2						1	1	1	1		2	5/5	4	6			5
'Taegat'																				1											3					
'TAESHAEN'			4											6		1		1			1		1		2						4		9/8			2
'Tolop'														1				1												1						
'Tongay'																																				
Do not know														4			2	3		5	2				4					5	5		3			

The results of the taro test support the conclusions drawn from the sweet potato identifications. Again, age does not seem to influence knowledge (the highest scorer with eight 'correct' identifications was a young married man and the lowest men's score of two went to two men, both of them in their early middle age; one married woman in her late twenties scored seven 'correct' identifications and an unmarried girl scored the lowest with none). Again the older persons were only average, the oldest man for instance scoring five 'correct' identifications. There is a noteworthy trend, though, for men of high status to score well in taro identifications (three men of the highest repute in the area scored seven each), which is understandable given that this crop sometimes figures in exchanges and it is an ability to excel in these which earns men renown (Sillitoe 1979a).

It is significant that again, as with sweet potato, some individuals with little idea repeated the name of a single cultivar a number of times in the absence of any firm knowledge (for instance one of the lowest scoring men repeated '*sizil*' and '*mond pombray*' twice and '*mond kongol*' three times, and the oldest man in the sample repeated '*sizil*' three times). Some individuals who were ignorant of cultivar names even invented new ones, which others denied referred to any cultivar at all (those who used these names could only point to the specimens they had 'incorrectly' named as evidence that these cultivars existed).

The results of these tests raise some difficult questions about the way the Wola order their crops. They have a number of cultivar names which, by and large, they hold in common, in that everyone agrees that a given list of names refers to the cultivars of a given crop, but when they apply these commonly held names to actual plants they only agree around fifty per cent of the time about which name goes with which plant. This casts the previous accounts of cultivar classifications in a questionable light. These represent the opinions of a small number of informants only, which the above tests indicate are quite possibly 'wrong' in some cases, in that the opinion of larger numbers of people might have changed them. Regardless of this and more importantly, they accurately document the principles by which identifications are made. Yet clearly such ordering differs in scope from familiar systems of classification. Why such absence of agreement?

Others have recently taken up the problem of people disagreeing over taxonomic issues (see Ellen 1975; 1979b; Gal 1973; Gardener 1976; Hays 1974, 1976);[10] although their concerns, unlike those above, centre on variations in the placing of organisms in categories equivalent to crops, and others at higher levels, both named and un-named (for instance, why should some individuals group slugs with molluscs, while others put them with annelids, and yet others with snakes and lizards?). The frequently used and straightforward common-feature or monothetic approach to classification, in which

[10] On other issues relating to informant variability see Heider (1972), Pelto and Pelto (1975), and Sankoff (1971).

a single shared attribute is sufficient for inclusion in a class (as the previous chapters order crops by their edible parts), generates little ambiguity, for something either has the diagnostic trait or does not, which offers little scope to account for these classing disagreements. The multiple-feature or polythetic approach (Needham 1975, 1979: 63–5; Sokal and Sneath 1963), in which members of a class are united by overlapping resemblances in a chain such that there is no centrally significant feature, gives an elasticity which may account for some of them, people linking organisms together differently according to varying shared attributes, any of them being right depending on the situation and their appraisal of it.

This may explain some apparent higher-level category confusions elsewhere, but not Wola disagreements over the identification of cultivars. Similarly, dualistic concepts deriving from Hegel's binary thought, which anthropologists currently use to advantage in contexts of symbolic classifications (discussed in the next chapter),[11] cannot cope with these disagreements, for the Wola recognise crop plants as wholes and not as things possessing a few identifying traits arranged systematically in pairs or any other way. Seeing crops like this parallels current thinking in theoretical biology concerning the apparently inherent misfit between the complex empirical discontinuities of nature and any system of phylogenetic categories. No clear boundaries can be drawn which take account of all possible varying criteria, even computers using modern numerical taxonomic techniques cannot reproduce on a piece of paper the complex network of interrelationships that snake back and forth between organisms in nature (Ellen 1979a: 2–10). The pre-literate Wola have never attempted any graphical representation of course, and so ironically in some senses come nearer to reality in their unexpressed ordering of crops. Their appreciation of these plants differs in some notable regards from any of the above classificatory approaches, which concern themselves with category membership as opposed to issues relating to identification (Ellen 1979b: 341).

POSSIBLE REASONS FOR CULTIVAR CONFUSIONS

It would be incorrect to suppose that, analogously to Western taxonomists, the Wola disagree only over the naming of plants which are very much alike, that the fine morphological distinctions (relating to the colour, size, shape and so on, of different parts of plants) are confused between markedly similar specimens. Indeed this can easily be done with some of the cultivars that are markedly alike, but the matrices show clearly that informants were not simply confusing one or two very similar plants with one another. They were mixing

[11] Also used extensively in kinship studies, were indigenous classifications, largely kin terminologies, have long been subject to systematic analysis (Needham 1971), leading to so-called componential analysis, which many have subsequently criticised as arid and unproductive.

up names which were applied by others to a wide variety of morphologically quite distinct plants. Their disagreements were more profound than straightforward taxonomic quibbles over whether this leaf is more lobed than that one, or this tuber's flesh is a creamier colour than that one.

Tearing some organisms out of context and presenting them for identification in a stuffed, preserved or otherwise tampered-with condition can seriously bias results, so that devising controlled and precise tests is difficult (this is especially so for animals where the absence of behavioural, vocal and other cues can mislead informants — see Hunn 1977: 21). But presenting the above crop specimens out of context did not confuse informants. Everyone knew that the plants occur in gardens only, and the specimens were presented in a fresh condition as they are seen growing.

Similarly, the disagreements are not the result of linguistic confusions, with informants using words from different dialects for the same plant (as Heider 1969 thinks may be the case among the Dani; see also Gardener 1976). All those co-operating in the tests came from the same region and spoke the same dialect of the Wola language. Also, aware of all synonyms (of which there are a number for sweet potato cultivars in particular), I standardised responses under single terms (there was no disagreement among informants over which names were synonymous with one another).

Neither was there any regular pattern in the differing responses which might indicate some social grid controlling them, that different cultural contexts prompt answers that vary consistently (Ellen 1975, 1979b; Gardener 1976). Nobody ever intimated that the naming of cultivars or crops varies in different situations according to cultural dictates; anyway all the identifications were made under the same conditions and in the same setting (the following chapter traces in detail the connection between crops and various social and ritual contexts).

Similarly, there was no observable behavioural patterning to individuals' differing identifications. In everyday contexts for instance they did not regularly differ on principle to score points off one another, they did not compete to demonstrate superior knowledge or establish some public image (the naming of plants is not a recognised context for such behaviour). A related sociological point, returned to in the following chapter, which somewhat begs the question, is that the propensity of individuals to differ reflects their acephalous social order, their refusal to acknowledge any authority, even in giving names (although they do concede that some are perhaps more knowledgeable about certain plants than others).

There is a marked correlation however, between the occurrence of different cultivars in gardens and agreement over their identification. The more often certain plants occur under cultivation the greater the consensus over their naming, as Table 25 shows with its listing of sweet potato cultivars according to the number of gardens in which they were seen growing and their

Table 25: The number and area of gardens in which different sweet potato cultivars were observed. Total number of gardens: 330. Total area of gardens: 390268 m^2

CULTIVAR NAME	NUMBER OF GARDENS	PERCENTAGE GARDENS	AREA (m^2) OF GARDENS	PERCENTAGE OF TOTAL GARDEN AREA
*'Sokol'**	190	57.6	236353	60.60
*'Kiyga'**	294	89.1	360350	92.30
*'Konoma'**	311	94.2	372342	95.40
*'Simbil'**	305	92.4	309706	79.40
*'Mendkauwa'**	49	14.8	60850	15.60
*'W^{e}nmuwn'**	297	90.0	357306	91.60
*'Olshor'**	49	14.8	58433	15.00
*'Mayow'**	99	30.0	108730	27.80
'Paeng'	14	4.2	25407	6.50
'Kinjuuwp'	43	13.0	85153	21.80
'Ma'	29	8.8	37279	9.60
*'Keret'**	1	0.3	293	0.08
'Hulma'	7	2.1	8841	2.30
'Mukba'	2	0.6	941	0.24
*'Mapopo'**	79	23.9	153510	39.30
'Waipg^{e}nk'	31	9.4	44319	11.40
'Komiya'	22	6.7	24227	6.20
*'Oma'**	19	5.8	31658	8.10
'Mundiyaem'	15	4.5	21834	5.60
'Tat'	8	2.4	17572	4.50
'Kuwliy'	6	1.8	13763	3.50
'Toriya'	4	1.2	9929	2.50
'Iybob'	2	0.6	6907	1.80
'Maegai'	2	0.6	4181	1.10
'Baegai'	1	0.3	255	0.07
'Bordorwiy'	1	0.3	178	0.05
*'Kogba'**	1	0.3	648	0.17
*'Maezayzuwla'**	1	0.3	857	0.22
*'Shorwat'**	1	0.3	941	0.24
'Taengiyael'	1	0.3	5653	1.40

* = arrived since European contact

combined areas.[12] The cultivars are arranged in order of the agreement over their identification in the test and those at the top of the list, where agreement was high, clearly occur more often in gardens. This is hardly surprising: it follows that the more often people see and cultivate any plant, the more they agree over its name.

Some cultivars are passing out of use, and others are only remembered as names — the plants have, as the Wola say, been 'lost'. The identification of plants on the way out, which occur less and less in cultivation until they fade

[12] This survey of the occurrence of sweet potato cultivars was rough and ready. One or two informants simply strolled about the gardens concerned calling out the names of the plants they saw (hence Table 25 is biased according to their idea of what cultivar names go with which plants).

away, is predictably poor. Other, recently introduced, cultivars on the other hand have become popular and occur frequently in gardens, and there is considerable agreement over their identification (as Table 25 indicates, with those sweet potato cultivars that have arrived since European penetration of the Wola region (marked with asterisks) tending to occur at the top of the list). One of the consequences of such change is inevitably some disagreement over naming, but it would be wrong to conclude that the sudden flood of new plants arriving with Europeans has sunk the Wola classification system at the cultivar level. No new taro plants for instance, nor hibiscus cultivars, have arrived since contact, yet disagreement is as high over them as it is over sweet potato with its significant number of new cultivars.

Plants come and plants go through natural hybridisation, genetic drift, preference shifts and the arrival of new strains from elsewhere (something that has accelerated considerably since contact with the outside world), and names change with the plants. This elasticity, which, among other things, readily manages change, comes over as an intrinsic aspect of Wola crop classification.

PRE-LITERATE ORDERING

This is a living and ever-changing system, which cannot be recorded at one instance in time — that is, committed to paper — without distortion. This is a critical shortcoming of this account (see Goody 1977; Ellen 1979a: 23–4). The literate reduction of the way the Wola classify their crops entails some crucial changes: it leads to an inevitable hardening and lifeless representation of their perceptions, so lifeless indeed that their ordering appears to verge on disorder; as the cultivar identification tests show clearly. But the apparent disorder, the lack of agreement over classes and the absence of names for them altogether at higher levels, which as pointed out is in a sense a truer reflection of nature, is not the result of error, it is an essential aspect of this oral system.

People inventing names, which others do not recognise, to cover plants they cannot otherwise identify, again shows the flexible and dynamic nature of this orally-transmitted ordering system. Some of these invented taro names for instance occur in spells which men own personally and keep secret (passing them on to their children), but others come to hear of them sometimes and apply them to plants. As generations get their identifications a bit 'wrong' compared with previous ones, and individuals mix up names and plants in this and other ways, they keep the system in flux. It is to be expected that the knowledge of individuals and their reflection upon it varies considerably (Ellen 1975: 220), a large part of it deriving not from any lore passed on systematically from generation to generation but from personal experience (which Morris 1976: 544 calls 'memorate knowledge'). Their vocabularies vary in extent (Hays, 1976: 491–5), so that when referring to the same range of organisms, those with narrower lexicons resort to lumping where those with wider ones discriminate.

In a literate society it is assumed that a classification can exist written down in an authoritative work regardless of the number of people out there who get it wrong. Indeed only a few specialists in such a society, for instance botanists, might know the classification while the majority live in ignorance of it. It is also granted that there may be some disagreement over identifications, in which event recourse to the acknowledged written authority settles the issue. These ideas are inapplicable in pre-literate societies, and to think that any permanent, one-off record can be definitive is to fail to understand their ordering. This qualifies the position of ethno-scientific accounts like this book. By recording the ideas of a few informants permanently they do not rank as reference works against which the identification of organisms may be checked. This study is not an authority (after all many Wola would disagree with parts of it), although it does accurately record the principles by which identifications are made.

Given the permissible absence of conformity in such systems of oral ordering, it is somewhat invidious to think in terms of correct and incorrect identifications, and to invent higher level categories that do not overtly exist. After all, it is recording them on paper that brings these inconsistencies to light. Yet in any classificatory system there has to be some consensus for it to exist, be it in people's heads or reference books. The data discussed here indicates that this consensus can be remarkably low: which raises a considerable problem for the study of orally transmitted classification systems, namely, what is the tolerable range of expected disagreement? How standardised must people's identifications of objects, and their collection of them into higher level groups be to allow a system of classification to work? Such a system, by definition, is one in which a number of people speaking the same language agree to arrange things into classes. An important word in the last sentence is *agree*, because without agreement it is difficult to conceive of classification; where individuals give different names to the same thing they have only an individual classification scheme known to themselves, not shared with others, and in such a situation they cannot communicate with one another, they are in effect speaking different languages.

The question is: when do we cross from classification to chaos? How many people have to disagree before the system ceases to exist — one per cent, ten per cent, fifty per cent? The matrices show that among the Wola a surprising amount of variation is permissible. In the identification of some cultivars there is a high level of agreement (over 90 per cent) while with others it is hardly meaningful to talk of a consensus (with as few as 12 per cent of those questioned agreeing over the identifications).

The Wola are similarly inconsistent in their identification of other natural phenomena such as marsupials, birds and other plants. And judging from the ethnographic literature they are not the ignorant exception but the pre-literate rule.[13] Yet, while it is acknowledged that writing has profoundly influenced European thought (Whorf 1956), sufficient cognisance is not always given to the fact that it is linked in some measure to our idea of the taxonomic hierarchy, so that the identification of levels in others' ordering of some aspects of their natural environment, and equation of these with the phylogenetic levels of the relevant biological science, is something which demands great care to avoid distortion.[14] There is an important distinction to be made, as Randall (1976) cogently argues, between the presence of a culturally sanctioned paradigm, like a taxonomic tree, and people displaying an ability to reason in this fashion from what they know and can remember when prodded by questions from someone who has such a model in mind. People may have the potential to reason this way and use transitive logic (as Gardener 1976 and Hays 1976 demonstrate), even when presented with phenomena strange to them, but the important question is: do they normally classify like this for themselves? Similarly, while informants' knowledge inevitably varies and no one knows everything, the achievement of an authoritative record comparable with the knowledge of an '"omniscient" native speaker—hearer' (Werner 1969: 333), which the equation of indigenous terms with the closely defined labels of Western science implies, assumes a fair degree of consensus over the identification of the things concerned. The situation with Wola crops questions this assumption. Only when there is almost unanimity over identifications (there will almost inevitably be some dissenting voices), as there is with the Wola as the crop level of naming, is it really legitimate to make any explicit equation.

GOOD TO EAT IS GOOD TO CLASSIFY

Confusion is inevitable without some mnemonic aid to record an elaborate classification outside the head (even given the impressive memories of preliterate people like the Wola, who regularly put the ethnographer to shame with their powers to recall). So, why have the Wola, and other pre-literate people, zealously over-stepped the capacity of their memories in their labelling of natural phenomena? After all, they could lump similar organisms together under single names, so reducing the scale of their classifications and achieving higher levels of agreement over identification. Why are the Wola for example,

[13] For further ethnographic examples of similar 'confusions' of identification see Hunn (1977: 36—7, 210) on disagreements between Mayan informants over the naming of mice in the genus *Peromyscus*, and Berlin, Breedlove and Raven (1966; 1974: 58) on their differences over identifying various plants; and Bulmer and Tyler (1968: 375, 381) and Bulmer and Menzies (1973: 101—2) on the lack of consistent correspondence between Karam frog and marsupial identifications and those of zoological science. Nearer to the interests of this book, see also Heider's (1969: 81—2) account of an experiment with Dani women in identifying sweet potato tubers, which produced similar conflicting results to the ones reported here.

[14] Others have recently pointed to the dangers of reifying the notion of the phylogenetic hierarchy (Ellen 1975, 1979a: 12—13; Friedberg 1968, 1970, 1979; Hunn 1976; Randall 1976).

not satisfied with their crop level of classification, going on in a muddled fashion to distinguish between cultivars? As Heider (1969: 78) points out, distinguishing between many tens of sweet potato cultivars does seem somewhat exorbitant.[15]

In their study of Mayan plant taxonomy, Berlin, Breedlove and Raven (1966; 1974: 96—103) compare the extent to which plants are classified with their importance in subsistence and everyday life. They come up with a positive correlation and conclude that the larger the part plants play in people's subsistence then the more elaborately they will differentiate between them. The Paiute classification of flora and fauna (Fowler and Leland 1967), based on the usefulness of the things concerned, is an even more explicit re-affirmation of the so called Herskovits—Nida—Conklin hypothesis, that the vocabulary of fields relates to their practical importance (Ellen 1979b: 350).[16] These and other writers bridle at Lévi-Strauss's (1966: 1—16) and Douglas's (1966) confusion of the study of *concrete classification* with the different process of *abstract association* (see Morris 1976, 1979; Bloch 1977; and also comments in the next chapter). The latter Lévi-Strauss equates with totemism, naming and so on, and castigates Malinowski (among others) for explaining the association of natural organisms with such social phenomena in terms of their importance to the people concerned, notably the part they play in their diet. In so doing he somehow implies that the detailed classification of related natural phenomena is likewise not associated with their importance to the livelihood of the people concerned, which is manifestly not so.

The Wola have developed their elaborate crop classification because these plants are important to them in a primary sense — they need them to survive. They think *about* them and not *with* them because they take up so much of their time and are essential for their subsistence (although they do use them sometimes as symbols in abstract thought, but this is an entirely different issue). This is not to argue that the Wola classify these plants elaborately because they are good to eat, it is more that their importance in their livelihood and their ensuing constant presence somewhat compels these people to distinguish between them and classify them.

A test conducted on the identification of sweet potato tubers by taste alone indicates that their discrimination does not rest on facile gustatory grounds. In this experiment several different sweet potato cultivars were cooked and pieces of the tubers fed to men who were asked to identify them. As Table 26 shows they consistently identified accurately some tubers, such as '*keret*' and '*kogba*', which have distinctive coloured flesh. But with others

[15] Although compared with the Irish potato classification of the Aymara Indians of Bolivia, who distinguish between over 209 named cultivars (La Barre 1947), the sweet potato classifications of New Guinea Highlanders pales a little (with Strathern's (1969: 193) ninety Melpa names standing as the current record).

[16] *Contra* Heider's (1969: 78) consignment of these explanations to 'the facile anecdotage of undergraduate teaching'.

Table 26: The identification of sweet potato cultivars by taste

CULTIVARS	CORRECT	INCORRECT	'DON'T KNOW'
'Bizuw'	4	1	0
'Boraynonk'	0	5	0
'Bordorwiy'	2	2	3
'Daebayda'	3	3	1
'Hagen hokay'	3	1	1
'Iybob'	1	4	0
'Keret'	5	0	0
'Kinjuwp'	0	4	2
'Kiruel'	0	3	3
'Kiyga'	0	5	2
'Kogba'	4	0	1
'Konoma'	2	2	3
'Kuwl'	2	3	0
'Kuwliy'	3	0	2
'Ma'	2	2	1
'Mapopo'	0	2	4
'Mayow'	2	4	0
'Mendkauwa'	4	0	2
'Mormuwn'	4	0	2
'Mukba'	2	2	1
'Mundiyaem'	1	3	4
'Olshor'	5	1	0
'Shorwat'	3	1	3
'Sigiliy'	0	5	0
'Simbil'	3	2	1
'Toriya'	2	1	2
'Tuwgiy'	2	3	0
'Waipg^{e}nk'	1	2	3
'W^{e}nmuwn'	6	0	1
No name	0	0	5
Totals	65 (37%)	62 (36%)	48 (27%)

they were not so accurate.[17] Indeed, overall they identified correctly only 37 per cent of the 175 pieces of cooked tuber given to them, which is poor even by Wola standards of agreement over the identification of entire plants. There is no discernible pattern either to the responses in terms of age, status and so on. So the Wola do not simply classify their crops into cultivars for practical reasons related to differing flavours and preference for the taste of some over others. They do so because of the importance of these plants in their livelihood and almost constant presence in their minds and lives.

Comments made by the Wola also support this contention. Although sweet potato tubers vary according to the hardness, consistency and dryness of their flesh, they claimed no marked preference for the taste and texture of certain cultivars over others.[18] They maintain that it is not only the cultivar type which determines this, but also the condition of the soil in which the tuber

[17] In this experiment all the participants agreed beforehand on the identity of the cultivars to be cooked, so there was no confusion here with them giving different names to the same tuber.

[18] They found amusing the ethnographer's finicky demands for soft and powdery tubers.

grows (a soft wet earth for instance, producing tender and moist tubers). When asked about their changing preferences, from some traditional cultivars to others arriving since contact, informants maintained that it was not the flavour they preferred but the speed with which the introduced plants grow and the overall larger tubers which they yield. But neither speed of growth nor size of crop vary with sufficient predictability to distinguish between cultivars. They vary markedly according to the conditions in the garden where the plants are cultivated, the fertility of the soil, the weather during their maturation, and so on. Also, as identification does not rest on these fluctuating criteria, the existence of the cultivar level of classification cannot be related to them in some functional sense (as Heider 1969: 83–4 suggests with his insurance and long-term harvesting hypotheses).

The extent to which the Wola divide up all their crops into cultivars parallels closely their relative importance in the diet. This is a further reflection of the concrete nature of the system, of the importance of these plants as food, not 'bits' of thought. Table 27 compares the number of cultivars each crop is divided into with their percentage occurrence by weight in the Wola diet.[19] It shows a clear association between the numbers of cultivars and importance of crops in the diet. The sweet potato for instance, constitutes 74 per cent of the total diet and has the largest number of cultivars at 64, whereas sixteen crops, each with only one cultivar, together make up less than 0.4 per cent.

There are some anomalies in the table which demand comment because they apparently contradict the argument advanced here; these concern taro, the screw-pine and pumpkin. Taro has a large number of cultivars but makes up only a little over 1 per cent of the diet. The probable explanation for this is that it is a classificatory hangover from a previous age when this crop was considerably more important. This is plausible because the available evidence suggests that this ancient crop was, prior to the arrival of the sweet potato, the mainstay of the diet of the ancestors of the Wola (see Bulmer 1964; Powell *et al.* 1975).

The periodic nature of the nut harvest from the *karuga* screw-pine accounts for its apparent contradiction of the above argument. During the three-month period when data on food consumption was collected, these nuts never came into season, hence their low occurrence on the table. Some of these trees yield a few nuts throughout the year but when large numbers of them come into season (as happens at intervals), the consumption of their nuts rises markedly (largely at the expense of sweet potato), and at these times they constitute the greater part of many meals. Hence their detailed classification into cultivars.

[19] The data on crop consumption, from which these figures are calculated, were collected in a survey of the food eaten by members of twelve homesteads over a three-month period (see Chapter 12). The weights used in the calculations include pods, inedible skins and other waste.

Table 27: Number of cultivars compared with importance of crops in diet (by weight). Crops scoring 0% were not eaten during the survey of food consumed from which these figures are calculated. These figures are calculated from the raw unprepared weights of the plants concerned

CROPS	NUMBER OF CULTIVARS	PERCENTAGE IN DIET
Acanth greens (*omok*)	1	0
Acanth greens (*shombay*)	3	0.710
Amaranth greens (*komb*)	5	0.011
Amaranth greens (*mbolin komb*)	1	0
Amaranth greens (*paluw*)	2	0.282
Bamboo	1	0.008
Bananas	10	0.923
Banana stem heart		0.044
Beans, common	4	} 0.594
Beans, hyacinth	4	
Beans, winged	1	0
Cabbage	5	0.233
Chinese cabbage	2	0.009
Choko	1	0.068
Climbing cucurbit	1	0.007
Crucifer greens	6	0.159
Cucumber	2	0.171
Dye plant	1	n.a.
Fig	1	0.004
Ginger	3	0.026
Gourd	3	0.244
Hibiscus greens	5	0.179
Highland breadfruit	1	0
Highland *pitpit*	9	7.407
Irish potato	2	0.009
Kudzu	2	0
Maize	2	0.978
Onion	1	0.023
Palm lily	25	n.a.
Paper mulberry	1	n.a.
Parsley	2	0.221
Passion fruit	1	0.002
Pea	1	0
Pumpkin: fruit	1	10.193
Pumpkin: young leaves		0.491
Screw-pine (*aenk*)	45	0.651
Screw-pine (*wabel*)	4	0.077
Sedge	1	n.a.
She-oak	1	n.a.
Spiderwort	1	0.002
Sugar cane	12	1.213
Sweet potato	64	73.841
Tannia	1	0
Taro: tubers	43	0.944
Taro: leaves		0.203
Tobacco	6	n.a.
Tomato	1	0.018
Watercress	1	0.055
Yam	1	0

The other anomalous crop is the pumpkin, the situation with which is the reverse of the above two crops. Pumpkins make up a larger part of the Wola diet by weight than any other crop except sweet potato and yet there is only one pumpkin cultivar. The absence of cultivars cannot be explained by this crop's recent arrival among the Wola because they have divided up into several cultivars other equally new arrivals (such as the common bean and cabbage). These plants have morphological and other discontinuities that make this possible, whereas pumpkins do not. The Wola say that it would simply be too difficult to distinguish between different plants on morphological or other grounds because a single plant can produce a variety of fruits differing in shape, size and colour (presumably fertilisation of flowers by pollen from different plants can produce both round and long fruits, all differing in size and colour, on the same plant). They cannot discriminate between different pumpkin plants on morphological or other evidence when they have no predictable natural discontinuities to allow it.

Allowing for the above aberrations, Table 27 supports the contention that the number of cultivars into which any crop is divided parallels the plant's importance in the livelihood of the Wola. Although not as painstaking as their ordering of crops, their classification of other natural phenomena supports this observation. To give an example: women and children eat a number of different insects, all of which are named. The Wola also name other insects which are inedible, and either are large and frequently seen or feature in their life (for example, as pests). Some other common insects they collect together under single terms (for example, they give a single collective name to all ants, regardless of their differing size, colouring and habitat). To the remaining insects they give no names at all, saying they are *imbiy na wiy* (lit: name not is) or *elel biy sem aengora* (lit: around do family only).

There are a number of plausible explanations for this lumping together of different creatures under a single name, or the absence of any distinguishing name for them at all. One relates to the nature of the pre-literate classification under discussion. There are far more species of animals and plants in the Wola region than they could possibly discriminate between and give names, and lumping organisms together, and simply omitting to name them, are ways of coping with this overload.

This comes back to a question left hanging in the air at the beginning of the chapter: why do the Wola not collect crops together into higher-level labelled groupings and ultimately into a single named supra-category? This relates to the common and, for the proponents of the taxonomic hierarchy's universal applicability, the worrisome absence throughout the world of so-called 'unique beginners' (Berlin *et al.* 1973), that is words equivalent to the English kingdom categories of plant and animal (Hays 1979: 255). The evidence presented here suggests an explanation for this omission, connected to the important pragmatic function of any classification. Whatever the French social philosophers and American cognitive analysts argue, people like the Wola name and order organisms to facilitate communication, and to store and pass on information so that individuals can make decisions and act with maximum effect in an unpredictable world (Randall 1976: 552). They use category words for identification in daily speech, to point things out in order to say something about them (Needham 1979: 3) not to reflect overtly on the world in either metaphysical or nascent scientific terms (regardless of what may be going on at a symbolic level): hence this study's effort to relate what they think about crops to wider everyday issues concerning these plants.

Put another way the problem is why people use categories at all instead of simply referring to organisms by their individual names. A plausible explanation relating to the foregoing is that more inclusive higher-level terms facilitate communication when a more specific name is either unknown (for example, referring to a strange ligneous plant as '*iysh*' (lit: tree) not beech, oak, pine or whatever), or the exact identity of the thing is not certain (for example, seeing a marsupial briefly up a tree a person may not be clear which kind it is and say '*sab beray*' (lit: large-furry-creature sits), unable to refer to it specifically as a Ringtail, Spotted Cuscus, Striped Possum or whatever), or there is disagreement over the use of a more precise term (for example, if one person thinks a dog is a pug and another a bulldog, they can compromise on ugly dog — see Gal (1973)). Communication is thus made possible, although less precise in denotation (Hays 1976: 497), and for this a complex ascending hierarchy of category terms is not necessary; consequently, as Glick (1964: 275) pointed out some time ago in a New Guinea context, folk taxonomies are usually shallow, often of two levels only (corresponding largely to the 'life-form' and 'specific' levels of Berlin *et al.* 1973). While 'tree', 'marsupial' and other more inclusive 'life-form' terms are less precise, they facilitate meaningful communication, where supra-categories equivalent to plant or animal would be redundant, for these are too general to have any signifying value (for example, to say 'I'll meet you near the plant at such-and-such a place' adds little or nothing to the injunction 'I'll meet you at such-and-such a place', whereas 'I'll meet you under the big tree at such-and-such a place' is notably more precise).

So it is with the Wola and their crops. They have no need for an overall term equivalent to 'crops' because everyone can accurately identify these plants at the species level. Indeterminancy at the cultivar level, though, prompts the use of 'specific' names as surrogate 'life-form' terms; for instance, if confusion arises over the use of a sweet potato cultivar name like '*konoma*', people can refer simply to *hokay* and get most of their message across. In other words, the hierarchical features which may be detected in Wola crop classification, and in their taxonomic arrangement of other natural phenomena too, come from the existence of category terms that may be substituted for more precise names when for some reason their exact meaning is not clear. The ethnographic literature abounds with apparent taxonomic oddities that support this line of reasoning; for example, the Karam and Nuaulu can afford to show indifference to or mix up their higher-level categorisation of the flightless cassowary because it is a unique creature never confused with anything else, unlike marsupials which they classify with care (Bulmer 1967, Ellen 1975).[20]

The contexts in which the Wola use different-level terms is a significant consideration. They refer to cultivars frequently in domestic contexts where considerable agreement occurs over which names apply to which plants, particularly between parents and children, the former teaching the latter their versions. Spouses and co-wives may differ more, especially if they grew up in settlements some distance apart between which considerable differences in naming occur; although in time they may learn how each applies terms to plants, a man coming to know for example that when his wife talks about the '*kwiydol*' hibiscus cultivar she refers to what he calls '*shumbuwhond*', and *vice versa.* Thus, between those using cultivar terms frequently, who share the same gardens and jointly consume their produce, there occurs the highest congruence in naming. Beyond these family groupings there is less agreement and people more commonly have recourse to higher-level crop terms. While accounting in some measure for the continued existence of otherwise confused cultivar naming, this does not explain the absence of wider standardisation.

CONFUSING CLASSIFICATION

Crops are too important for lumping or non-labelling and this accounts for their over-zealous classification, which outstrips the capacity of the human memory. It would seem that with the particularly important things in their world (and there is nothing more important than their crops for their continued existence), the Wola almost spontaneously as individuals classify and name without, in all cases, following an agreed code. Without pretending to address itself squarely to the philosophical issues relating to such classification, this chapter sounds an empirical warning shot across the bows of those who do steam on this course without sufficient regard to the significant differences that exist between cultures.

The distorting mirror effect, as Lévi-Strauss (1966) aptly puts it, where we see familiarity together with strangeness in any foreign culture, plagues any discussion of exotic classification. On the one hand, to what extent are the distortions with a familiar feel to them part of the alien cognitive system described and hence, as Lévi-Strauss (1966) argues, evidence of some common bedrock upon which rests all human psyches ('savage' and 'civilised')? On the other hand, to what extent do these familiarities result from the ethnographer

[20] The work of certain universal theorists also paradoxically lends support to this straightforward explanation; Brown's (1977: 330—2) correlation of 'life-form' terms with species diversity for example, shows after a fashion that as the latter decreases and confusions become less likely so the number of all-embracing names declines, and there is Berlin's (1972: 83) observation that urban dwellers rely heavily on 'life-form' terms as their knowledge of particular names decays with their separation from nature.

interpreting what he sees and thinks he hears into his own cognitive system, so imparting to them an inevitable feeling of similarity, somewhat distorted like our own image in a fairground mirror? Committing spoken and memorised ideas to paper compounds the problem, and demands great sympathy if any meaningful appreciation is to be achieved without doing unnecessary violence to the indigenes' thinking and using a travesty of it to promulgate the observer's preconceptions.

Section II: Crops discussed

Chapter 9
THINKING WITH CROPS: THEIR ABSTRACT ASSOCIATIONS

While the utilitarian or consumable value of organisms, as argued in the previous chapter, correlates with the elaborateness of their classification, this does not imply a crude material rationale at the root of this cognitive process. The Wola for instance, classify many things in their natural environment which are useless, although not in the painstaking detail they do crops. One possible reason they name some things and not others, related to the foregoing utility argument, is the frequency with which they occur, things seen often demanding more attention. For instance, it is not enough to call a frequently-seen woody shrub just 'a tree', it has to be the 'such-and-such tree'.

This classification of non-utilitarian natural phenomena raises starkly the issue of why the Wola, and by extension others, classify anything at all. Social philosophers addressing themselves to this question relate their speculations to society, as opposed to the bio-psychological functioning of the brain. Two notable anthropological studies in this tradition are Durkheim and Mauss (1963) and Lévi-Strauss (1966). These concern themselves with the abstract association of things that physically are invariably dissimilar (often tracing their cognitive connections in a binary fashion, for example putting a natural species with a certain social group). This approach, as pointed out, differs markedly from that of the last chapter which attempts to explain the logic behind the differentiation and ordering of like things (usually the taxonomic ordering of natural phenomena).

Some anthropologists (Morris 1976, 1979; Bloch 1977), responding largely to Lévi-Strauss's (1966: 138–9) and Douglas's (1966: 112) proposal that technologically primitive societies have undifferentiated views integrated through all-embracing systems of symbolic logic, have recently warned against the confounding of concrete classification with abstract association, or non-ritual with ritual communication (Bloch 1977: 285), or 'formalism' with 'social constructionism' (Ellen 1974: 4).[1] Although they have a point, at least in some ethnographic contexts, it is necessary to guard against a swing too far in the opposite direction, taking the distinction between the concrete and the abstract too rigidly. Each relates to and may influence the other in a number of ways (Ellen 1979a: 20–2); after all, many of the same principles operate in both cognitive domains (Needham 1979: 62).

This chapter turns to look at the abstract associations of crops with social phenomena in an attempt to understand further how the Wola conceive of these plants and order them. Following on from the last chapter, it proceeds on the assumption, advocated by Friedberg (1979: 84), that an appreciation of their use in cultural contexts depends on seeing their place in the overall classificatory scheme, that it is questionable to isolate restricted taxonomic points to account for the cultural roles of particular species (Douglas 1966). Although it has no significant contribution to make to the debate why humans have the facility to classify, this chapter underlines some pedestrian points that are not always given their due regard, so serving as an empirical counterweight to high-flying abstract philosophies that sometimes appear to have lost touch with the ground where we find human beings behaving and expressing their thoughts.

CROPS AS TOTEMS

According to Lévi-Strauss (1966; 1969), totemism is particularly informative about the workings of the 'savage' mind.[2] Totemism is the association of a social group with some natural phenomenon, usually in the sense that members of the group in some way believe they descend from this thing. It is common for its 'descendants' to observe some kind of avoidance of the thing concerned (in the case of edibles, usually avoiding eating them).

The Wola have totemic-like associations which fall into this broad category. They associate the founding of their *semonda*, the land-controlling groups, with the actions of some animal or plant in the mythical past. They relate these actions in stories which they call *ol maerizor injiy* (see Sillitoe 1979a: 43–5 for two such myths and further details). Of the fifty-four *semonda* that fall within the social universe of Haelaelinja (for a list see Sillitoe 1979a: 35), four claim descent from an edible crop and one from an inedible crop; the remainder claim to descend from birds, insects, reptiles, trees and grasses.

The four *semonda* claiming descent from an edible crop think that they come from sugar cane, and in their origin story they specify the cultivar '*ar*'

[1] See Needham (1979: 3) for a straightforward definition of class and symbol: the grouping of like things together according to shared traits, and the using of one thing to stand for another.

[2] Regardless of his dubious claim to demonstrate that it does not exist, after all he retains the word throughout his work prefixed occasionally with a disdainful 'so-called'.

as their 'ancestor'.[3] According to the story some '*ar*' sugar cane impregnated an unmarried girl who then gave birth to two sons who founded the present-day *semonda* of Yarol (centred on a place called Ndidiym) and Wolol (centred on a place called Kwaray; literally translated Wolol means 'sugar cane men'). After this, in the historical past, some people left the latter place to found two new *semonda* called Wolol (centred on Dayowtinja) and Wolharok (centred on Shariyp).

The members of the territorial group that claims descent from an inedible crop point to the palm lily, from which men obtain their buttock coverings, as their 'ancestor'. In their origin story they specify the cultivar called '*diy*' and relate how it procreated a son who founded the present day *semonda* of Shuwma (which is centred on a place called Shuwma-howma).

A point that should perhaps be made clear in the light of the previous chapter's argument is that the utilitarian argument of totemism, which Lévi-Strauss rightly ridicules, does not apply to the Wola. In the first place the range of organisms involved is too restricted to give this explanation credance. Furthermore, from a subsistence point of view, some of them are useless, while other very important things are omitted. Again, although the Wola pay lip-service to the idea that people ought perhaps to avoid the organism from which their group descends, only those descended from inedibles even pretend to follow this dictum, the others certainly do not (the men of Wolol *semonda* for instance, grow and chew sugar cane like everyone else). So the totemic beliefs are not acting to conserve food resources or control their allocation in any way.

Adopting a compromise position between the undifferentiated world-view approach and a too rigid distinction between pragmatic and symbolic classification, Bulmer (1978, 1979) argues that the Karam use as totems, and mark ritually in other contexts, those creatures which are salient, prominent representations of their classes. This is so for the Wola with the sugar cane cultivar '*ar*', which grows the largest cane and is pre-eminent in this class, but it is not so for the palm lily cultivar '*diy*' which, with its small green leaf, is only ordinary compared to the valued red and yellow variegated plants. There is also the knotty problem of why only these two plants are totemically marked, and not other more important ones. This matches the situation with other organisms the Wola believe they descend from, which appear as a random collection displaying no regular correlation between their everyday classification and totemic standing (see Worsley 1967: 153–6).

The extent to which the Wola are using natural phenomena to think about and order their social environment is not clear either. In everyday conversation they rarely refer to the organisms concerned. They refer to *semonda* by using their names or the place name where most of their members live.

[3] Warner (1962: 405) notes the irony of some people in New Guinea having legends recounting their origin from sugar cane, when the reverse is the case, these people being the first in the world to domesticate this plant (see Chapter 4).

Indeed I have never heard anyone refer to them by using the names of their 'totemic' organisms. Moreover, given the relatively small part which these territorial groups play in Wola social organisation anyway, it seems doubtful to suggest that they use their totemic ideas to think about them to any extent (Sillitoe 1979a: 30–46).

Indeed the nature of Wola ideas on this topic do not fall in readily with any explanation; they are, for me, an enigma. But the purpose of this study is not to account for Wola 'totemism', it is to see how the association of crops with land-controlling groups furthers out understanding of these plants. Neither of the crops concerned (the palm lily or sugar cane) holds a special place which accounts for their selection as ancestral emblems:[4] although the Wola believe a few *semonda* originated from these crops in the mythical past, this does little to advance our understanding of their conception of crops in general.

CROPS IN RITUALS

A similar negative conclusion is reached over crops occurring in certain rituals. Here, some characteristic of the plants selected makes them apt symbols for something in the rites concerned. A few examples follow.

During the *porot* ritual, which men perform to increase the flow of wealth to themselves (see Sillitoe 1979a: 160–4 for further details), participants eat a diet of bananas and meat only. Informants explained that they eat bananas because these curved fruits resemble the crescent shape of pearl shells, and they grow prolifically in pendulous great bunches, in the same way that participants in the ritual anticipate receiving bundles of shells, not just one or two here and there.

When people dream that they are going to die, or see an omen to this effect, they prompt the performance of a ritual called *aenk way boi* (lit: screw-pine crown-cutting plant). In the course of this rite a small screw-pine crown (together with some other plant cuttings) is planted at the threatened person's feet. The crown of this tree is already living (unlike a seed) and grows vigorously when planted, so (informants explained) the person fearing death is again firmly planted in life and encouraged to grow quickly and strongly like the screw-pine.

Following a death attributed to a malicious ancestral spirit attacking and 'eating' someone's vital organs, relatives of the deceased stage an elaborate series of rituals to protect themselves from attack and banish the spirit responsible. Taro figures in both the rites and the spells recited during them. The reason, a ritual officiator explained, was that the chafer beetle, which

[4] The erect nature of the plants concerned (indeed the phallic-like appearance of a length of sugar cane) might explain why they were chosen as procreative agents. This observation bears on the discussion of the next chapter. But not all totemic organisms are erect.

the Wola call *twenj* (a Scarabaeidae, *Papuana* sp.), sometimes attacks taro corms and eats out their insides. In the same way ghosts attack and 'eat out' the vital organs of their victims. Roasting a healthy taro tuber in a fierce fire 'drives out' any lingering ghost. Scraping off the burnt outside while reciting a spell, and finally sharing the tuber with the deceased's relatives, 'strengthens' them and ensures that their insides remain whole and uneaten.

The Wola use other crops in rites because they have a specific effect. For example, people who believe that they have been poisoned must take an emetic to expel the toxic substance from their stomach. This potion contains, among other ingredients to induce vomiting, some diced up ginger rhizomes. The hot and burning sensation which they cause, people say, helps to produce the desired stomach contractions.

In all the above cases the crops have some property which makes their use apt in the rites concerned, but this relates to the rituals in question and not the plants themselves.[5] The crops are simply contrivances of the rituals which, although they tell us something about the properties seen in the crops by the Wola, are not instructive on how they conceive of these plants as a category.

NAMES OF CROPS

Perhaps the words used to label crops and cultivars can further our understanding of this botanical category.

Some of them have unambiguous meanings which the Wola can relate. For example, they explain that they call two sweet potato cultivars '*hagen hokay*' (lit: Hagen sweet potato) and '*tari hokay*' (lit: Tari sweet potato) because they come from the direction of the administrative centres of Mount Hagen and Tari, where in their opinion they originated. Both of these plants are recent arrivals; they were first planted in the Haelaelinja area some time in 1975 and people can remember the first women to obtain and plant cuttings (in the case of '*tari hokay*' for instance, it was Mayka Kot's wife Lenday, who obtained them from her natal place, which is located towards Margarima patrol post in the direction of Tari). The Wola can explain the names of other recently arrived crops and cultivars in similar prosaic terms. The newly arrived tannia plant they call *mbolin ma*, meaning 'whiteman's taro', after its similarity with indigenous taro plants; and the 'Dwarf Cavendish' banana cultivar, which arrived in the mid-1970s, they call '*banana*', after the Pidgin word for it.

The majority of crop and cultivar names that are amenable to such straightforward explanations apply to recently arrived plants,[6] although there are

[5] For some comparative ethnographic data see Girard (1957) on the Buang people of Morobe Province, New Guinea.

[6] See Barrau (1979: 143) on the coining of names for new plants and how these can help in understanding an exotic botanical taxonomy, although the way the Wola cope with new crops questions his optimism about the placing of the unknown revealing the operational logic of a classification system.

some traditional ones that can be explained in this way too. One such for example, is the sweet potato cultivar '*oba*', a pre-contact plant that people say has that name because it is found growing predominantly in a place called Oba (about three miles east of Haelaelinja). A few plants even have names that are a mixture of recent and traditional idioms. The '*mendkauwa*' sweet potato cultivar, for instance, is an introduced plant which came, people say, from the Mendi direction — hence the '*mend*' prefix — but no one can explain why it is called '*kauwa*', for it bears no resemblance to the traditional sweet potato cultivar of this name.

The term '*mendkauwa*' falls on another dividing line too, in that part of it can be explained while the other cannot. The names of many plants similarly have no explanation. When asked what they mean the Wola shrug their shoulders and say that they are just names that refer to certain plants. Other than this they say that they have no other meaning. Sometimes, though, they will point out that the same name is associated with two crops because they have something in common. This is the case with tannia, the 'whiteman's taro' mentioned above. Similarly there is both a sweet potato cultivar and a taro cultivar called '*kogba*' or '*muwt*'. The tubers of both plants have purple veins running through their flesh, hence the use of the same name. Not all associations are entirely consistent. Another synonym for the purple-veined sweet potato cultivar is '*oma*', but this does not apply to the taro plant.

On the other hand there are plants that have the same name but nothing in common; for example there are both sweet potato and taro cultivars called '*keret*' with nothing apparently in common. Interestingly the name '*keret*' is a synonym for '*tari hokay*', the recently arrived sweet potato cultivar 'from' Tari. While informants can explain the '*tari hokay*' name as described, they cannot account for the '*keret*' one. It is, they say, just an alternative name given to that plant, one which they think came with it and so originated elsewhere. Clearly such associations of names do not explain their meaning. Why call taro *ma* in the first place, or purple-veined tubers '*kogba*' and '*muwt*'?

Sometimes it is possible to speculate on the possible origin of a name. The name '*keret*' for example, might come from a species of screw-pine called *keret* which has red thistle-like fruits, the redness of which parallels the orangey-red colour of the flesh of the sweet potato tuber.[7] Few names are amenable to speculation of this kind though. Resort may then be made to the literal meanings of the words concerned (where they exist). The meaning of the following sweet potato cultivar names might be sought in their literal translations: '*denshor*' (lit: herb leaf), '*olshor*' (lit: man leaf), '*shorhiy*' (lit: leaf short), '*waipg^{e}nk*' (lit: carving small). But the Wola meet with a blank look any suggestion that the meanings of cultivar names are related in any way to their literal translations. They agree for instance, that *den shor* means

[7] No Wola ever made this association: it is the product of the author's imagination.

'the leaf of a herb', but insist that when '*denshor*' is applied to a sweet potato plant it is simply a homonym of the former phrase.

A possible next step is to split up the words used for crops and cultivars and look for some pattern of meaning in their syllables. This would make even less sense to the Wola than searching for origins in the literal translations of plant names. Imagine the reverse situation, of a New Guinean anthropologist trying to convince an Englishman that the meaning of potato lies in the phrase 'pot a toe' — what he would ask, having no knowledge that this is a loan word of Haitian origin, do toes have to do with pots and tubers? Given the understandable refusal of the people concerned to go along consciously with such reasoning, recourse may be made to the subconscious. Plant names can be either literally translated into phrases or split up into syllables and then grouped according to recurring elements, from the patterning of which the association of these words with certain plants could be speculated upon. For instance all Wola crop terms with the syllable *iy* in them could be listed and some metaphorical pattern looked for relating to the meaning of the word *iy*, which is faeces.

This linguistically founded approach may have some relevance on an unconscious level in certain societies (as Gell 1975, 1979 argues for the Umeda people of the Western Sepik), but it does not help in understanding how the Wola conceive of their crops or the part these plants play in their culture at large. Without going into the questionable methodology of this approach (which blurs the lines between ethnography and interpretation, and the writer's imagination and the people's unconscious — see Lewis 1980), there is, as far as I can see, no detectable patterning in either the literal translations or the syllables of Wola crop and cultivar terms.

An isolated and stable culture is required, Gell (1979) argues, for such patterning, or 'language poetry' as he calls it, to develop. The incorporation of loan words from neighbouring languages damages the pattern irreparably. This may be one reason for the absence of any pattern in the syllables of Wola crop names. These people are in contact with those who speak a number of other languages (Sillitoe 1979a: 9; 24—8), and as already mentioned, they have received crops from them together with their names; as in the above example of the '*keret*' sweet potato cultivar of Huli origin. Clearly, the Wola cannot be expected to give an exegesis of these words which complies with any hidden syllabic pattern that may exist. In all probability the original word has changed passing from mouth to mouth with the plant anyway, so that even if the place of its origin could be found, people living there might not recognise the name in its altered form. Similarly plant names probably change over time. Someone may have found a plant mutation in the dim and distant past and christened it according to something they associated with it. Then, as the generations passed, not only might the name itself undergo changes but also the association which prompted it be forgotten.

It appears that the words themselves, their patterning and possible associations do not further any understanding of how the Wola either think of their crops or the part they play in their life.

NAMES FROM CROPS

While the names of crops are not instructive above how the Wola conceive of these plants, perhaps the equation of human names with crop terms will prove to be so. In his provocative treatise dealing with 'savage' classificatory thought, Lévi-Strauss (1966: 200–16) equates after a fashion the proper names of human beings with the terms used for other natural species.[8] The Wola explicitly make this equation too in the sense that they sometimes use the terms that label certain natural phenomena to name individuals.[9]

They sometimes name both males and females after crops and cultivars. Eight men, for example, out of a sample of seventy-six (that is 10.5 per cent), had names which were also the terms used for crops or cultivars.[10] Sometimes individuals are given such names because of some event in their life; for example at Andwariy there lives a young man whose father called him Ar, after the sugar cane cultivar of that name, because while he was still a baby his mother ran off deserting him, and his father had to feed him by giving him sugar cane to suck *in lieu* of his mother's breast milk. People insist that other crop-term names have no such history behind their bestowal on individuals, they are simply what springs to mind when the time comes for a parent to name an infant at the age of a few months.

Even when there is no reason behind the choice of a name, those that are the same as plant terms are not simply homonyms. This is illustrated by people's behaviour when for some reason it is taboo for them to mention an individual's name (usually because of a marriage, the ensuing affinal connections of which prompt name avoidance). When the tabooed name is also the term of a crop and there exists a synonym for that plant, they switch to use this alternative. So, for example, men called Kot (which is a

[8] In his discussion of naming and its relation to classification and 'savage' cognitive processes, Lévi-Strauss (1966) is concerned to establish a cyclic and undifferentiated pattern on an abstract structural level. Like the Penan and others discussed by Lévi-Strauss, who prohibit saying the names of the dead, the Wola maintain that individuals should not bear the same names as their deceased relatives because saying them would attract the attention of the deceased's ghost and provoke it to attack and cause illness. Each person demands a new name. Over many generations it may be supposed that people forget their distant ancestors' names and these may be used again; forgotten they re-enter the pool of potential names. But the Wola are unaware of this, they have no conception of a cycle, or conceptual structure analogous to their classification of natural phenomena, an empirical discord which Lévi-Strauss accomodates by pitching his analysis at the level of the subconscious structure.

[9] There is no evidence of the reverse, of giving individuals' names to plants, as European flower growers do when christening new blooms (Lévi-Strauss 1966: 211–13).

[10] See Panoff (1969: 27–8) on the way the Maenge of New Britain use crop names for human beings too.

fairly common Wola name) may be called Pombiy by those who may not say their real name, or a woman called Ebel may alternatively be called Diyr.[11]

There is a significant difference though between proper names and crop terms. Individual human beings are unique and their names refer to them alone.[12] The Wola do not think of all males called Kot as a category, they share nothing in common (other than their name) for them to do so. Whereas when they call an individual plant *kot* they do not think that it is inimitable; it is one of a kind, similar to unnumerable other plants in that category collected together under the term *kot*.

This suggests a modification of Durkheim's and Mauss's (1963) hypothesis that logical thought had social origins (regardless of their puzzling claim to concede the prior existence of technological classifications — 1963: 81n). Today the Wola name some individuals after crops, cultivars and other natural phenomena; as do many other people (witness the number of girls' names in English which are also the names of flowers). In the primeval past, the reverse could plausibly have been the case. Classification may have started with the naming of individuals. It is reasonable to argue that humans distinguished between members of their own species first and so named them, as toddlers do their parents and close relatives (contra Durkheim and Mauss 1963: 86). If this was the first cognitive step towards classification, the second might conceivably have been the ordering of other organisms, giving them names like kin (possibly the same ones). The crucial point is that even at this stage each human being would have been seen as unique, whereas any other organism would have been conceived of as a member of a category bearing the same name. Giving any non-human organism a name predicated the conception of a category. There is no need to presuppose certain social formations for their conception. The level of ordering is significant too. To think of and even name similar kinds of grass as a category does not presuppose the cognitive gathering of all grass-like plants together. Alternatively, conceiving of higher level groups does not assume distinctions within them. But this is to speculate on the evolution of hierarchical taxonomies, the universal applicability of which the previous chapter has queried.

This naming hypothesis, which relates to the classification of subsistence items, both complements and challenges that suggested by Durkheim and Mauss. When they argued that society served in the distant past as the model which generated the first non-technical classifications, they were referring to social classifications, symbolism and abstract associations, but their theory implies that the taxonomic ordering of natural phenomena relates to this, if it

[11] See Ryan (1958) for an account of naming in the Mendi Valley where the situation is similar.

[12] Something which Lévi-Strauss (1966: 191—216), so far as I can understand him, questions in the light of necronyms, tekonyms and classificatory kin terms, which in some societies may be substituted for proper names.

is not determined by it (1963: 4, 83–4).[13] There is a parallel in discriminating between social groups (lineages, clans, tribes, or whatever), and the placing of all trees, for instance, in one 'tribe' and dividing them into 'clans' according to leaf shape, size, colour and so on. The ideas outlined above, focusing on the individual, accord with the values that underpin the Wola social constitution (Sillitoe 1979a). Organised groups play a small role here and, as pointed out, the association of crops with territorial groups as totems is not instructive about the ordering of these plants. It is arguable that the apparently indeterminate nature of Wola hierarchical ideas also relates to this, supporting Durkheim's and Mauss's contention that there is a social imperative behind any classificatory scheme. Lacking the social analogy (of corporate lineages, clans, tribes and so on), and living in an acephalous society in which there is no established political hierarchy, they have not developed an elaborate grouping of organisms into 'tribes' and so on. While this is true, at least linguistically if not cognitively, of crops, it does not entirely hold for all organisms, some of which the Wola do lump together under higher-level terms. Similarly, the notable absence of consensus over the identification of some cultivars might be related somewhat to a sociological imperative; that the value placed on the sovereignty of the individual is more likely to lead to disagreements over the codes to be followed in naming, particularly when these become complex.

These sociological correlations are not very persuasive, though, and possibly not even significant. However, while the nature of the relationship between society and classification is not clear (whether, for instance, the existence of particular social hierarchies presupposes classification in taxonomic hierarchies), it is noteworthy that the models used to elucidate others' classificatory schemes parallel those employed in social contexts. Perhaps, cynics may argue, this tells us more about the intellectual conditioning of the writers concerned, their almost reflexive resort to social models, than it does about any relationship between classification and society. This analogy is striking from a New Guinea viewpoint, where the usefulness of hierarchical descent concepts has been debated for nearly twenty years. In a search for alternatives the network model has featured prominently, notably elsewhere (Sillitoe 1979a). Similarly, where others have found inapplicable the group-orientated notion of the hierarchical taxonomy, which assumes a fictitious consensus that does not allow for individual variation, they have tried to depict classifications as networks or webs (see for instance, Friedberg 1968, 1970, 1974 and Ellen 1979b).

While the above speculative hypothesis founded on the individual accords better with the Wola social situation, it leaves numerous questions unanswered.

[13] It verges on the otiose otherwise, arguing that humans needed social groupings before they could conceive or represent them symbolically. Also, the implication that technological schemes do not have the same logical origins as scientific ones is incomprehensible (1963: 81).

A classification may in a sense express a society but will not be conditioned by it alone. It is logically doubtful and dissatisfying to argue that the human classifying facility relates simply to the patterning of categories of phenomena according to the ordering of groups in society.[14] While it is beyond doubt that individuals cannot order things according to the complex classificatory schemes evolved by their culture over numerous generations without education in the categories concerned, it is questionable to assert that the human mind lacks the innate ability to classify. Indeed it is difficult to conceive of someone understanding a classification scheme without the innate cerebral capacity to perform the operations necessary to distinguish between and group classes according to the relevant criteria. After all many animals can perform the first stage in any classificatory process, that is discriminate between things, and experiments with some mammals (notably the apes) indicate that they can also associate like things with each other. Hence as Durkheim and Mauss assert, human beings have possibly always had what they call technological classifications, which developed before they achieved hominid status.[15]

Behind any classification there is a logical imperative, in addition to a possible social one. When relevant there will be material compulsion too, related to subsistence needs (as argued in the previous chapter). Any discussion that ventures on to the intellectual quicksand and speculates on the genesis of human ordering ability needs to consider all these possible forces. The current foundations of classificatory thought, and its development by our proto-hominid ancestors, depend on the simultaneous working of social, psychological and biological factors (their contributions and how they acted upon one another not being at all clear).

This somewhat reluctant speculation on the development of classificatory thought does not account for the complex classifications of extant cultures. These may serve practical ends; for instance grouping together plants that require certain conditions to flourish, so informing gardeners where they should cultivate them. But such functional ends are not necessary, which to some extent accounts perhaps for the disagreements over identification allowed in some systems. It is arguable that humans classify and order their environment because this is the first step in many cerebral processes, chaos is antithetical to thought and understanding. It is from this orderly base that abstract associations are made and ideas manipulated. And once started, classification might continue with no relation to practical ends. Classification for classification's sake is not irrational;[16] witness Linnaeus's initiation of

[14] As Needham (1963: xi—xxix) argues in his critical introduction to the English translation of Durkheim's and Mauss's book.

[15] This observation puts a further question mark over Durkheim's and Mauss's theory. If human beings could already classify, why the need of a social template? And who is to gainsay that they ordered their social groups, and made their first abstract associations, on the basis of this prior ordering of natural phenomena, passing from naming individuals, to naming and classifying other organisms, to naming and and classifying social groups?

[16] Although some fight shy of conceding it — see Heider (1969: 84).

present-day biological taxonomy some time before Darwin developed his evolutionary theory accounting for the differences upon which it rests. While people may intuitively divide up and classify things around them simply because they are different, before they reach that level of technological achievement that encourages so-called scientific reflection, they heed more those things which they use; which are edible, serve as a resource for making artefacts or are otherwise significant. In other words, they pay more attention to those important things that come to their notice more often.

This cursory look at the cognitive foundations of classificatory thought has not significantly advanced our understanding of how the Wola order and conceive of their crops. The part these plants play in their lives is not systematically related to the abstract associations and thoughts built on them, which echoes Morris's (1976, 1979) and Bloch's (1977) warnings about accepting, except perhaps in some ritual contexts, Lévi-Strauss's (1966) and Douglas's (1966) assertions that pre-literate people draw no line between their natural and social universes, thinking of them as integrated conceptual wholes. The situation with the Wola and their crops parallels that reported for the Hill Pandaram and Navaho (Morris 1976: 553, 1979: 127) whose on the one hand fragmented and on the other more integral natural taxonomies cover many organisms of no symbolic relevance and constitute domains largely independent of other aspects of their cultures.

This suggests a somewhat one-sided response to Friedberg's (1979: 96) question about the usefulness of comparing the cultural roles assigned to plants with their classificatory status. This may be informative about a crop's use in some social context, why it is apt there, but it will reveal relatively little about how people see it, except narrowly as having some feature that suits it to the symbolic role concerned. Regardless of these lukewarm observations, there is one association which is particularly important for understanding Wola thoughts about their crops, and this is their connection with one or other sex, something that markedly influences their cultivation.

Section II: Crops discussed

Chapter 10
THE GENDER OF CROPS

An intriguing aspect of Wola crop classification is their division of plants into male and female categories. For them, as described in earlier chapters, there are some crops which only men may plant and tend, and there are others which only women may cultivate, plus a third category which members of either sex may cultivate.

This gender division of crops is found in other Highland New Guinea societies too; it occurs, for instance, among the Melpa (Powell *et al.* 1975: 14), the Chimbu (Whiteman 1965: 307), the Maring (Rappaport 1968: 43, Clarke 1971: 124), the Kapauku (Pospisil 1963: 146), the Daribi (Wagner 1972: 96–7) and the Enga (Waddell 1972: 51). These people tend to put the same crops as the Wola into their male and female categories; so sugar cane, bananas and yams are male, while sweet potato, cucurbits and certain greens are female. The pattern is not entirely consistent: sometimes the same crop is classified as male by some and female by others; for example taro is a male crop for the Melpa but a female one for the Kapauku.

Why do the Wola, and by extension other people throughout the Highlands of New Guinea, categorise their crops in this way? When asked this question, the Wola themselves are unable to give an answer. They say simply that this is the way their ancestors did things, and so tradition dictates that they should do it. This chapter, in answering this question attempts to show how this association relates to Wola society at large.

THE GENDER OF CROPS

When classifying their crops the Wola distinguish between those which are unambiguously male or female, those which are more male than female or more female than male, and finally those which members of both sexes are equally entitled to plant and tend. Table 28 lists the crops grown by the Wola

Table 28: The gender ascribed to crops

MALE ONLY	PREDOMINANTLY MALE	BOTH SEXES	PREDOMINANTLY FEMALE	FEMALE ONLY
Amaranth greens (*komb*)	Amaranth greens (*paluw*)	Acanth greens (*omok*)	Cucumber	Acanth greens (*shombay*)
Bamboo	Climbing cucurbit	Beans	Irish potato	Dye plant
Bananas	Tobacco	Cabbage	Onion	Highland *pitpit*
Chinese cabbage		Gourd	Parsley	Sedge
Crucifer greens		Maize	Pumpkin	Spiderwort
Fig trees		Passion fruit		Sweet potato
Ginger		Pea		
Hibiscus greens		Tannia		
Highland breadfruit		Taro		
Kudzu		Tomato		
Palm lily				
Paper mulberry trees				
Screw-pines				
She-oak trees				
Sugar cane				
Watercress				
Yam				

according to these five classes. These order their horticultural practices.

It is very wrong for a member of one sex to meddle with the crops which are the sole domain of the other (that is those at either end of the continuum). If a man for example, found his wife cutting down his bananas or sugar cane he would be angry and chastise her, and arguments arising over such transgressions can lead to losses of temper and beatings. For a woman there is the threat of a physical sanction if she breaks the rules, whereas for a man there is not. Hence women observe the rules in a different spirit to men; for example they consider it smart if they can sneak a length of sugar cane out of the centre of a clump without their husbands finding out. But they never replant the top of such filched cane (or indeed ever plant any all-male crop) which would advertise their behaviour, instead they throw it away and destroy the evidence. It is different if a woman is in the company of a man who authorises her to cut down some sugar cane or bananas for them to share, but the man, never the woman, will do any necessary replanting. Planting is a significant criterion in the marking of male from female crops. (These rules cover only the planting, tending and harvesting of crops, not their consumption; members of both sexes are free to eat all crops.)

Similarly, men may harvest all-female crops, and they may do so both without a female present and without her permission. For men there is no fear of a physical sanction. It is their dignity and the fear of ridicule which prompts them to observe the rules. A man may for instance, harvest some sweet potatoes or Highland *pitpit* for himself at any time and indeed, if he is monogamously married and has no daughter old enough, nor any other female relative in his household to fill the breach, then he will do so for a few days each month during his wife's menses (a time when a woman is particularly polluting for a man and he cannot accept food from her — see Sillitoe 1979b). But a man will usually only harvest small amounts of such crops for his own consumption, he will not bring back a load to feed his family and pigs (except in emergencies, if his wife is too ill to do so for example, when he will receive sympathy, not ridicule from others). However he will never plant these crops, just as a woman will not plant those which are in the male domain. If a man did plant them then others would laugh at him, his position would be ridiculous.

The attitude of people living in some settlements on the Was river towards a man called Saemom, whose predicament they think was hilarious, illustrates clearly Wola feelings about men handling women's crops: In a garden which his wife Yaelten had recently planted, Saemom marked out an area as his exclusively, from which he would harvest sweet potato during her periods of menstruation (this is quite a common practice). One day he found Yaelten digging tubers from his area and flew into a rage. He pulled up a fence stake and beat her with it, one of his blows catching her on the neck and killing her. This murder cost Saemom and his relatives a considerable amount of wealth in compensation exchanges with Yaelten's relatives, and as a result no one was willing to contribute to another bridewealth so that he could remarry. As a man of only average ability in manipulating wealth in ceremonial exchanges, Saemom was unable to muster a bridewealth on his own and so remained a widower until he died. The irony of the situation, and for the Wola the pervesely funny side of it, was that he had killed his wife for digging tubers on his patch and as a result he had to plant his own sweet potato from then on. He had no sister to help him and his only daughter was about five years old at the time of the murder. Other female relatives (such as his elder brother's wife and daughter) helped him out, but still he had to do a considerable amount of planting himself until his daughter was old enough to take over responsibility for it. He died some twelve or so years later. This ignominious episode is remembered for Saemom having to get down on his knees and heap up sweet potato mounds. This was seen as ridiculous.

The moral is that members of the excluded sex should never plant those crops exclusively associated with the other. Those crops, on the other hand, associated predominantly with one sex may, on the odd occasion, be planted by members of the other without fear of ridicule and loss of face. But such occasions are rare, the predominant sex doing nearly all the planting and harvesting. Finally, those crops which fall in the middle of the continuum and which either sex may plant are subject to no restriction, both men and women are free to plant them.

These observations raise the following questions: what is it that qualifies any crop for inclusion in a certain category, and what is the point of this anyway?

'TO BE' A CROP

When talking about their crops, the Wola distinguish between them by using two forms of the verb 'to be'. There is a connection between this and the sexual categorisation of these plants which gives significant clues to understanding the nature of these gender distinctions.

The Wola language has two verbs which can be equated approximately with the English verb 'to be'. They are *wiy* and *hae*. The word *wiy* is used for things in a recumbent state, which are horizontal to the ground, whereas the word *hae* is used for things in an erect state, which are vertical to the ground. So, for example, when talking about a raincape the Wola say *saebort wiy* (lit: raincape is), whereas when talking about a tree they say *iysh hae* (lit: tree is). A human being too is *ol* (or *ten*) *hae* (lit: man (or woman) is). However, while both men and women are *hae*, this verb carries more male connotations than the *wiy* verb, which has more female implications. The Wola themselves explicitly make this equation (as do the Enga; see Brennan 1977: 21–26).[1] Things that stand erect are strong and masculine for them, whereas recumbent things are weaker and female.[2] And from a male point of view strength and erectness implies importance, so that important things are *hae*. For example, the two most important valuables of the Wola are *hae*, pigs being clearly erect but pearl shells less clearly so (although men prop them up in a vertical plane in ceremonial exchanges, and do not lay them on the ground).

Are female crops *wiy* and male crops *hae*? It is reasonable to ask whether there is such a correlation relating gender to the physical character of the plants concerned, with those standing erect as male and those lying recumbent as female, given the sexual associations of the two words *wiy* and *hae*. Table 29 lists the crops cultivated by the Wola and indicates their *wiy* or *hae* status and the sex of those who plant them. It reveals some correlation, which the pie-diagram of Fig. 36 shows more clearly. Men do plant considerably more *hae* crops than *wiy* ones, and women the reverse. In the middle range, that is those crops which both sexes may plant, *wiy* predominates, and this fits in with the pattern which is developing because, although either sex may plant these crops in theory, it is women who do most of the planting in practice.

Although the *wiy* and *hae* status of crops correlates with their gender category, there are exceptions to this pattern which demand explanation. To discriminate simply between plants as either erect or recumbent appears somewhat gross. A refinement of the features used to divide the plants into categories may explain the anomalies in Fig. 36. For instance, where do short

[1] According to Brennan (1977: 23), the Enga make this equation even more explicit than the Wola and use their equivalent of *wiy* (which is *petenge*) for women, and their equivalent of *hae* (which is *katenge*) for men.

[2] The sexual equation with the erect penis will also be clear, especially to those sympathetic with certain schools of psychological thinking — no Wola ever made this connection verbally to me though (possibly because the subject is a shameful one, to which they avoid making reference).

Table 29: Gender ascription and planting compared with *wiy/hae* status

COMMON NAME	WOLA NAME	*WIY*	*HAE*	MEN ONLY	MEN MAINLY	WOMEN ONLY	WOMEN MAINLY	BOTH SEXES
Acanth greens	*omok*	*						*
Acanth greens	*shombay*	*				*		
Amaranth greens	*komb*	*		*				
Amaranth greens	*mbolin komb*	*		*				
Amaranth greens	*paluw*	*			*			
Bamboo	*taembok*		*	*				
Banana	*diyr*		*	*				
Beans, common	*taeshaen pebway*	*						*
Beans, hyacinth	*sokol*	*						*
Beans, winged	*wolapat*	*						*
Cabbage	*cobaj*	*						*
Chinese cabbage	*kwa*	*		*				
Climbing cucurbit	*tat*		*		*			
Crucifer greens	*taguwt*	*		*				
Cucumber	*laek*	*					*	
Dye plant	*komnol*	*				*		
Fig	*poiz*		*	*				
Ginger	*shombiy*		*	*				
Gourd	*senem*	*						*
Hibiscus greens	*huwshiy*	*		*				
Highland breadfruit	*shuwat*		*	*				
Highland *pitpit*	*kot*	*				*		
Irish potato	*aspus*		*				*	
Kudzu	*horon*		*	*				
Maize	*kwaliyl*	*						*
Onion	^{e}*nyun*	*					*	
Palm lily	*aegop*		*	*				
Paper mulberry	*korael*		*	*				
Parsley (dropwort)	*taziy*	*					*	
Passion fruit	*ya iyl*		*					*
Pea	*mbin*	*						*
Pumpkin	*pompkin*	*					*	
Screw-pine	*aenk*		*	*				
Screw-pine	*wabel*		*	*				
Sedge	*hurinj*	*				*		
She-oak	*naep*		*	*				
Spiderwort	*hombiyhaem*	*				*		
Sugar cane	*wol*		*	*				
Sweet potato	*hokay*		*			*		
Tannia	*mbolin ma*		*					*
Taro	*ma*		*					*
Tobacco	*miyt*	*			*			
Tomato	*tomasow*	*						*
Watercress	*kuwmba*	*		*				
Yam	*bet*		*	*				

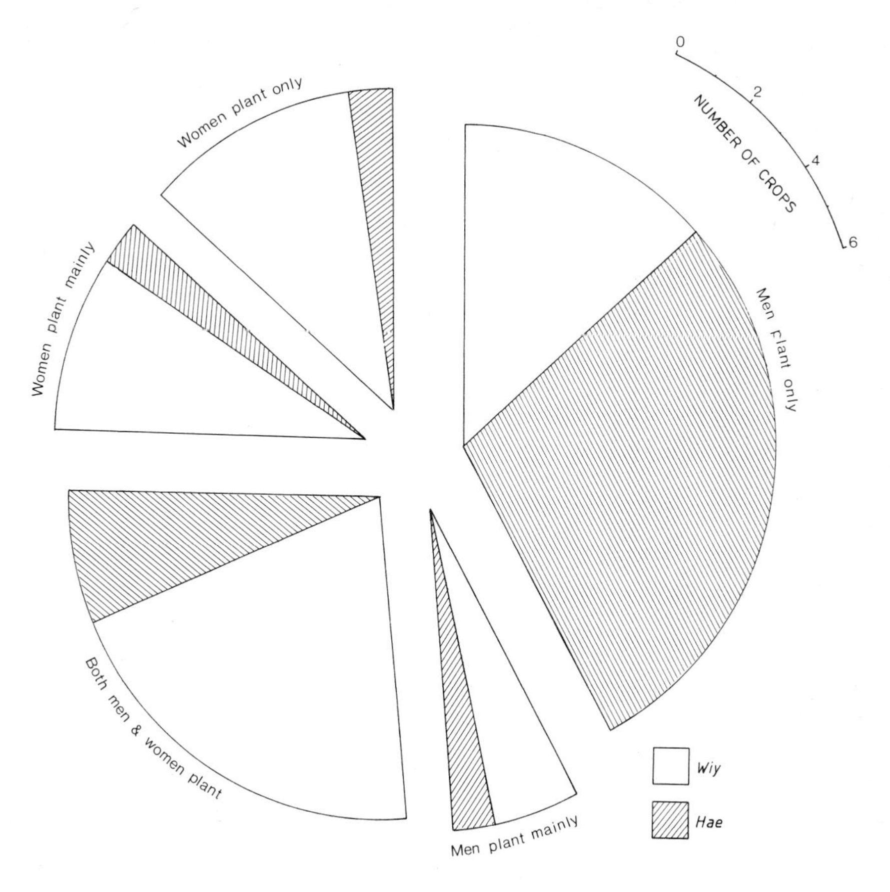

Fig. 36: The *wiy/hae* status of crops compared with sex of planters

erect plants fit in, which near to the earth are recumbent, and what about climbing ones which, without something to climb up, would trail and so pass from erect to recumbent? Possibly those plants which are ambiguous in terms of their erectness or recumbency fall into the middle category which either sex may plant. To test these possibilities, Fig. 37, in comparing the *wiy/hae* status of crops with the sex of those who plant them, discriminates between plants as either erect and over 50 cm high, or climbing, or erect and under 50 cm high, or creeping.

Erect plants over 50 cm high may be taken as those which ought to be unequivocably *hae* and so quintessentially male, while those which are creeping ought to be unambiguously *wiy* and female. Those plants which are climbing or short and erect fall mid-way between these, they are not clearly *wiy* or *hae*, male or female. Again Fig. 37 demonstrates a correlation along these lines, in that that men plant all the erect crops over 50 cm high and climbing ones, and women plant only creeping plants and erect ones under

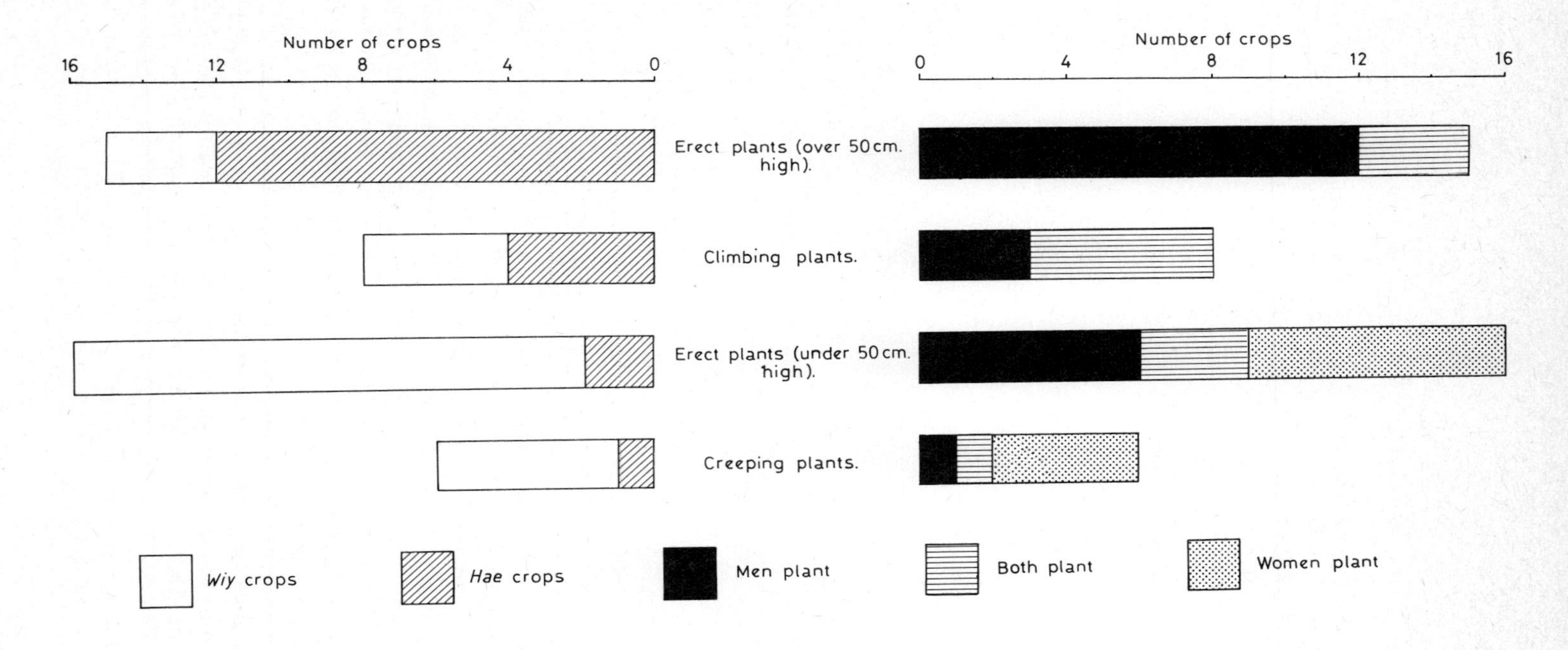

Fig. 37: Plant morphology correlated with *wiy/hae* status and the sex of planters

50 cm high. Similarly, significantly more climbing plants and erect ones over 50 cm high are *hae*, while significantly more of the creepers and erect plants under 50 cm are *wiy*. Although Fig. 37 demonstrates a more conclusive correlation between the *wiy/hae* ascription of a crop, the sex of those who plant it, and the physical appearance of the plant in question, than Fig. 36, it nevertheless still has some aberrations from the general pattern which require explanation.

GENDER CONFUSIONS

One possible source of confusion is that a number of new crops have arrived in the Wola area since European penetration of the Highlands of Papua New Guinea. Perhaps this influx of several new crops over a short period has confused the Wola into misclassifying some of them. This is feasible because they are not consciously aware of a direct correlation between the gender ascribed to a crop and its *wiy/hae* status; their classification depends on some intuitive recognition of the apparent connections.[3] So, is there any evidence that the ambiguities in the previous figures have resulted from the misplacing of recently introduced crops?

Table 30 compares the above four plant categories, their *wiy/hae* association, and the sex of those who plant them, with their status as either indigenous or introduced. This comparison indicates, in answer to the above question, that although a few of the recently introduced crops are misplaced according to the classificatory criteria suggested above, an almost equal number of the indigenous ones are also misplaced. What then are the reasons for these misplacements? In an attempt to answer this question Table 30, like Fig. 37 is arranged with the plants which are most clearly of *hae* status, and therefore male, at the top, and those of unambiguous *wiy* status, and therefore female, at the bottom. The climbing plants, although more male than female, are less clearly *hae* than the self-standing and high erect ones, while the short erect plants represent the zone of transition from male to female crops (being almost equally represented on either side). The apparently misplaced plants are also identified in Table 30.

In the table's top row there are three high erect male plants in the *wiy* class, not the expected *hae* one. Two of these are indigenous and one introduced. Both of the indigenous ones, hibiscus and tobacco, are among the shortest in this erect category and so fall near the border with erect plants under 50 cm high which, with two exceptions, the Wola classify as *wiy*: hence the *wiy*-ness of hibiscus and tobacco. However, men plant both, which accords with the category in which they occur. Both sexes on the other hand plant maize, the introduced tall and erect plant of *wiy* status. A possible

[3] This relates to Lévi-Strauss's (1966: 16—36) elusive concept of the cultural *bricoleur*.

Table 30: Plant gender and status correlated with morphology and origin. The left-hand colemn under each heading contains indigenous plants, the right-hand column introduced plants

	HAE		*WIY*		MEN PLANT		BOTH PLANT		WOMEN PLANT	
Erect plants (over 50 cm)	11	1	2 Hibiscus Tobacco	1 Maize	12		1 Taro	2 Tannia Maize		
Climbing plants	3	1	2 Hyacinth beans Winged beans	2 Common beans Peas	3		2 Hyacinth beans Winged beans	3 Passion fruit Common beans Peas		
Erect plants (under 50 cm)	1 Ginger	1 Irish potato	8	6	3	3	1	2	5	2
Creeping plants	1 Sweet potato		3	2		1 Watercress	1 Gourd		3	1

explanation for the designation of this crop as *wiy* is its propagation from large seeds which, in size and dibble method of planting, are similar to those of the pulses and the gourd — crops also planted by both sexes. Furthermore, as discussed below, seeds have an intrinsically female character.

The other two plants in the erect over 50 cm high category which both sexes plant are taro and tannia. The former is indigenous and the latter introduced, which explains why tannia occurs here: similar to taro in many respects, it was classified with it as a *hae* crop planted by both sexes. The question is why taro was so classified in the first place. On currently available evidence it is reasonable to assume that prior to the arrival of the sweet potato, taro was the staple of the Highlanders' ancestors (see Bulmer 1964, Powell *et al.* 1975: 50, Yen and Wheeler 1968). That both sexes may plant the crop nowadays could be, it is fair to suggest, a throw-back to bygone days, a legacy from the past that does not tally with the rationale of the current classificatory system and which has never been 'up-dated'.

In the climbing crop category (the second row on Table 30), it is the pulses which are apparently misplaced as *wiy* crops planted by both men and women. Nearly all seed-propagated crops, though, like beans are *wiy*,[4] as Table 31 shows, comparing the way crops are propagated with their *wiy/hae*

Table 31: Crop gender and propagation

METHOD OF PROPAGATION	*HAE*	*WIY*	MEN PLANT	BOTH PLANT	WOMEN PLANT
Cutting	11	9	11	3	6
Seed	2	15	7	8	2
Seedling	6		6		
Lateral shoot	2	3	2	1	2
Budded rootstock	2	1	1		2

status and the sex of those who plant them. This table suggests why the pulses are *wiy* and planted by both sexes. The two exceptions shown, of *hae* status seed-propagated crops, are the passion fruit and climbing cucurbit. The former, as Table 30 shows, may be planted by either sex, while the latter is planted largely by men. Both climb high up trees and so are noticeably erect. The climbing cucurbit is further unusual in that it is more generally propagated by transplanting a seedling found growing wild than from a seed and, as Table 31 shows, all seedling-propagated crops are unambiguously *hae* and male; hence the sometimes seed-propagated cucurbit's *hae* status.

The question raised by the pulses, though, and the propagation data in general, is why are seed crops *wiy*, while neither sex predominates in their

[4] Tobacco, one of the ambiguously classified plants in the previous category, is also propagated by seed, which goes further to explain its *wiy* status.

planting? This may have something to do with the intrinsic *wiy* nature of a seed as something laid down in a hole and not erect in any sense; which contrasts with the intrinsic erectness of a seedling. Table 32 supports this suggestion. Its comparison of the parts of crops eaten with their *wiy/hae* status and the sex of the planter shows that all seed-bearing crops are *wiy* too.

Table 32: Gender compared with parts of plant eaten

PARTS EATEN	*HAE*	*WIY*	MEN PLANT	BOTH PLANT	WOMEN PLANT
Leaves	2	14	9	2	5
Fruits	6	4	5	3	2
Seeds		5		5	
Tubers	7		3	2	2
Stems	1	1	1		1
Shoots	2		2		

This correlation, and the previous one, indicates that it is not only the physical appearance of a plant, its erectness or recumbency, that determines its classification. The nature of the food it yields and the parts planted also play a part. Using several such cross-cutting criteria, which occasionally conflict, results, not surprisingly, in anomalous assignments sometimes when considered from the viewpoint of a single criterion. The pulse crops illustrate this. As climbers they stand erect, but without supports, they would be creepers, the epitome of recumbent plants, and this resemblance, together with the fact that they are seed-propagated and seed-producing plants, makes them, on balance, *wiy*. The crop categorisation of the Wola is therefore a more complex process than initially supposed, in which a number of factors, some of them possibly conflicting between categories, are simultaneously weighed one against another.

The *wiy/hae* status of erect plants under 50 cm high and the planting rules covering them (recorded in the third row of Table 30), illustrate further how these sometimes conflicting factors are balanced. The proportion of indigenous to introduced crops in each class is almost equal in this row, as it was too in the previous one of climbing plants. Men and women also plant these crops in almost equal proportions; they span the boundary between male and female crops (or conversely, they are the boundary, where a grey and blurred transition occurs based on the ambiguous and conflicting features of the plants). Nevertheless, all but two of them are *wiy*, so on balance they are more female than male. Considering the two *hae* crop exceptions, it is understandable that the introduced Irish potato was equated with the sweet potato, a *hae* plant — although there was more compulsion behind this equivalence than mere likeness, for, as Table 32 shows, all tuber-producing plants are *hae*. This accounts for the otherwise anomalous *hae* status of the ginger plant too, the rhizomes of which are eaten.

Why are all the tuber-bearing crops *hae*? One of the anomalous crops on the final line of Table 30, which details the creeping plants, offers a clue. Creeping plants epitomise the recumbent *wiy* state and are the female crops *par excellance*. Yet the one tuber-producing crop which has creeping foliage is *hae*. This is the sweet potato, the staple of the Wola diet. This apparent misplacement of sweet potato as a *hae* crop cannot be explained on the grounds that it is a recent introduction with confusing characteristics, some of which link it with other *hae* crops. Sweet potato is, for those Wola now living, an indigenous crop which they have always grown.[5] Its categorisation is not confused, either by the recent sudden influx of many new crops or anything else. It is part of the system and no slip.

The designation of an unambiguously female and creeping crop like sweet potato as an erect *hae* one is in direct contradiction to the classificatory system. It is blatantly wrong, and as such it serves (as do many such inversions) to emphasize and point out sweet potato. It is arguable that the great importance of this crop as the mainstay of the Wola diet is signalled by its anomalous *hae* status. It is so important that it must be *hae*, and this patently inverted use of the *hae* concept serves to underline its importance. Before sweet potato, the staple of Wola ancestors was taro. This is, and presumably was, *hae* too. This explains perhaps why all tuber-bearing crops are of *hae* status. The tuber has always made up, since antiquity, the main part of Wola subsistence.

The sweet potato is also the only female crop that is *hae* (except for the recently introduced Irish potato which, as pointed out, is *hae* too because of its similarity with sweet potato). All creeping plants, as shown by the bottom row of Table 30, are planted largely by women. They are intrinsically female. The only exception is the very recently introduced crop of watercress, the recent arrival of which accounts in part for its misplacement. The reason for this is its appearance. Its leaves look like those of the indigenous crucifer spinach the Wola call *taguwt*, indeed they explicitly equate them (one of their names for it is whiteman's *taguwt*). Men plant these greens, hence they plant watercress too as another kind of *taguwt*.

THE GENDER MESSAGE

The question remains: why do the Wola make these gender distinctions anyway? Why have male and female crops, and equate this with the erectness or recumbency of plants?

[5] There is, as noted in an earlier chapter, currently some debate about the antiquity of the sweet potato in New Guinea (see Brookfield and White 1968, Yen 1974), but it certainly pre-dated European penetration of the Wola region by several generations.

Clearly, the categorisation of crops into those planted by men and those by women is not simply founded on something to do with the crops *per se* — either their physical appearance, method of propagation, or parts eaten. Rather, it is as if these latter features are amalgamated in some way to support the *wiy/hae* distinctions, which themselves in turn serve to reinforce the distinctions made between those crops planted by men and those planted by women. In other words, the *wiy/hae* status and its attendant considerations gives added meaning and strength to the sexual planting distinction. Here male crops are *hae*, erect and strong, while female crops are *wiy*, recumbent and weak. The overall morphology of the gender distinctions made between crops does match this *wiy/hae* differentiation (the anomalies, as shown, can be accounted for in terms of conflicting criteria).

If its rationale is not centred upon the plants themselves, it is necessary to look elsewhere to explain why the Wola have male and female crops. The apparently deliberate misplacement of sweet potato in the above scheme points the way to a possible explanation. This relates to the importance of this crop in the Wola diet, and the place of others in it too. This is something not considered so far, but it is undoubtedly important because, after all, these plants are cultivated ultimately for consumption.

The pie-diagram in Fig. 38 follows this line. It compares the percentage weight contribution to the Wola diet of various classes of crop, differentiated according to the sex of those who plant them and their *wiy/hae* status.[6] All the hatched *hae* in the 'women plant only' segment of the circle (which amounts to about 78 per cent of all the plant food eaten by the Wola) is the inverted sweet potato. Otherwise women clearly plant the *wiy* crops, and the inversion of the creeping sweet potato's status to *hae* stands out markedly.[7] Men on the other hand, are clearly responsible on this diagram for the *hae* crops. And those plants which both sexes can plant are a mix of *wiy* and *hae* plants.

This figure illustrates clearly why the sweet potato should have an inverted *hae* status to underline its unique position. In the male-dominated society of the Wola it is things masculine that are important, which are erect, strong and epitomised in the concept of *hae*. Now it is obvious to anyone on the Wola diet that sweet potato is very important to their livelihood, so that to say that it is *wiy* (weak and unimportant) would be a mockery of the *wiy/hae* distinction. It has to be a *hae* status crop, but it remains a creeping plant cultivated by women only — it is an aberration, a deliberate inversion to make an important point which relates to the importance of the crop in the Wola diet.

[6] This data on crop consumption was collected in a survey of the food eaten by the members of twelve homesteads over a three-month period (see Chapter 12 for details).

[7] The *hae*-status Irish potato figures to such a small extent in the Wola diet that, on the scale used, it does not show up in the 'women plant mainly' segment of the circle.

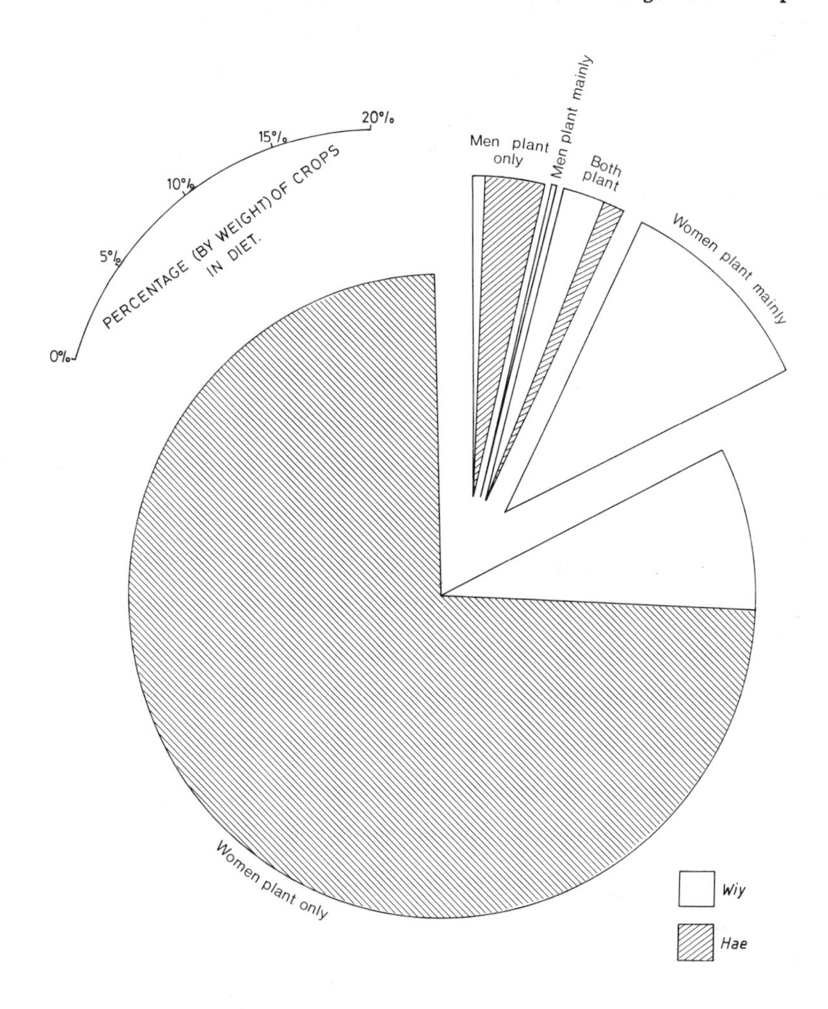

Fig. 38: The gender of crops correlated with their importance in the diet

The truly startling fact revealed by the pie-diagram though is the overwhelming importance of female crops in the Wola diet. Those planted by men count for little. Without women planting crops the Wola could not exist, whereas they could if men's crops disappeared (albeit on a less varied, more boring and nutritionally poorer diet). It is here, I think, that we may be coming close to an explanation of why the Wola distinguish between male and female crops, and why they use the opposed *wiy* and *hae* statuses to reinforce and underline it. In reality, as the pie-diagram shows graphically, it is the women who plant and produce the food. What men add is no more than a luxury, an addition of some variety.

This is how the Wola understand the situation: the women produce the 'bread and butter' (or sweet potato) of the diet, and men the appetisers and treats. All the male crops can be seen in this light; sugar cane and bananas, for instance, are relished and eaten with delight, they are the foods a man will offer a visiting friend as a treat to mark the visit. But people do not eat these male crops often, unlike the female ones which they consume daily (in fact, for the Wola a meal is incomplete without sweet potato — even when bloated they will say they are 'hungry' if the meal just eaten did not include some of these tubers).[8]

Clearly, the women are the producers in Wola society, and the division into male and female crops, together with the associated *wiy/hae* ascriptions, relates to this. The Wola, I argue, receive a 'message' from these distinctions. They are telling them that women are the producers and men the transactors. This is a crucial point in this society, indeed it is one upon which its very existence depends. Its constitution stands on the exchange of wealth (see Sillitoe 1979a), it is essential that men continually participate in exchange transactions and pass valuables to one another. They are the transactors, women are the producers. The danger is that men, whose reputations depend on handling wealth, may be tempted to produce wealth; for instance, to grow tubers earnestly and build up their pig herds to large sizes, or to grow surplus food and try to trade it for wealth. Such behaviour would be antithetical to the ceremonial exchange system, which requires them to obtain valuables through transaction, not production, and then to give the items received away again in other exchange transactions. The whole system depends on exchange, production is out.

In a truly acephalous society like that of the Wola, which is also preliterate, there are no enforceable laws to control behaviour nor offices endowed with authority to see that people observed them if they did exist. In such societies other forces regulate and direct behaviour. One such force for the Wola is the distinction between male and female crops. This ensures that men keep exchanging wealth, and do not turn their minds to its production; not by institutionalised and coercive sanctions, but through subtle pressures involving ridicule and shame. No man wants to find himself in Saemom's ridiculous position, which is where he would be if he started planting women's crops in an attempt to boost his household's production, with the intention of diverting this surplus into the exchange system in some way. This is just not possible, nor tolerable, in Wola society.

There are other ways in which the production of women is contrasted with the transactions of men in Wola society, so underlining the point that the two shall never meet. The relations of reproduction between men and women are hedged around with taboos which also give this 'message' (see

[8] As my wife and I found out the first time we had some people to eat with us; after filling them with plates of fish and rice (the most relished of introduced foods) we were surprised by their exclamations of hunger and rush to bake some sweet potato tubers in our fire.

Sillitoe 1979b). Interestingly, aspects of these reproductive taboos extend to the planting of crops too. A menstruating woman, for example, should not enter a garden to do any planting, neither should anyone who has indulged in sexual intercourse in the previous two or three days.[9] The implication is that too much productive effort is bad; one kind cancelling out another. The moral is that production must be controlled and not allowed to expand beyond certain limits. Women, as the producers, should not be expected to work hard to keep production at high levels. For men to compete through their womenfolk by goading them to increase production would be as harmful as them entering the sphere of production themselves. Unlike the rationale behind Western capitalist economies, that behind the Wola requires a certain level of production and no more.[10]

As a postscript, it is noteworthy to repeat that in many other Highland New Guinea societies people categorise their crops into male and female plants; for instance, the equation of sugar cane with men and sweet potato with women is very common. It seems reasonable to suggest therefore that some of the ideas developed here may have a pan-Highlands relevance (for example, Clarke 1971: 124, writes that the Maring equate their tall crops with males and their shorter ones with females).

[9] The Enga (Waddell 1972: 51) also fear that menstruating women will pollute male crops, and the Goodenough Islanders (Young 1971: 149) say that sexual intercourse contaminates the hands and through them the crops, which then exude a scent that attracts crop mould and white ants.

[10] An interesting footnote with regard to Wola ideas relating to production and crops is their apparent unawareness of how plants reproduce. For the Wola they simply do so, some having flowers (*paepuwliym*) that turn to seeds (*iyl*) which they can plant, while others seemingly do not.

Section II: Crops discussed

Chapter 11
THE OCCURRENCE OF CROPS

A pertinent point related to the separation of production from exchange is the relation between men's status and the crops they grow. Men sometimes give crops such as bananas, screw-pine nuts and taro in exchanges, and they also cultivate them (women produce the staples). Do men of renown, the *ol howma* of Wola society, grow more of these valued crops than others, or a wider variety in general?

The short answer to this question is no. Although men's reputations and renown depend on the amounts of wealth they handle in ceremonial exchanges (Sillitoe 1979a: 111–23), they do not compete in the cultivation of these crops. The ethos of productive moderation extends to them too, it is exchanging that counts and not growing. The Wola themselves deny that men of higher status and renown work hard to grow more of these crops; they say that all men, regardless of their exchange achievements and commitments, grow about the same amounts.[1]

A survey of the crops occurring in men's gardens compared with the status of the owners supports these assertions. Figure 39 compares the number of gardens in which men have various crops growing with their renown as established in an opinion poll (for details of the poll results see Sillitoe 1979a: 117; in this figure the men are put into groups of eight, ranging from group 1 with the eight men of unequivocally high, or *ol howma*, status down to group 10 comprising youths of low social status). It shows clearly that while men of higher status have larger areas under cultivation, there is no proportional difference in the crops they and others grow.

Successful men — not surprisingly, given their larger area of garden — produce more of all crops, but compared proportionally with the number of

[1] These crops feature rarely in ceremonial exchanges, compared to other valuables such as pearl and cowrie shells, pigs, decorating oil, and so on (Sillitoe 1979a: 144–5).

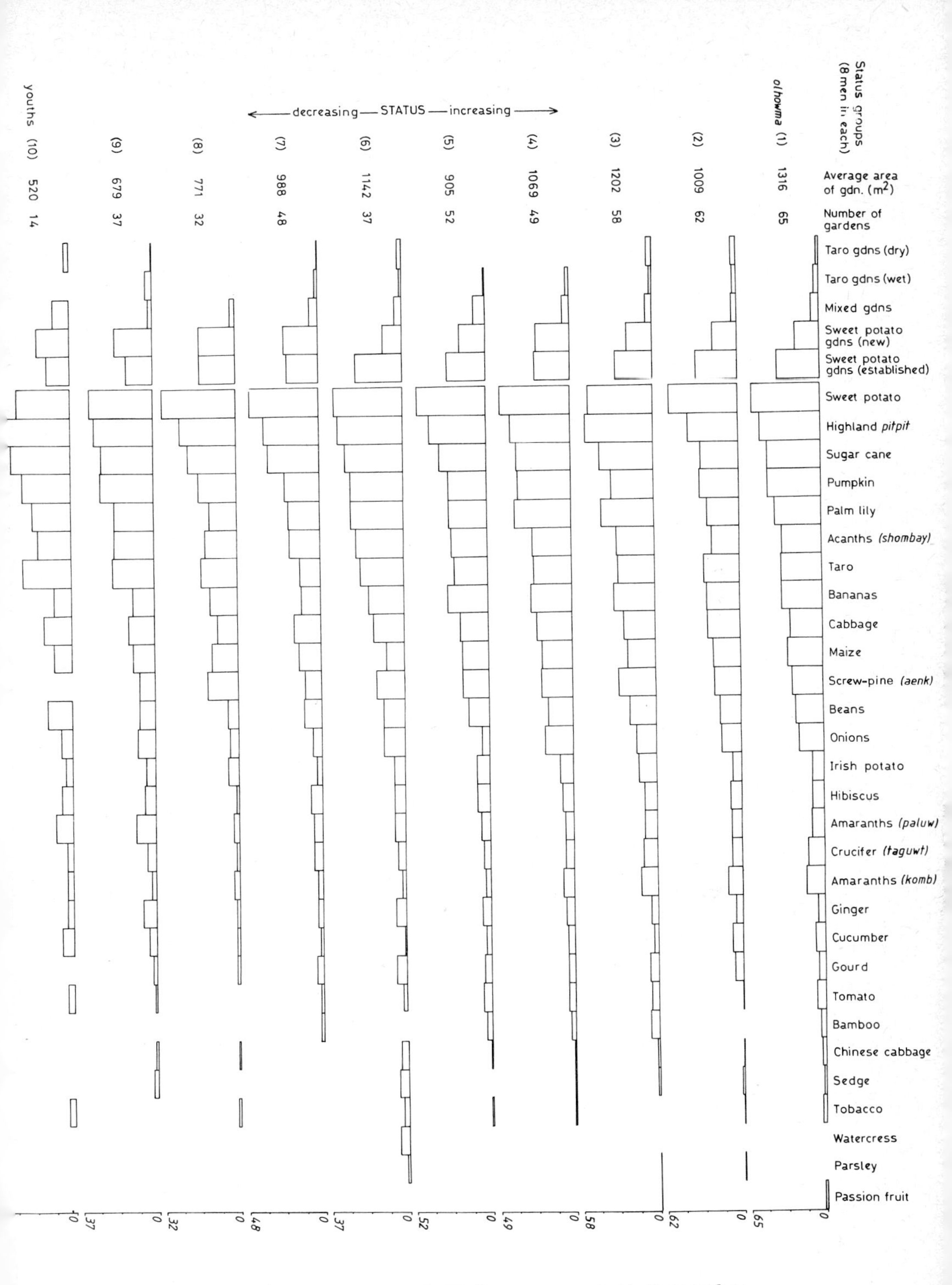

Fig. 39: Men's status compared with the crops occurring in their gardens

gardens they have under cultivation all men grow all crops in about the same amounts. In gross terms, growing more of all crops, *ol howma* do cultivate more exchangable ones, but their increased cultivation of these is actually very small in relative terms because they are infrequently grown compared to the staples. So men of high status do not cultivate noticeably more male crops to accord with, or bolster, their positions (although their households' production of female crops is markedly higher).[2] And their marginally higher proportion of them, an increase which accords with their larger cultivated areas, follows, since their renown brings more visitors to them to discuss exchange matters. They need more luxury foods to be hospitable, and also to stage a few more of the infrequent exchanges involving crops.

It is significant that the balance of crops grown by men of high status does not differ from that of ordinary men. They do not as a result of their position, or as a means to its achievement, noticeably increase their efforts to grow more of certain luxury crops. The opinions of the Wola fully concur with these findings. Fig. 40 compares indigenous judgements about the occurrence of crops in a sample of gardens with the status of the gardeners (in this figure men are again arranged in ten groups of eight according to their status). The variety of crops occurring in the gardens surveyed increases from the right to the left of the figure, and the words in the Wola phrases for the seven categories are as follows: *hombuwnja* means 'everything' (that is, in this context, 'all crops'), *onduwp* means 'a lot', *g^{e}nk* means 'a few', the *mon ... sha* construction means '... ish' (*mon onduwp sha* meaning 'many-ish'), and *ora* means 'very' or 'only'.[3] This figure shows that the proportions of crops grown remain more or less constant regardless of status, and that the number of crops occurring in gardens is usually few anyway, consisting largely of staples.[4]

The important point is that the balance of crops grown by a man and his household does not change with an increase in his status. All households grow the same proportions of all crops, although the larger ones of more successful men produce more overall. But then they have larger families to support and more mouths to feed, yet their diet is identical to that of everyone else.

[2] They are usually polygynously married and so have more females in their households to cultivate their gardens, itself an indication of their ability to manipulate wealth because each wife will have demanded a bridewealth payment in the first place (see Sillitoe 1979a: 134).

[3] For further comments on the way the Wola qualify and rank things (in the context of men's status) see Sillitoe 1979a: 116.

[4] The gardens surveyed for both Figs. 39 and 40 varied in age and the time that had elapsed since they were planted, two things that significantly influence the occurrence of crops.

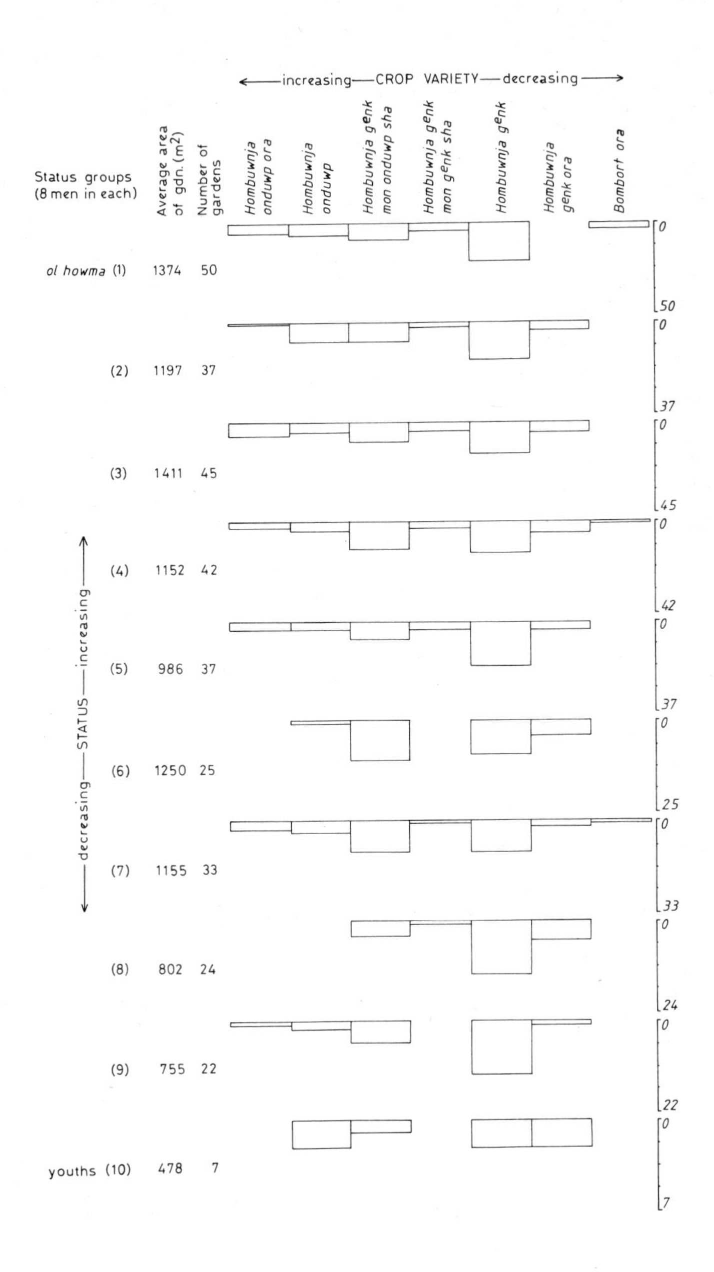

Fig. 40: Indigenous judgements of crop variety compared with gardeners' status

CONSTRAINTS ON CROP CULTIVATION

The foregoing discussion of the gender of crops shifts the emphasis from what the Wola think of their crops to where and how they cultivate them. This is logically the next step for the so-called ethno-sciences as a whole (see Johnson 1974),[5] to see how peoples' cognitive ordering and ideas comply with and influence their behaviour towards the natural phenomena in question.

The earlier floristic chapters describe Wola cultivation practices for each crop, where these people think plants grow best and why, and so on. These accounts convey clearly the foundation of this knowledge in practical experience, showing that the extensive gardening lore of the Wola gives them a shrewd appreciation of the conditions under which various crops flourish. They also have some ideas, again rooted in generation of experiment and accumulated knowledge, about the influence crops have on soil fertility.

While the Wola are aware of nature's limitations — that they cannot grow crops where she will not permit (such as sweet potatoes in water-logged places) — these practical constraints are not the only ones on their cultivation practices. Cultural precepts are also significant. For example, Wola men will grow tobacco only in gardens adjacent to their houses, usually in beds worked under the eaves, not because it would not flourish elsewhere but because of all crops it is the one most often stolen. Cultivating it almost 'in the house' reduces the chances of theft (although this still occurs regularly, prompting irate owners to tear up and display plants which have been tampered with atop poles called *showaip* in their houseyards to 'shame' the thief and let vent their anger). The reverse is the case with spiderwort. Women plant this around the edge of gardens away from homesteads because it is really a 'wild thing' and not properly speaking a cultivated plant at all. It continues to flourish after a garden is abandoned, and grows wild along the edge of paths and in other cleared areas, so it would not only be out of place in a garden adjacent to a house, but given its prolific growth might also choke it. Both these practices are the result of man-made, not natural, rules. Tobacco and spiderwort could grow in the right locations in the gardens from which they are barred, but Wola custom prevents it.

Broadly speaking, there are two principal factors constraining the cultivation of crops. These are the kind of garden concerned and its age. Firstly, the kind of garden. This depends on the interplay of a number of factors, such as the physical nature of the piece of ground in question (the steepness of slope, the nature of the soil, and suchlike), and the distance of the area from the homestead, and so on. The Wola distinguish three kinds of garden, and they sub-divide each category into two.[6] They are as follows:

[5] Dealing with Brazilian share-croppers Johnson correlates the people's ideas about the conditions preferred by their crops with actual planting practices.

[6] See Waddell 1972: 39—57 for a description of the different kinds of garden of the Enga, whose classification is similar to that of the Wola.

Garden type	*Sub-types*
(1) Sweet potato gardens (*hokay em*)	(*a*) newly-cleared (*ka*) (*b*) established and planted more than once
(2) Taro gardens (*ma em*)	(*c*) wet site (*suw pa*) (*d*) dry site (*iyb na wiy*)
(3) Mixed gardens (*em g^{e}mb*)	(*e*) abandoned house sites (*em aenda munk wiy*) (*f*) cleared site

The topography of garden sites acts as a constraining influence on the cultivation of crops in several ways. The Wola are well aware for example, that soil slips down to the bottom of slopes (especially on sites denuded of vegetation for gardens) and that crops such as greens, brassicas and beans flourish on the deeper layers of top soil occurring there and in surface folds. The nearer the clay sub-soil to the surface (for example, at the top of slopes, from which rain has denuded the top soil) the lower the yields, they say, for a variety of crops. Some established gardens planted several times over have pockets of deep top soil (often in surface folds) and here a variety of crops may be planted repeatedly, not just sweet potato, which dominates elsewhere in gardens of this type.

The second principal constraint on crop cultivation is the age of the garden. This relates primarily to the time that has passed since the garden was planted, although in the case of sweet potato gardens it also relates to their absolute age (upon which their division into sub-types depends). Different crops take varying lengths of time to reach maturity, and when ready the period of which they yield also varies. Fig. 41 shows the time the crops grown by the Wola take to reach maturity and the period for which they yield. They are listed according to the time taken to ripen, with the rapid growing spinaches at the top and the slow-growing tree crops at the bottom. Some crops, notably beans, maize, gourds, cucumbers and some greens, have a short productive life, whereas others go on yielding for many months, in some cases continuing to do so when gardens are abandoned.

The time elapsing since the planting of gardens is a significant factor to consider when thinking of constraints on the occurrence of crops because the Wola, in common with subsistence gardeners in other parts of the world, do not follow the mono-crop practices of mechanised agriculture, cultivating a single crop in a specific area. They mix up crops throughout gardens, dotting plants about in the spots which suit them. So a three-month-old taro garden for instance, with a profusion of various crops, looks considerably different

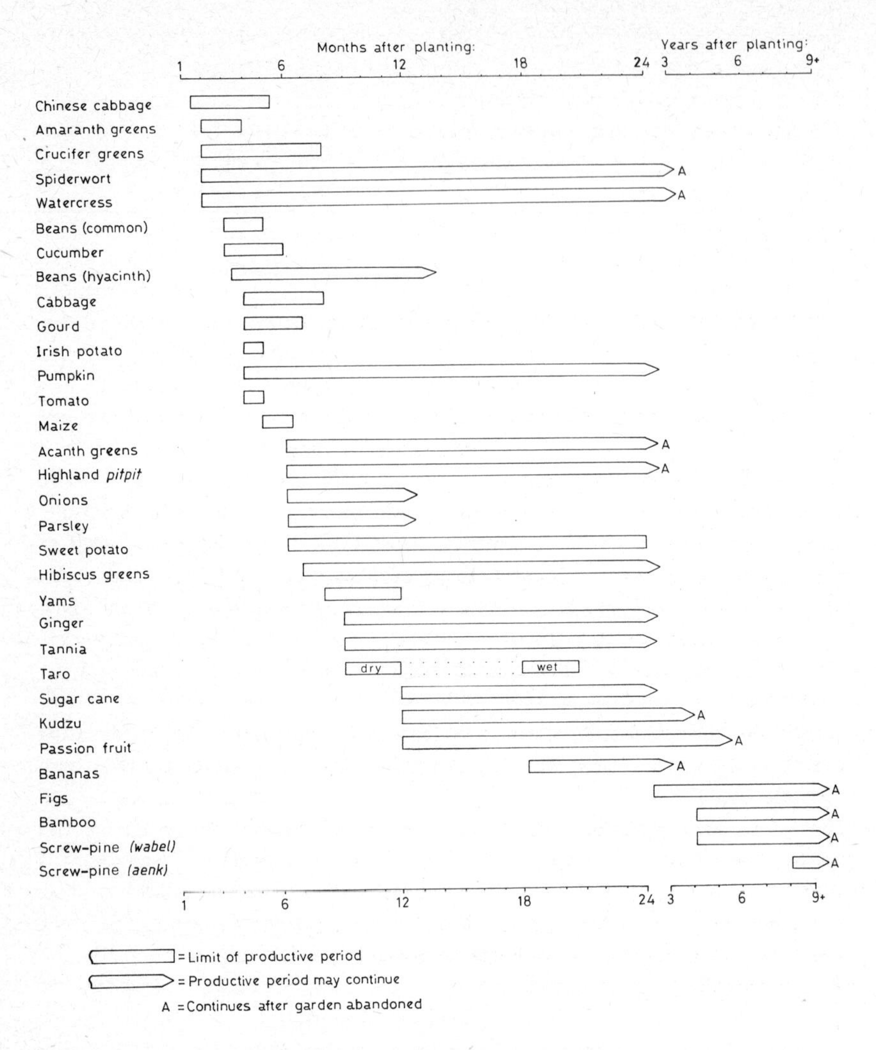

Fig. 41: The time for which crops yield

to one nine months older, in which taro will predominate to the virtual exclusion of any other crop.

All newly-planted gardens, with the exception of established sweet potato ones (where the variety of crops cultivated is usually markedly less than in others), have a similar appearance, although sweet potato predominates in sweet potato gardens and taro in taro ones. The appearance is one of confusion, with plants of all shapes and sizes growing intermingled with one another. Clarke (1971: 76—8) describes in detail the plants occurring in a short traverse of a garden, an experience which he says 'is to wade into a green sea' (1971: 76).[7] Discussing the intermingling of crops in such gardens, Rappaport (1972: 349—50) suggests four benefits: (1) the vertical stratification allows the plants to make maximum use of the surface area and so achieve a high rate of photosynthesis for any given area; (2) it discourages plant-specific insect pests which would otherwise spread easily where the same plants all grow in close proximity; (3) it allows the gardener to exploit to the full slight variations of habitat in the garden; and (4) it protects the otherwise exposed soil more effectively, the rapidly growing plants springing up to cover the entire area until the slower growing ones have established themselves and take over. Also, in this intermingled planting there is little competition between different species of plants because they are ready for harvesting at different times (Waddell 1972: 53).

This mixed cultivation of crops throughout gardens makes a reliable quantitative assessment of their occurrence difficult. Nothing short of the detailed recording of the actual plants occurring within specific measured areas in a statistically significant sample of gardens of different types and ages would achieve this. A less accurate approach is simply to record the presence or absence of crops in gardens. The results of such a survey of Wola gardens is recorded in Fig. 42 and 43.[8] These show the number of gardens in which different crops were present, and their overall area (both expressed as a percentage of the total number or area of the gardens in the sample). Both sets of data compare broadly one with the other showing an expected correlation between the occurrence of crops and their importance in the Wola diet (as established in the next chapter). In terms of the actual areas under different crops though, the two graphs are misleading — something like 75 per cent of the area under cultivation supporting sweet potato.[9] This information on the occurrence of crops was collected by walking round the edge of gardens, and traversing them at random, where the crops growing could not be seen clearly; thus, compared to the detailed study of quadrants, this was something of a casual survey, in which some small plants were undoubtedly missed, though the general trend of the results is reliable.

THE CROPS OCCURRING IN SOME GARDENS

To remedy the deficiencies of these graphs, there follows a series of accounts detailing the crops actually growing in a sample of gardens surveyed closely. These descriptions are supported by diagrams of the areas

[7] Geertz (1963) describes the confusion of similar gardens in Indonesia as duplicating the structure of the rain forest from which they are cleared.

[8] Compare these with Clarke's (1971: 79) figure, the data for which he collected by recording in detail the plants occurring in 383 three-foot-square plots in a sample of gardens; see also Pospisil (1963: 436).

[9] Again, the precise measurement of this would be difficult because of the intermingling of crops. Also, these graphs show the crops occurring in a sample of gardens at one instance in time only, they indiscriminantly lump gardens planted for different periods — some of the older ones would have supported a wider variety of crops than shown a few months earlier.

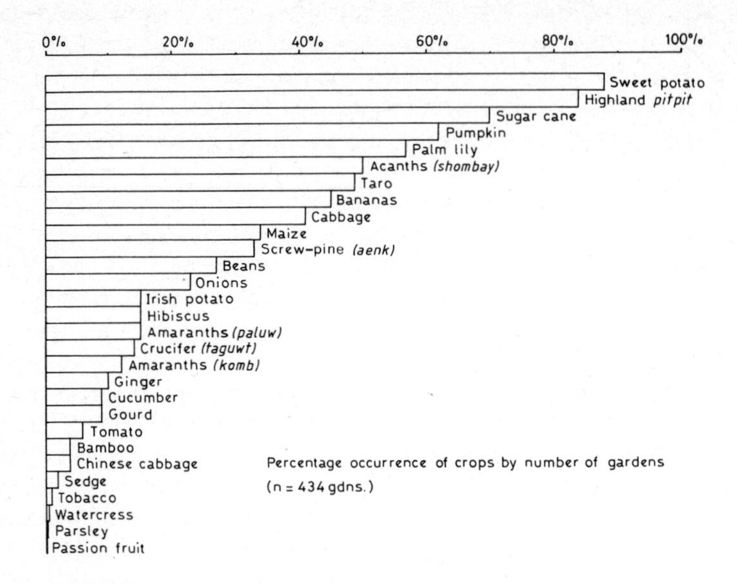

Fig. 42: Percentage of gardens in which different crops occurred

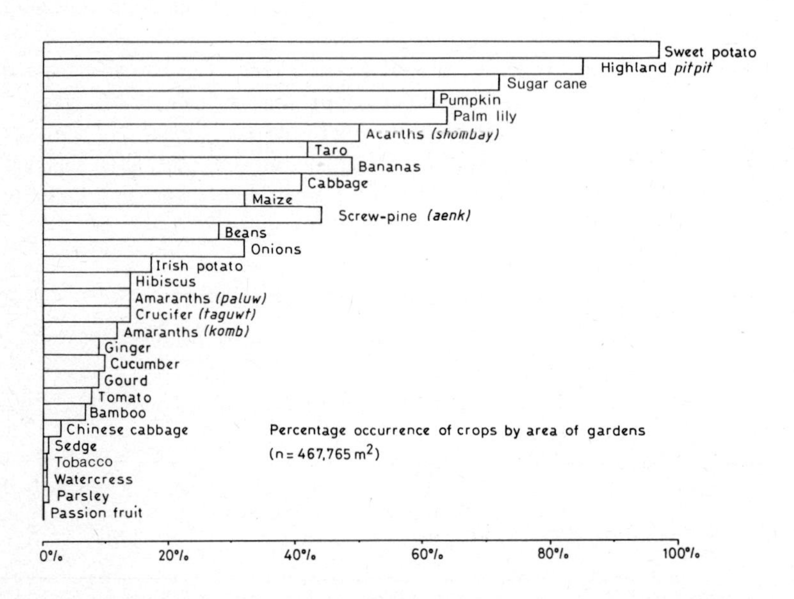

Fig. 43: Percentage of area cultivated in which different crops occurred

concerned, showing the location of the crops.[10] They illustrate the crop pattern of the three principal kinds of gardens, and how these change and vary over time, notably in long-established areas under sweet potato.

The first gardens in this series are mixed vegetable ones (*em g^{e}mb*). Those surveyed and illustrated in Fig. 44 had been planted for one, four and five months respectively. Such small gardens usually carry a wide variety of crops intermingled with one another, as illustrated by the four-month-old area (unstippled in Fig. 44). There are a number of reasons for this. Firstly, they

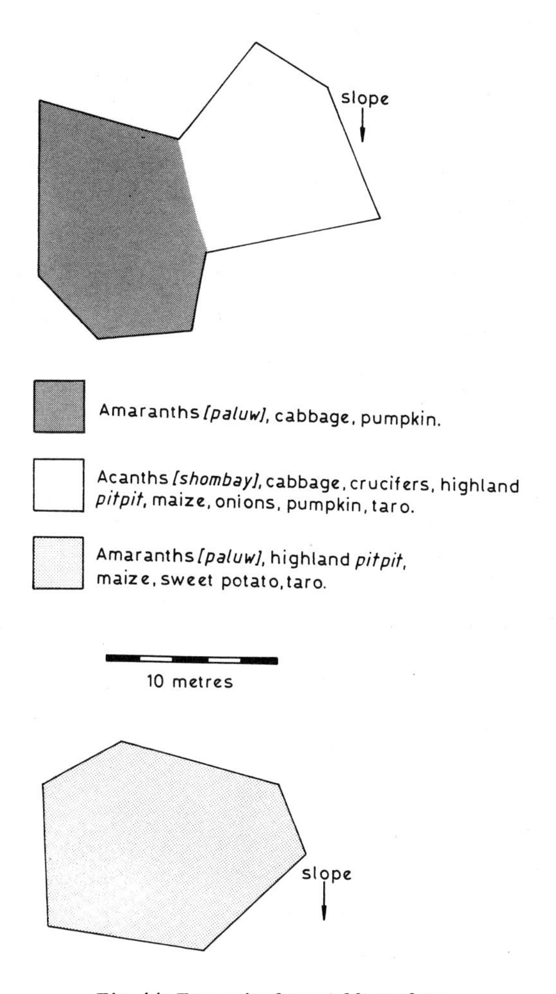

Fig. 44: Two mixed vegetable gardens

[10] For comparison, see Panoff's (1972: 378—9) diagram showing the occurrence of crops in a Maenge garden and Pospisil's (1963: 108) showing their location in a Kapauku one.

are usually located adjacent to houses, supplying a ready-to-hand selection of crops to vary the predominantly sweet potato diet. Secondly, their location discourages stealing, a significant consideration when it is recalled that some of these crops are eaten infrequently, as something of a treat, and so, like tobacco but not mundane sweet potato, tempt thieves. Thirdly, as the previous crop accounts note time and time again, many of the crops cultivated by the Wola thrive on newly-cleared areas only, and small gardens are a popular way of securing areas of virgin soil for their cultivation, because they demand less work to clear and fence than more extensive taro or sweet potato gardens. Hence the utilisation of ready-cleared and particularly fertile abandoned house sites for such gardens. The Wola are well aware, as the previous chapters again point out several times, of the good yields obtainable from the cultivation of such locations, something which they attribute to the rotting refuse of human habitation, and in the case of women's houses the accumulation of pig faeces.

The Wola do not always plant a wide variety of crops in small mixed vegetable gardens though. The area planted a month (small stipples on Fig. 44) does not have the expected wide selection usually occurring in these gardens. This demonstrates that the occurrence of crops follows no rules, only general patterns; a point made a number of times in the coming accounts. When asked in these cases why they are not cultivating a wider variety of crops, people usually reply that they had no seeds or cuttings to hand when planting, and no one would help by giving them any. Sometimes they mention laziness too.

As time passes and the fast-growing plants are harvested, the crops remaining in such mixed vegetable gardens become sparser and less varied. In the five-month-old garden for example (large stipples on Fig. 44), only sweet potato, taro, and Highland *pitpit* remained after another month had passed. Sometimes small vegetable gardens are planted with only a selection of the quicker-growing crops and are abandoned within this time, that is after half a year. In this event they usually have a temporary fence only, made of *Miscanthus* cane. Mixed vegetable gardens are usually cultivated once only and then abandoned, for they cannot support a wide variety of crops a second time round (not even the extra-fertile old house site beds will go beyond this). Occasionally men extend the area of an exhausted mixed vegetable garden, establishing an extensive sweet potato one encompassing it. At other times they plant banana suckers on the cleared area and establish a stand of these plants next to their houses, a popular location for them because they are also highly attractive to thieves.

The next two gardens in this series are taro ones (*ma em*), one of the wet type and the other dry (see Figs. 45 and 46). Both were newly planted, the waterlogged one with fewer crops, which is usual because fewer plants are tolerant of wet conditions. The owners of this garden planted some greens and maize on the drier edges, away from the stream filtering through the

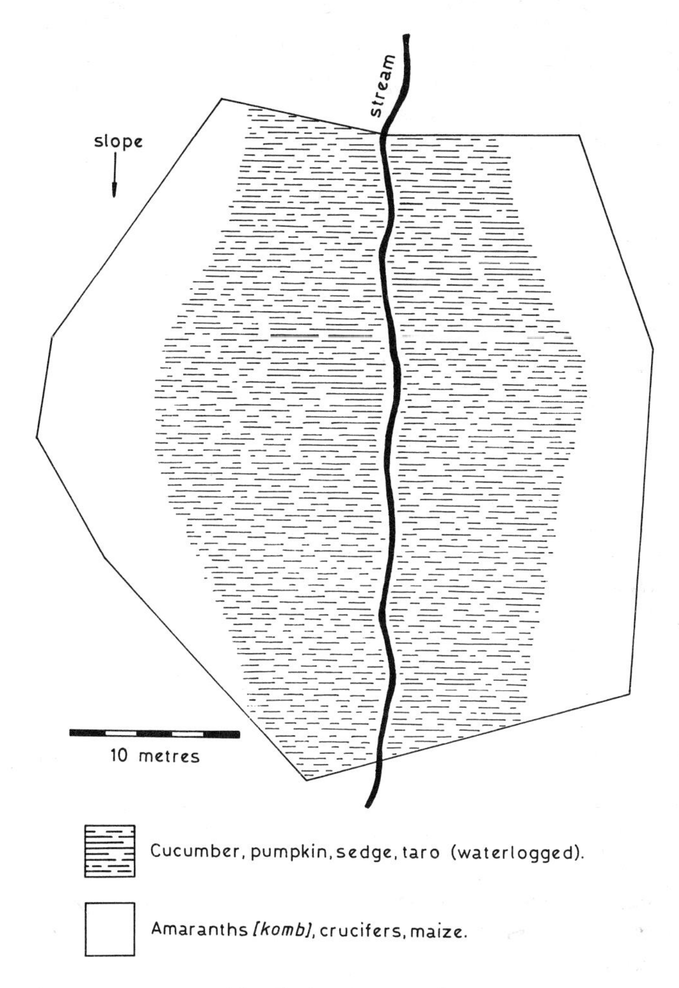

Fig. 45: A wet taro garden

centre and its waterlogged margins, where they planted taro, cucurbits and sedge. The crops in the dry garden on the other hand, like those in the previous mixed vegetable ones, were planted intermingled with one another; indeed it is only their size and the preponderance of taro in them that distinguishes such dry gardens from mixed ones. When the owner of the dry garden excavated his taro five months later, the following crops were still growing: Highland *pitpit*, pumpkin, acanth greens and hibiscus bushes. Eighteen months later, when the gardeners of the wet area dug up their taro crop and staged an exchange of corms, these were the only remaining plants in the garden, together with sedge used for skirt-making.

Following the excavation of taro (which men usually do on a single day, staging an exchange of the tubers), these gardens are effectively abandoned,

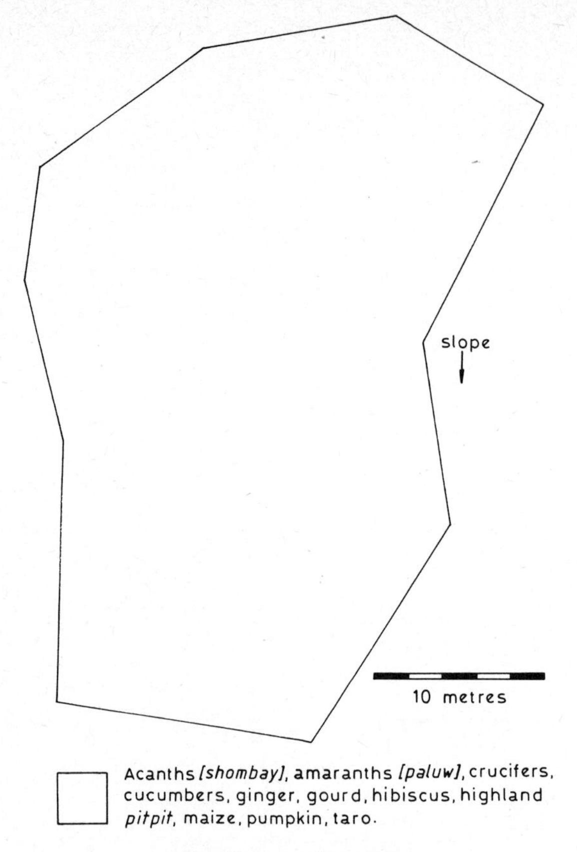

Fig. 46: A dry taro garden

although women will return occasionally to harvest what crops remain, until the regeneration of secondary regrowth renders this a waste of time. The Wola never attempt to grow a second taro crop in a garden because it would, they say, be a failure. Nor do they often plant some other crop on the site; in the case of wet ones because there are few alternative plants suitable, and on dry ones, because taro, they say, takes so much out of the soil that not even sweet potato can grow tolerably well on it. Another unvoiced reason is that taro requires a ritual and spells to ensure its good growth, so there is something improper about turning a taro garden over a mundane sweet potato; although some are not above such impropriety and, if they think it will thrive, occasionally plant a second crop of sweet potato on a site.

The next eight gardens in this series are sweet potato ones (*hokay em*) of varying ages and at differing times since planting (see Figs. 47–54). The first of these gardens was newly cleared and planted for one month (Fig. 47). It is usual for gardens at this stage to carry a large variety of crops intermingled with sweet potato (the unhatched areas on the figure). Women plant

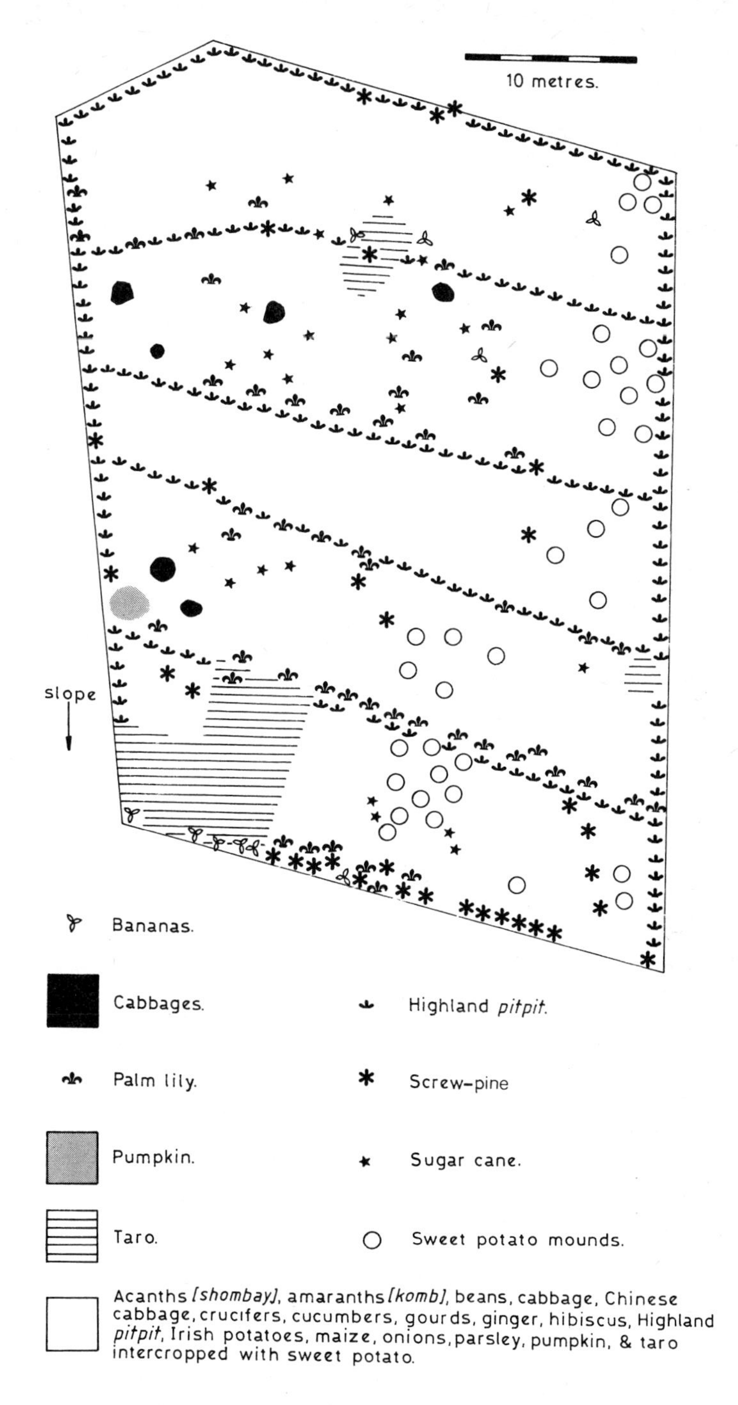

Fig. 47: A newly-cleared and planted sweet potato garden

sweet potato, which dominates, straight into the ground largely (called *suwl* planting) because it is naturally soft when first cleared of vegetation. In places where they think the newly cleared soil is too moist and compact for tubers to grow well if *suwl* planted, they break it up and build mounds (called *mond* planting). This is the case on the right-hand side of the garden, where *Miscanthus* cane stood prior to clearing. It often leaves a dense soil.

The planting of Highland *pitpit* plants and palm lily shrubs along the fence and internal boundaries between women's areas, to mark them clearly, is a common practice (the five women's areas in this garden are obvious from these plants). So is the cultivation of screw-pines along the fence line (the ones growing within the garden were standing on the site prior to its clearance), which will help to identify the location of the site in the next generation when it is gardened again. Bananas are also located largely along fence lines because when established their large leaves, like those of screw-pines too, cast considerable shade which prevent other crops flourishing under them. It is usual to site pumpkins in one or two places on the edge of a garden, as shown, for similar reasons: to keep these trailing plants from spreading out and covering too many others. Conversely, tall sugar cane plants, with relatively little foliage to cast shade, are cultivated here and there throughout a garden, as Fig. 47 shows.

This garden also shows how the Wola exploit the micro-variations of a site. The two smaller pockets of taro were established in slight depressions in which water accumulated and where the soil as a result, was too wet for sweet potato. The large area of taro at the bottom of the slope occupied the shadiest part of the garden, which, receiving less sunlight, was notably damp and so unsuitable for sweet potato too. When the time comes, a year or so later, to replant such new gardens, these initially damp areas have frequently dried out through exposure to the sun and can then take sweet potato. All the brassicas in the garden shown were located on the sites of large fires, where they would thrive on the nutrients released from the vegetation burnt during clearing.

The next garden is also a newly-cleared one, a year after planting (Fig. 48). This garden differs from the previous one, and is atypical, in the large number of sweet potato mounds it contains. The explanation is that this site was damp with a moist, heavy soil that demanded breaking up and mounding (except for the area on the left, in which the sweet potato vines were *suwl* planted straight into the unprepared soil). There was a level patch in the bottom left-hand corner of this site where water collected and the soil was consequently wet: hence the planting of taro here.

The palm lily shrubs running down the left-hand fence were not planted by the gardener but were mature plants marking the fence line of a previous garden on the site owned by a relative. In this garden the initial crop flush, of greens, beans and so on, was over and only sweet potato remained together with taro, slow growing sugar cane, perennial palm lilies and a single screw-pine. The mounds in this garden had been harvested from a number of times

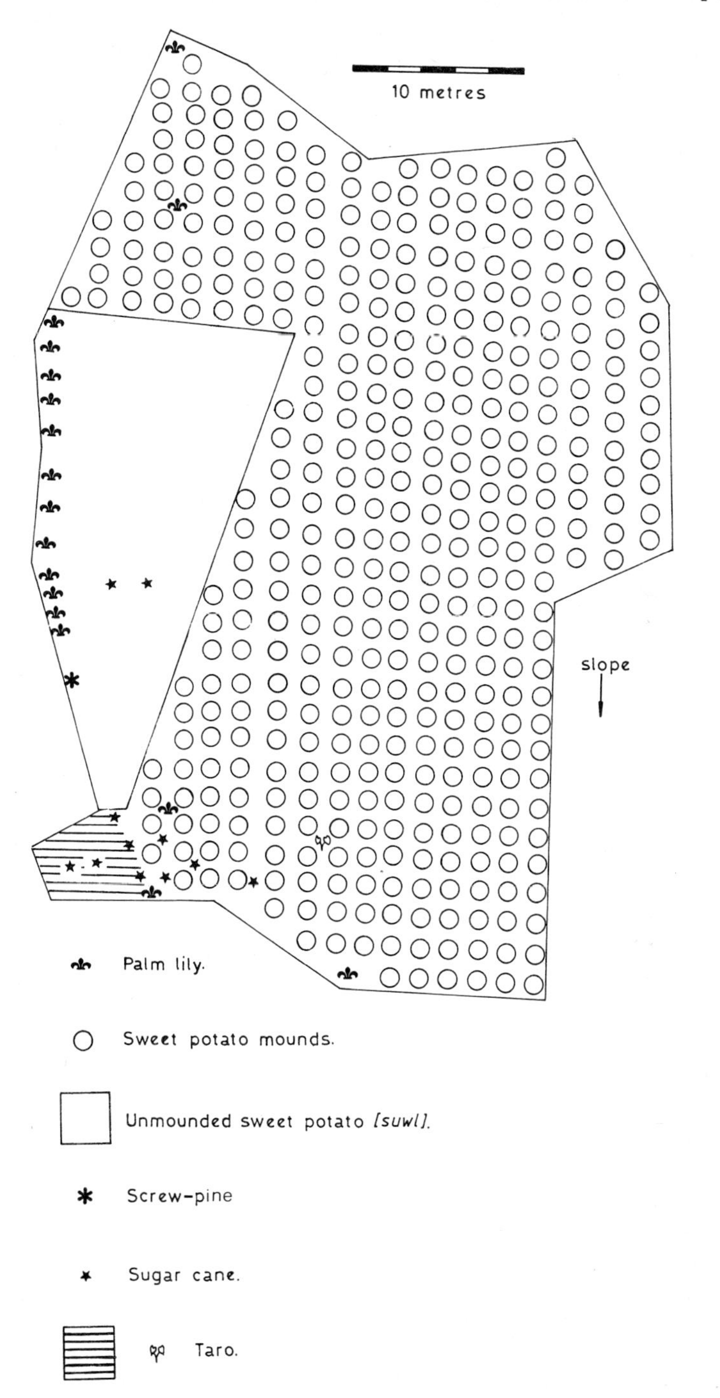

Fig. 48: A newly-cleared sweet potato garden

prior to its survey and their yield had fallen off to such an extent that it was due for re-working. The women responsible for it said that they would mound it all when they replanted.

Fig. 49 also shows a newly-cleared garden about a year after planting, but it is something of a joker illustrating that although Wola planting practices tend to follow a pattern, there are exceptions. In this case social, not natural

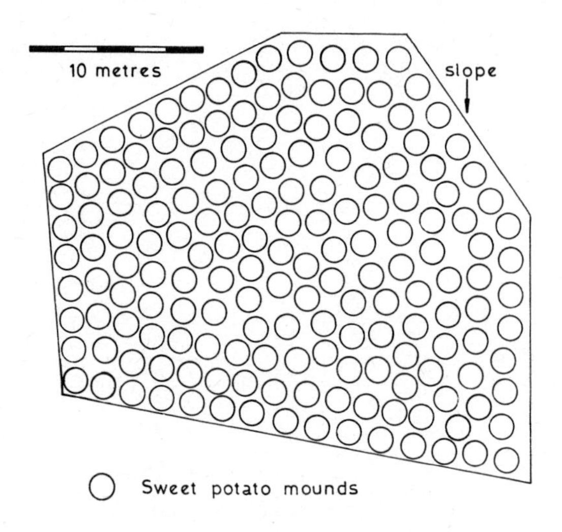

Fig. 49: An aberrant newly-cleared sweet potato garden

conditions influenced the pattern and prompted a bending of the general rules. Rights to the land on which the garden stood were under dispute, and the current gardener cleared it only five or six years after his rival claimant had abandoned a garden there (he claimed a prior right to it through his paternal ancestors, which his rival had usurped by marrying one of his distant relatives).

Clearing the area of cane regrowth so soon after the last period of cultivation to prove a point meant that the gardener could hardly expect high crop returns. Indeed, one of the few crops that can produce any kind of yield on such tired soil is sweet potato, and this was the only crop planted in this garden. The soil was 'strong' because of its recent use and so demanded breaking up and mounding to soften it. The higgledy-piggledy pattern of the mounds is unusual, though; the practice of women is to arrange them in straight lines called *tiyt* running down the slope of a garden (see other figures). In this one, the young unmarried girl who was responsible for it demonstrated her disgust, at having to waste her time cultivating such tired soil to score a point for her belligerent father, by scraping up mounds here and there in no pattern.

The fourth sweet potato garden in this series shows an established garden

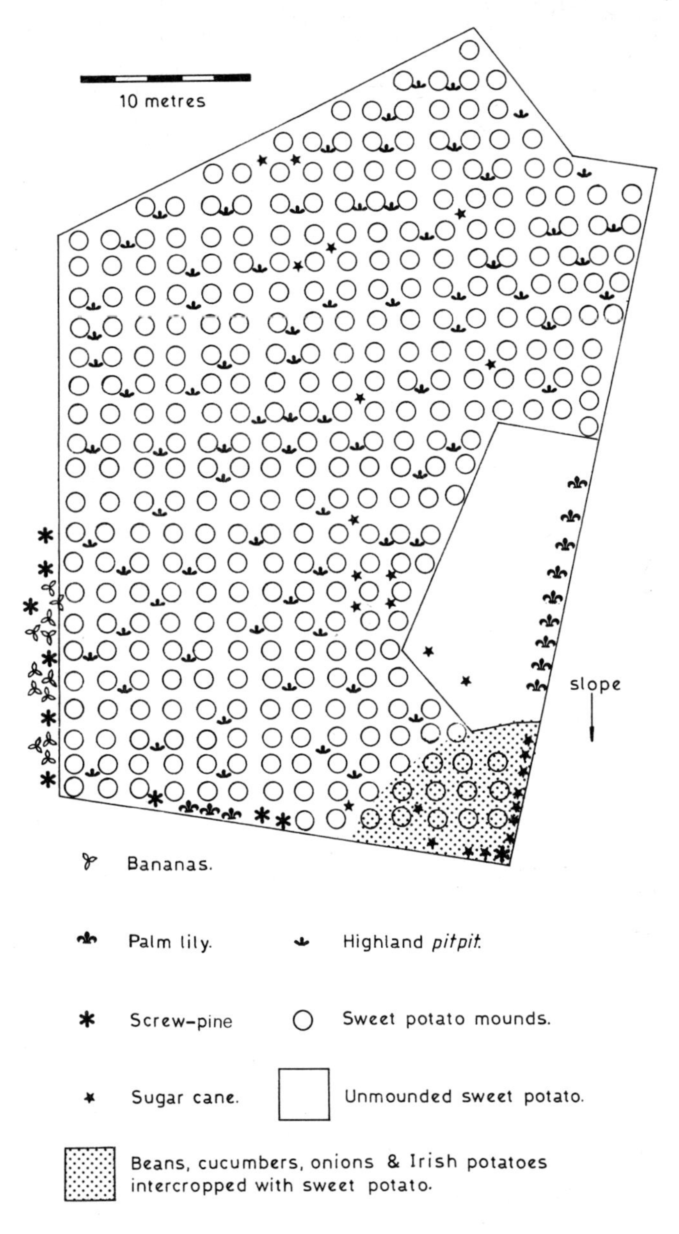

Fig. 50: A sweet potato garden established about three years

about three years old and nearing the end of its second planting (Fig. 50). The rows or *tiyt* of sweet potato mounds are typical, as are the clumps of Highland *pitpit* dotted about between them and the occasional stands of sugar cane. Again screw-pines and bananas are located around the edge of the garden where they cast less shade on the crops growing within it; the small grove on the left was adjacent to the gardener's house. Palm lilies once more demarcate part of the boundary, giving way in one place to a row of sugar cane, which is not commonly used as a border marker in this manner.

This garden illustrates the common practice of extending areas under cultivation, so introducing anomalous crop patterns, both new and old together. Two such extensions can be seen on the right-hand side of the figure. In the lower corner a variety of crops grow intermingled with sweet potato on a patch planted only a month (stippled on figure). Above this is an area of unmounded or *suwl* planted sweet potato, about six months old, with its initial flush of various crops finished and only sweet potato, together with palm lilies and sugar cane, remaining. Men commonly enlarge established gardens in this manner if there is room within the area enclosed, but they infrequently extend fences out to incorporate such extensions.[11]

The next figure (Fig. 51) also depicts a garden about three years old. The lower part, with the mounds, is newly planted for the third time, the upper part being what remained of the second planting, where women had recently obliterated the mounds in a final harvest of sweet potato tubers prior to remounding. The lower, newly-mounded, area displays the characteristic crop pattern of older established gardens with the virtual monoculture of sweet potato, the mounds also having clumps of Highland *pitpit* planted here and there between them (acanth greens, pumpkins, sugar cane, bananas and screw-pines are commonly dotted about gardens of this age too). Around three sides of the garden, up to the fence, are patches of long grass (this is the first stage of regrowth, which the Wola call *taengbiyp*). These grassy areas were planted when the garden was first cleared, but have been left uncultivated since then, which is not unusual. Sometimes, after a number of years have passed, women pull up the grass and plant such areas again, on occasion with a variety of crops (after their rest from cultivation), giving patches like the extensions in the previous garden. At other times they never bother with them after their initial clearance, and they develop into areas of cane or woody regrowth hiding the fence.

Fig. 52 shows a garden about three and a half years old, planted with new mounds down the centre for the fourth time. As in the previous garden, sweet potato predominates, with a few Highland *pitpit* plants dotted about between mounds. The three areas from which the tubers have all been harvested consist of overturned earth mixed with sweet potato vines. In the two on the right women have obliterated the mounds in their final harvest of the tubers they planted when they reworked this garden for a third time. Whereas the larger excavated area on the left was a newly-cleared extension added a year previously, with sweet potato cultivated unmounded or *suwl*, together with

[11] Previously they said this was the way they established large sweet potato gardens in stone age times (see Sillitoe 1979c).

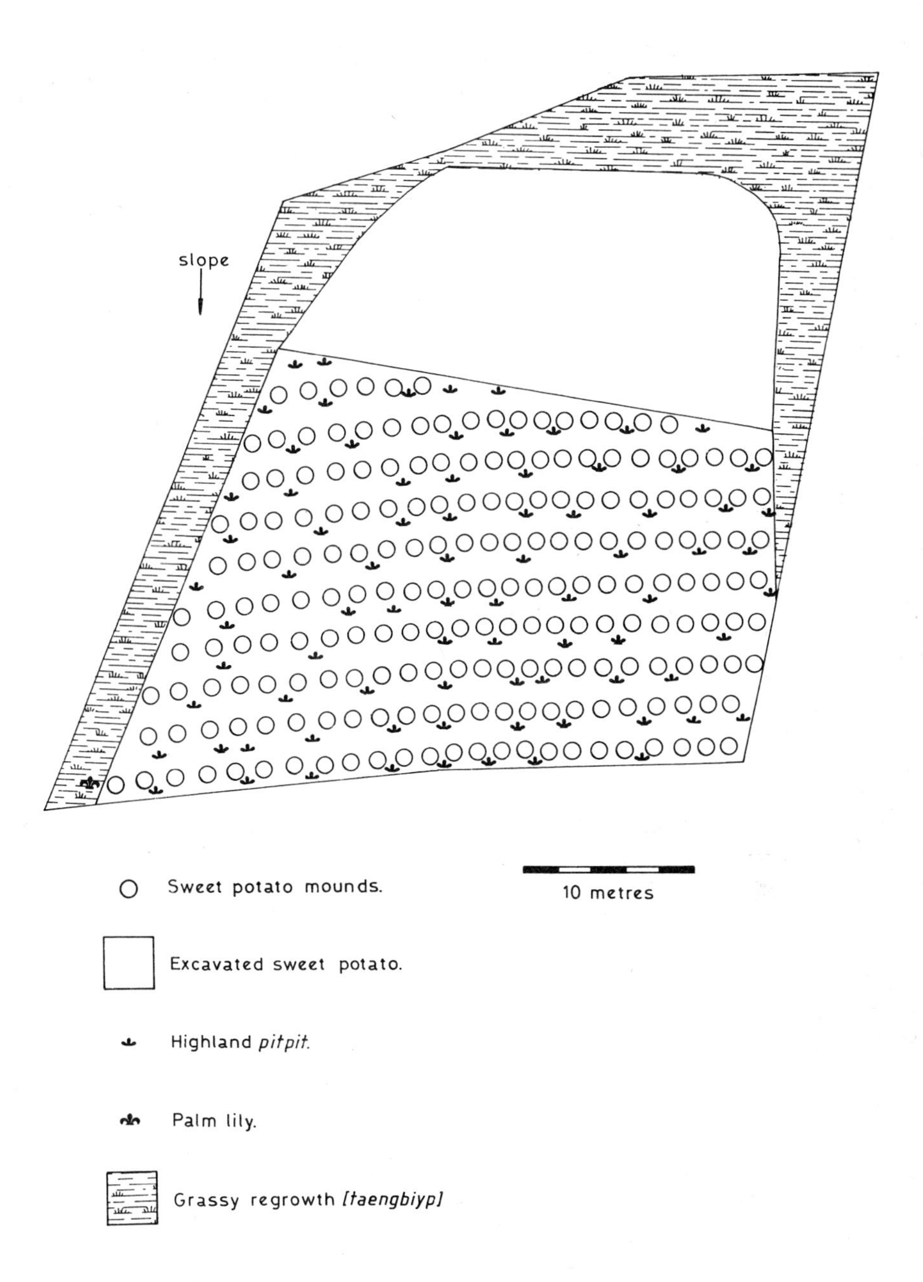

Fig. 51: A sweet potato garden established about three years and newly planted in part

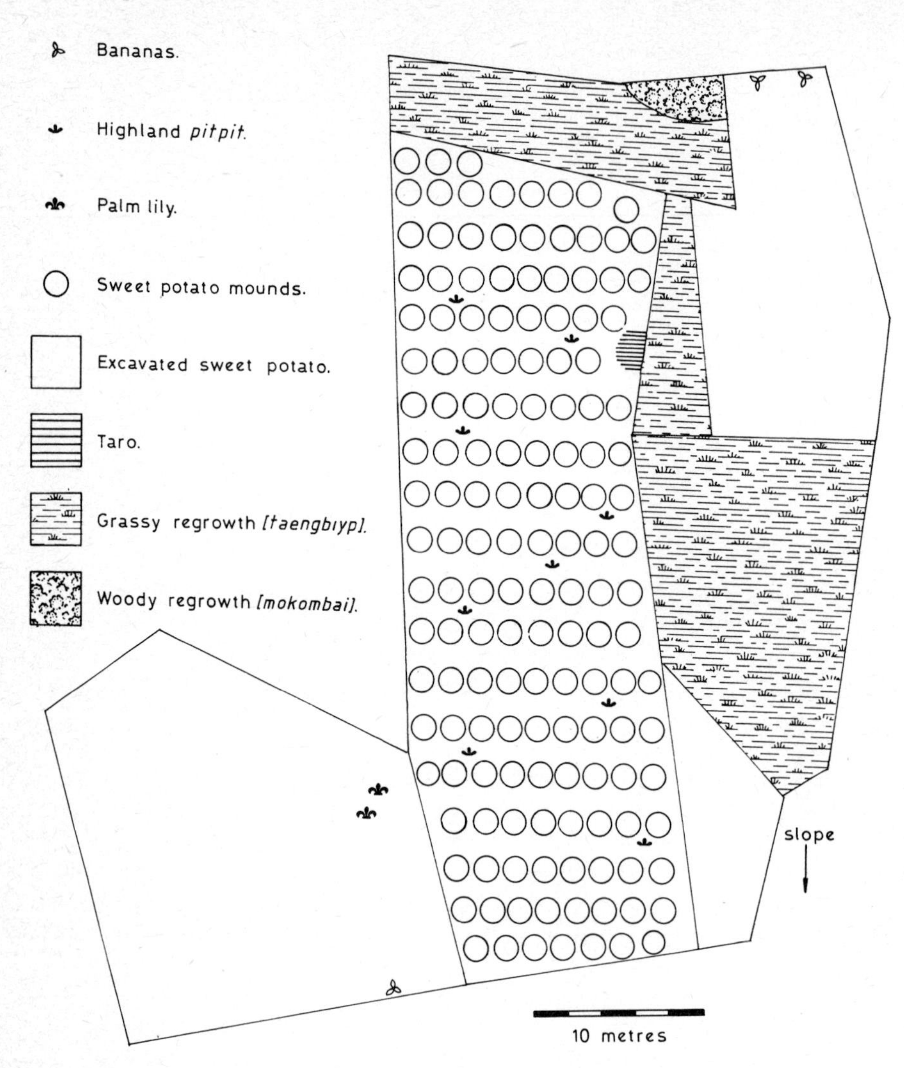

Fig. 52: A partially replanted sweet potato garden about three and a half years old

a profusion of various other crops. At the time of the survey this addition looked the same as the older excavated areas on the right, and when mounded and replanted it was indistinguishable from them in terms of the occurrence of crops. The small patch of taro illustrates again the exploitation of micro-environmental variations, standing in a dip of wet soil which had lain fallow since the first planting of the garden. Considerable areas of grass regrowth occur within the garden too, like the previous one, and in the top right-hand corner cane grass and tree saplings are beginning to establish themselves (this is the stage of regrowth following grass, which the Wola call *mokombai*).

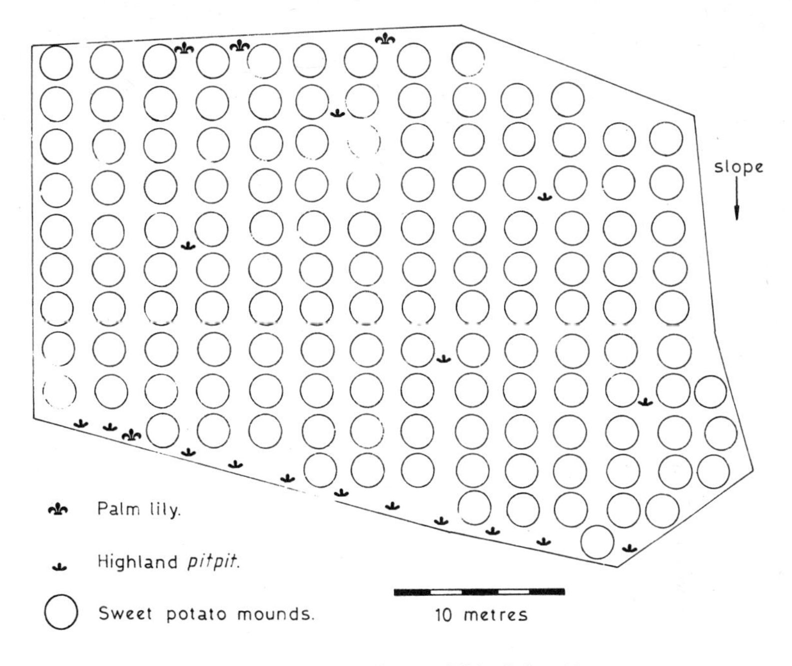

Fig. 53: A sweet potato garden established about ten years

The seventh garden in this sweet potato series was approximately ten years old, and the woman responsible for it thought that she had planted it thirteen times (Fig. 53). It was surveyed a month after this last planting. The garden follows the general crop pattern of established cultivations, with sweet potato predominating, and clumps of Highland *pitpit* scattered between the mounds and in places around the edge, together with a few palm lilies. The woman who gardened this area said that her tuber yields were always good, even improving with the years as she broke up the soil time and again, giving her the kind of fine tilth in which sweet potatoes flourish.

The final garden in this series illustrates further this unexpected aspect of Wola shifting cultivation, that some yields remain constant or improve with time. Instead of suffering a decline with each subsequent crop, the garden depicted in Fig. 54 has experienced a gradual increase. This garden was over twelve years old when surveyed and had been under continuous cultivation, such that the old woman responsible for it was unable to remember how many times she had replanted it when asked and could only reply 'many, many times'.[12]

In parts this garden follows the typical sweet potato-predominating-pattern of established cultivations, while in others it supports a variety of crops, something like a newly-cleared area; demonstrating again that the occurrence

[12] *'Onduwp ora, kau'* (lit: many very, (I) say).

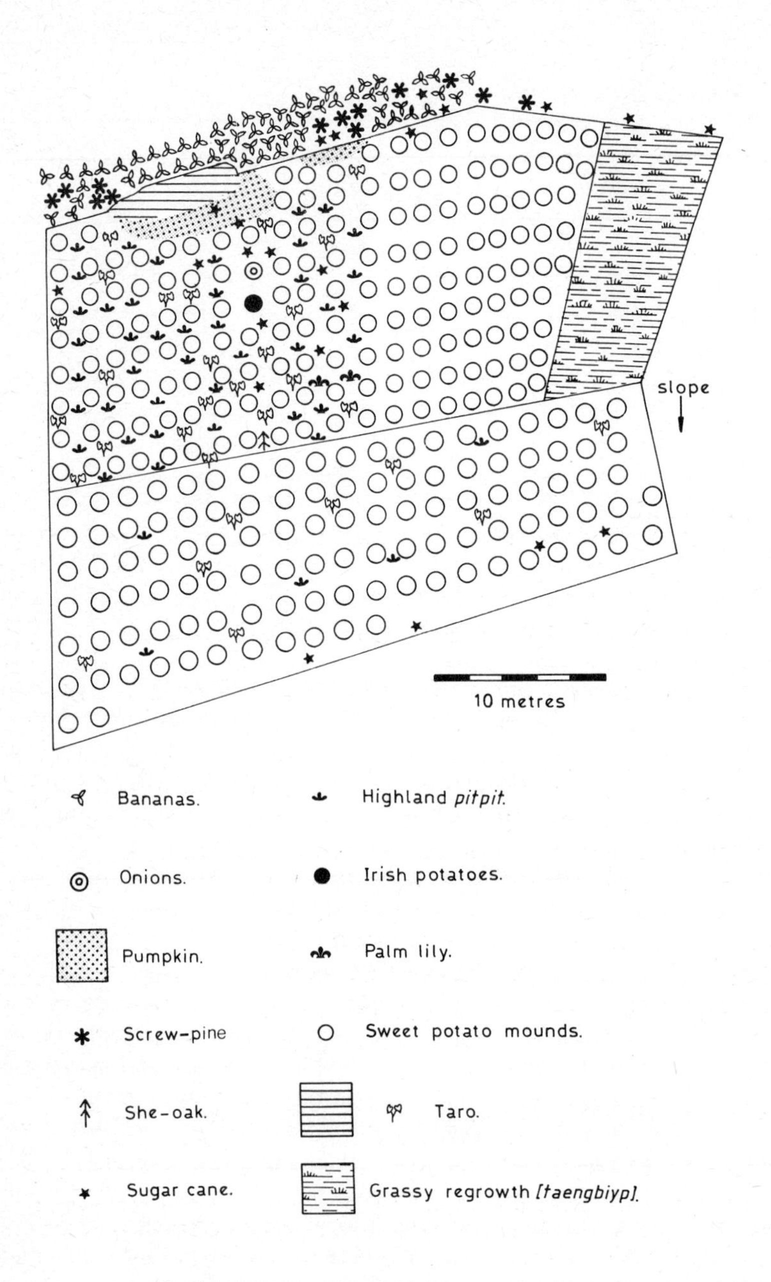

Fig. 54: A sweet potato garden over twelve years old

of crops follows flexible patterns not rigid rules. The lower part is more typical: planted for about nine months, and harvested from for about three, its mounds still intact (it had not reached the stage of the final tuber harvest), with clumps of Highland *pitpit*, stands of sugar cane and a few taro plants dotted about. The presence of taro, which does not generally grow well on land replanted for a second or subsequent time, in a cultivation of such age, is somewhat anomalous, but it is in the upper half that this garden deviates noticeably from the general pattern.

Part of this area is typical too: on the right a patch under grass, which the woman responsible said she intended to pull up in the next week or so, and mound and replant; and in the centre an area of sweet potato mounds conforming, like the lower part, with the standard monocrop pattern of established gardens. It is the upper left-hand quadrant that contradicts the established pattern with its variety of crops. Several taro plants are dotted about here and a stand of taro too in a flat area of wet soil on the top edge. Below this, and also to the right, are areas of pumpkin, which, together with the Highland *pitpit*, sugar cane and palm lily plants dotted around the garden, are not atypical in established sweet potato gardens, except in such density. The mounds planted with Irish potatoes and onions though, located towards the centre of the garden, are like the taro, unusual. The grove of bananas along the top edge, situated on a former house site and mixed vegetable garden, although not exceptional, is of an unexpected size; the screw-pines growing there, planted when the area was cleared and now mature nutbearing trees, are typical.

Why is there this variety in an established garden of such age? The Wola explain that some places prove particularly fertile and productive over many years, although they are unable to predict which ones will prove to be like this beforehand. Some sites they say improve with recurrent planting, especially with regard to sweet potato which flourishes on the well broken up and finely crumbled soil that results from several replantings and repeated re-working. In the case of the garden represented in Fig. 54, just beyond the grove of bananas and screw-pines stood the gardener's current homestead, the close proximity of which further encouraged the continued planting of a variety of crops (for reasons noted earlier in the discussion of mixed vegetable gardens).

REGULARITIES IN THE OCCURRENCE OF CROPS

While it is difficult to plot the occurrence of crops accurately, their intermingled planting making difficult the measurement of areas under different plants, it is possible to indicate regularities. Fig. 55 does this by showing how often various crops were seen growing in different kinds of gardens. This approach though overlooks the other principal factor influencing the occurrence of plants, namely the time that has elapsed since planting. This figure lumps gardens together regardless of their stage in the cultivation cycle

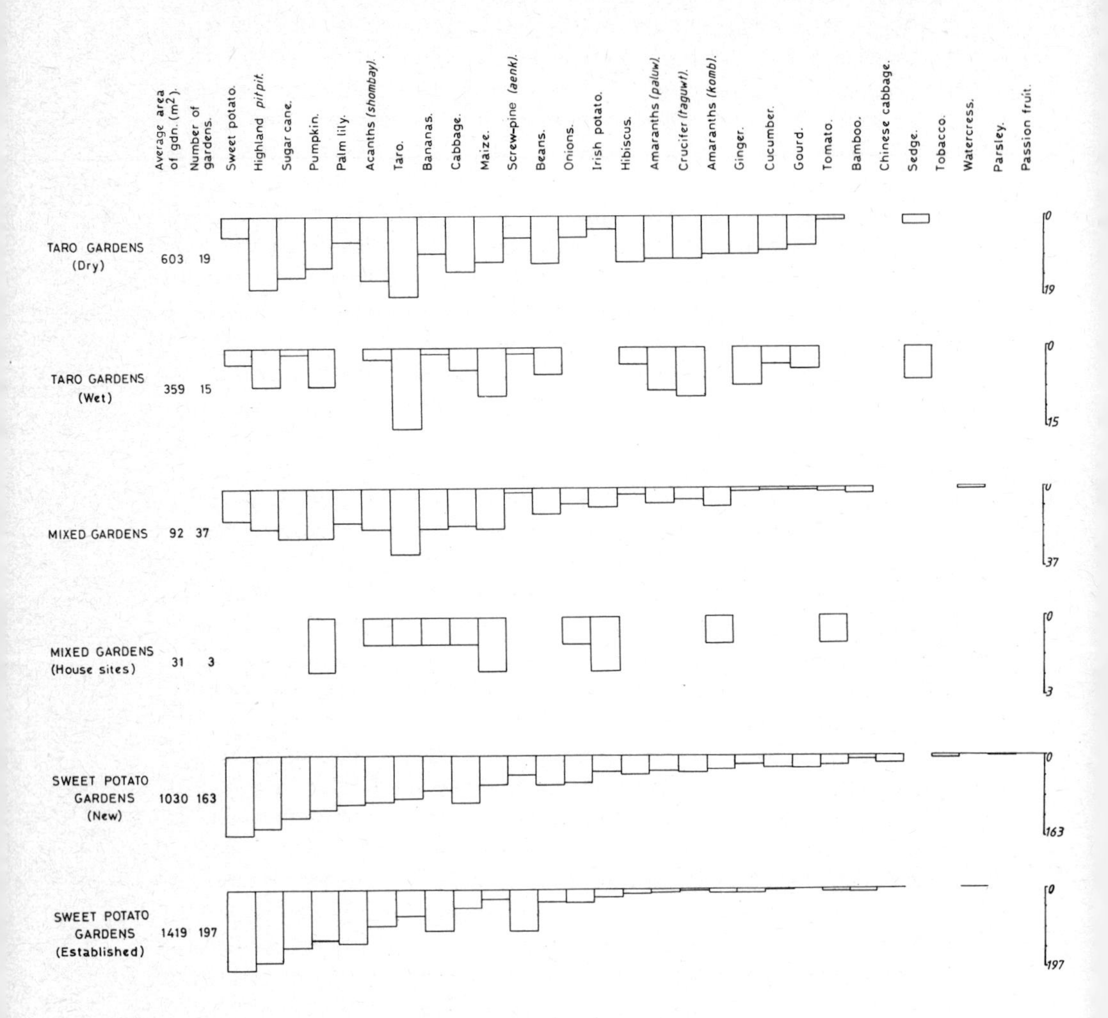

Fig. 55: Occurrence of crops correlated with garden type

(from those newly planted through to those approaching abandonment). Hence it indicates the overall occurrence of crops in all the various gardens of a region at one instant in time. A graph showing only newly-planted gardens would present a different picture, with the lower-scoring crops increasing in occurrence; conversely, one of the older gardens would depress the score of these plants.

In general, the figure underlines the crop occurrence pattern established in the foregoing series of garden accounts. The dry type of taro garden supports a larger range of crops than the wet one, where the moist and sometimes waterlogged soil is inimical to the growth of several plants (those dry-soil-loving crops recorded as occurring in these gardens — such as sweet potato, maize, greens and beans — were planted on a few higher and drier spots, never numbering many plants). Taro, predictably, dominates in all these gardens. The mixed gardens have a wide range of crops, tailing off towards those occurring infrequently overall. The newly-cleared sweet potato gardens also support a range of crops like mixed ones, although here the staple sweet potato predominates, and to a lesser extent Highland *pitpit*, pumpkin and sugar cane. In the established sweet potato gardens these crops dominate even more, others tailing off noticeably, although not disappearing entirely — those gardens with folds of rich soil and other fertile patches, like the final one in the above series, continuing to support a variety of crops, albeit on very small areas compared to those under sweet potato.

The differing occurrence of crops reflects Wola botanical understanding, and shows how it influences their cultivation practices. They have a wealth of traditional knowledge accumulated through the experiments of previous generations; for example, that different crops flourish on soils of varying wetness and 'strength', are sometimes incompatible with regard to physical location, and so on. Everyone knows that some plants thrive markedly when cultivated on the fire sites of recently-cleared gardens, where they benefit from the recently-released nutrients and high carbon content of the soil, and conversely, that only a limited number of plants can tolerate long-standing sites that have dried out through prolonged exposure to the sun. They are aware though that crops need to 'see' the sun to thrive, and they try to reduce the shade cast on a garden by keeping banana plants and screw-pines on the edge, together with the prolific and creeping pumpkin, whose broad leaves will shade out any low-growing plant beneath them. But the Wola, as some of the previous cases show, are not hidebound by this traditional lore; for instance, if a fertile pocket in a garden maintains a tolerable moisture level, then they will experiment on it, trying a wide variety of suitable crops until low returns prompt them to cease. Their knowledge is pre-eminently of a practical kind.

A few features have recurred regularly throughout this discussion and may be seen as the major principles influencing the occurrence of crops. These can be isolated independently of either garden type or the time elapsing since planting, taken earlier as the principal factors constraining crop location, although they are integral aspects of both. These features relate to the absolute age of gardens and the state of their top-soil, with regard to moisture and structure, as shown by Fig. 56. They overlap, as indicated, countering one another to an extent. The figure lists crops according to the influence these features exert on their location, although, as stressed, they do so in a flexible manner and not as rigid rules. They do no more than prompt regular trends.

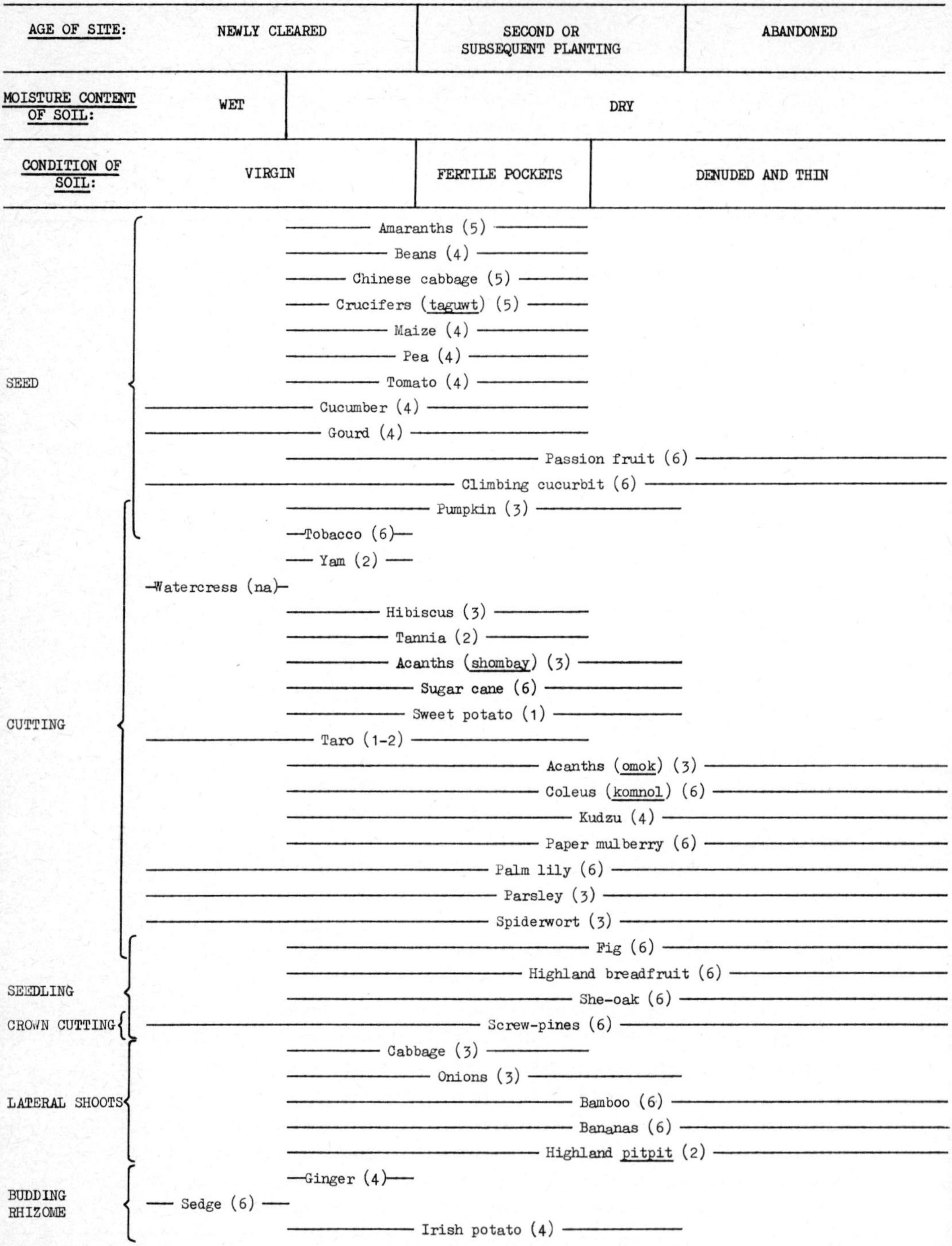

AGE OF SITE:	NEWLY CLEARED	SECOND OR SUBSEQUENT PLANTING	ABANDONED
MOISTURE CONTENT OF SOIL:	WET	DRY	
CONDITION OF SOIL:	VIRGIN	FERTILE POCKETS	DENUDED AND THIN

SEED
- Amaranths (5)
- Beans (4)
- Chinese cabbage (5)
- Crucifers (taguwt) (5)
- Maize (4)
- Pea (4)
- Tomato (4)
- Cucumber (4)
- Gourd (4)
- Passion fruit (6)
- Climbing cucurbit (6)
- Pumpkin (3)
- Tobacco (6)

CUTTING
- Pumpkin (3)
- Tobacco (6)
- Yam (2)
- Watercress (na)
- Hibiscus (3)
- Tannia (2)
- Acanths (shombay) (3)
- Sugar cane (6)
- Sweet potato (1)
- Taro (1-2)
- Acanths (omok) (3)
- Coleus (komnol) (6)
- Kudzu (4)
- Paper mulberry (6)
- Palm lily (6)
- Parsley (3)
- Spiderwort (3)

SEEDLING
- Fig (6)
- Highland breadfruit (6)
- She-oak (6)

CROWN CUTTING
- Screw-pines (6)

LATERAL SHOOTS
- Cabbage (3)
- Onions (3)
- Bamboo (6)
- Bananas (6)
- Highland pitpit (2)

BUDDING RHIZOME
- Ginger (4)
- Sedge (6)
- Irish potato (4)

[1] The numbers in brackets (from 1 to 6) show the crops' place in the planting sequence of a garden (those shown as 6th may be planted sometime after a garden is established).

Fig. 56: Factors constraining the occurrence of crops

Fig. 56 lists crops according to how they are planted, as bracketed together in the left-hand column. This reveals no correlation between their occurrence, as affected by the features adumbrated, and their propagation. Method of planting is unrelated to occurrence. Similarly, there is no correlation between the location of plants and their place in the sequence followed when planting gardens. It is worth noting in passing, though, that there is some correlation between propagation and place in planting sequence. Cuttings of various kinds tend to precede seeds, the smaller seeds coming later. Large plants and trees go in last, often some months after the establishment of a garden and its yielding of some early crops, when it is less crowded and easier to move about without damaging crops, and also when these large plants will not cast shade over smaller ones struggling to root themselves, so killing them off.

Although the Wola have a well-developed lore regarding the conditions that suit their various crops best, they do not appear to have any notions of sequential cultivation, of growing plants one after the other on the same piece of land, or in conjunction with one another, so that the specific qualities of some might benefit the others. Nobody ever hinted either that the differing requirements of plants suggests that they draw upon different qualities in the soil. Their intermingled planting indicates the absence of any ideas relating to crop rotation; planting them jumbled up ensures that they draw simultaneously on the soil's nutrients.

Similarly, the Wola have poorly-developed ideas about some crops changing the soil by putting something back into it.[13] For them, most cultivated plants simply use what is available in the soil until it is dry, hard and exhausted, when yields fall and they abandon the site to regenerate under regrowth. This abandonment has prompted the apt description of such cultivation as one in which land is rotated and not crops. Some crops, notably sweet potato and Highland *pitpit*, can draw on the soil for longer; indeed in some old gardens, which only occasionally grass over, they seem to be able to find an almost inexhaustable supply of the nutrients they require. But only one crop, according to the Wola, improves soil fertility: the she-oak.

Men often plant these trees in and around gardens as described previously in Chapter 6, and when they clear a site on which one stands they will not fell it, unless it is old or they have a specific use for it, but leave it and pollard its low branches to reduce the shadow it casts (as with the she-oak standing in the last garden described in the above series). When asked why these trees benefit the soil and make it more productive, the Wola point to the bed of dead needles underneath them and say that when women work these into the soil as they break it up they make it softer and improve its growing qualities. They are probably half right as these needles, and the tree's cone-like fruits, are reported to be high in nitrogen (Waddell 1972: 143). She-oaks are one of the few non-leguminous angiosperms which have nodules on their roots

[13] See Clarke and Street (1967) on soil fertility and gardening practices.

containing bacteria of the *Rhizobium* species, which can fix atmospheric nitrogen in the soil (see Mowry 1933).[14] While the Wola however, are aware of the soil-enriching capacities of the casuarina, they strangely have no apparent knowledge of the soil-improving qualities of their pulse crops (again through the excretion of aspartic acid produced by bacteria in root nodules), possibly because they cultivate them relatively infrequently, for brief periods and mixed up with other crops, so that their benefits pass unnoticed (unlike she-oaks, which are popular and grow alone for long periods).

In conclusion, it is noteworthy that while the Wola are fully aware of the constraints that restrict the occurrence of crops to certain locations and, as this chapter makes clear, relate their cultivation practices to them, this knowledge is not overtly reflected in their classification of these plants. They do not categorise crops according to their location, although they realise that this is patterned, as reflected in their gardening lore. Their classification is not a straightforward functional one, related to cultivation or use. The situation is analogous to the structuring we have discussed in the earlier chapters on plant morphology, notably parts eaten, which the Wola can apprehend, although they do not order crops this way. This absence of a direct correlation between cultivation and classification contrasts with the relation between the frequency with which crops occur growing and their ordering. This relates to consumption, for the more frequently crops occur under cultivation the more often they are eaten, and, as argued earlier, there is a connection here with classification; it is to this topic, of consumption and nutrition, that the next chapter turns.

[14] It is not known whether the species of she-oak cultivated by the Wola actually fixes atmospheric nitrogen. It is not included in Bond's (1976) list of twenty-one casuarina species (out of the total of forty-five accredited to this genus) known to have root nodules, although Waddell (1972: 144) records that nodules have been found on *Casuarina oligodon* (while not committing himself to its nitrogen-fixing status).

Section II: Crops discussed

Chapter 12
YIELDS AND CONSUMPTION

This chapter documents the yields obtained by the Wola from their gardens and investigates the part the crops harvested play in their diet. It rests largely on the analysis of quantitative data, discussed in the light of cultural practices influencing yields and consumption, like food taboos, the use of plants on specific occasions, their preparation, the etiquette surrounding their sharing, and so on. It also deals in a somewhat cursory manner, for I am unqualified to attempt anything more thorough, with the nutritional value of the diet which consists largely of the plants studied here. This predominantly vegetable diet, low in high-quality protein, is problematic, especially with regard to infants who may, as a result, suffer malnutrition.[1]

YIELDS

The results of a survey of the crops harvested from a series of sample garden areas are given in Table 33.

The method used to collect this data was as follows: I selected a number of gardens ranging in age from newly-cleared through to about fourteen years old, and in each one I measured out an area, marking its boundary clearly with wooden pegs, and arranged with the woman, and her relatives, who harvested from it, in return for a reward,[2] to bring everything gathered to my house for weighing. I have every reason to believe that those involved co-operated fully and did not cheat. When someone arrived with a load of food, each crop harvested was weighed in a string bag from a simple spring

[1] This problem may worsen though when a people's traditional subsistence base is no longer intact, and they live on an unbalanced diet of processed food.

[2] Consisting of a meal of fish and rice, some money and an item of clothing (usually a blouse). Here I record my thanks to them for their ready co-operation and help: *chay tenow hombuwn*.

Table 33: The crops yielded by a sample of gardens (weights in kg)

Area index letter	Name of woman harvesting from area	Location of garden (name of place)	Age of garden	Area (m^2)	Length of time (in days) from first harvest to last harvest	Number of days on which crops were actually harvested from area	Number of times pigs broke in & damaged garden during survey	Number of sweet potato mounds in area	Sweet potato	Highland[2] pitpit	Pumpkin	Onions	Irish potatoes	Taro	Chinese cabbage	Crucifer greens	Cabbage	Amaranths (paluw)	Maize[3]	Gourd	Cucumber	Beans[3]	Acanths (shombay)	Parsley	Sugar cane	Ginger
A	Leda[1]	Ganonkiyba	Newly cleared	344	237	44	4	13[4]	165.169	45.162		0.283	0.340	4.763	2.041	10.887	3.080		1.134		2.495	3.004	8.496			0.567
B	Pombray[1]	Ganonkiyba	Newly cleared	332	244	47	6	5[4]	174.075	36.285	3.629		0.907	1.814	6.123	15.667	9.753		0.794	0.340	4.083	3.628	1.417	0.113		0.227
C	Horshiyow[1]	Ganonkiyba	Newly cleared	237	241	22	3	4[4]	154.850	24.494		0.227	0.280		4.762	1.396	2.721		0.340		0.851	2.722	0.227			
D	Kwalten[1]	Ganonkiyba	Newly cleared	309	235	24	4	9[4]	195.956	23.414				1.361	5.670	2.189	7.994				0.492	2.041	0.680		0.511	0.113
E	Twaen[1]	Ganonkiyba	Newly cleared	355	244	30	3	4[4]	162.904	30.559		0.113		5.897	1.134	7.371	6.237		0.113		0.340	3.288	0.170		0.397	0.057
TOTAL OF NEWLY CLEARED GARDEN AREAS:				1577		167	20	35	852.954	159.914	3.629	0.623	1.275	13.835	19.730	37.510	29.785		2.381	0.340	8.261	14.683	13.655	0.113	0.908	0.964
F	Kombaem	Maenshiyow	Established 1 year	158	92	6	3	24	52.164																	
G	Oliyn	Honael	Established 3 years	156	159	18	6	37	115.329	1.134																
H	Kaelenj	Honael	Established 5 years	137	113	22	1	35	245.114	4.196																
I	Wenya	Honael	Established 8 years	151	170	12	1	32	119.409																	
J	Nonk[1]	Senz	Established 10+years	170	212	14	0	22	72.292	11.793	29.087	0.454	0.454	0.454	0.057	0.170	0.454	4.536	1.814							
K	Wa[1]	Tomb	Established 14+years	132	176	19	0	27	241.882	1.927		0.454														
TOTAL OF ESTABLISHED GARDEN AREAS:				904		91	11	177	846.190	19.050	29.087	0.908	0.454	0.454	0.057	0.170	0.454	4.536	1.814							
GRAND TOTAL OF NEW & ESTABLISHED GARDEN AREAS:				2481		258	31	212	1699.144	178.964	32.716	1.531	1.729	14.289	19.787	37.680	30.239	4.536	4.195	0.340	8.261	14.683	13.655	0.113	0.908	0.964

[1] These areas were harvested from again after I had completed the survey and left the field, although the yield would have been negligible compared with those recorded here because the gardens were approaching the end of their productive periods.

[2] The weights here are of unshucked stems without their leaves (humans eat about1/3 of this weight, and pigs the remaining 2/3).

[3] The maize and bean seeds planted in the Ganonkiyba garden areas were eaten largely by rats before they germinated, hence the yields are abnormally low.

[4] The sweet potato in these areas was largely planted straight into the unprepared ground (suwl), and not in mounds.

balance.[3] Each area had a record sheet, on which a note was made of the weights, the date, the number of people and animals to be fed from the produce, the state of the garden, and, if relevant, the number of sweet potato mounds excavated.

Unfortunately these yield records are incomplete for some gardens because I left the field before they were fully harvested (as indicated on Table 33).[4] However, the amounts harvested subsequent to my departure would have been small compared to those weighed in the survey. Prior to leaving, in my last week, I had the areas still in production excavated of all mature tubers and harvested of all other ripe and near-ripe crops. The gardeners said that it would be something like six months before they could harvest from the denuded areas again and that the yields would be small.

Another problem concerns the expression of the survey's results. The Wola practice of planting different crops mixed up in one garden, as described in the previous chapter, makes difficult the estimation of yields according to the standard agricultural method of so many kilograms to a hectare, which assumes mono-cropping. On the basis of test beds, and from areas where people practise mono-crop agriculture, some writers record yields for the crops discussed here in this standard agronomic fashion (for instance see Massal and Barrau 1956; Purseglove 1968, 1972; and Powell 1976), but to do so for the Wola would be distorting. It would require the establishment of test-beds, which could not replicate Wola gardens in evéry respect. The intermingled growing of several crops for instance affects yields, for competing with a diverse range of plants is markedly different to doing so with other members of the same species only — different crops draw varyingly on the nutrients in the soil, compete for sunlight in differing ways, are ready for harvesting at diverse times, and so on, all of which factors influence yields. Mono-cropping would influence yields not only by creating artificial growing conditions but also by imposing artificial cultivating conditions on the native gardeners. They would require direction in cultivating the crop, which would modify their traditional methods. Another problem would be arranging the harvest. A single crop assumes one harvest reaping it all, whereas the Wola leave their crops growing like a living larder, harvesting them daily over several months as required.

Another hindrance to direct comparison is that agronomists look for the highest yields possible under experimental conditions; for example Powell (1976) talks about average sweet potato yields of 26,000 to 38,000 kg per hectare, acanth greens (*Rungia klossii*) yielding 20,000 to 40,000 kg per hectare, and Highland *pitpit* (edible portion) yielding 8,000 to 48,000 kg per hectare. Under experimental mono-crop conditions these yields might be feasible in some Highland regions, but they do not reflect the returns of

[3] Of the type fishermen commonly use to record the weight of their catch.
[4] Rappaport 1968: 252—5 had similar problems.

indigenous horticulture. Nor do they necessarily imply that it is inefficient; the Wola after all grow sufficient to maintain a vigorous and growing population, and also feed large numbers of pigs.

While intercropping renders awkward the precise calculation of single crop yields per unit area, it is preferable to work within this frame and avoid introducing any unnecessary bias. The data presented here give intermixed yields, that is according to the Wola horticultural practices. Thus, for example, a sweet potato yield may be expressed as so many kilograms per square metre or hectare when planted in certain proportions with specified other crops like Highland *pitpit*, pumpkin, sugar cane and various greens, which yield so much for the same area.

The results of the survey show clearly that yields from different gardens vary considerably; for example, the most productive area yielded six times more food by weight than the least productive. This poses problems for the calculation of average yields, compounded by the difficulties mentioned above surrounding the computation of single crop yields. An enormous sample would be necessary to achieve any degree of accuracy, and even then fluctuations from the average would make it little more than an academic figure. The pie-graph in Fig. 57 illustrates proportionally the extent of these fluctuations for the areas surveyed. It contrasts the average yield with the largest and smallest recorded, all expressed in kilograms per hectare.

This figure also shows the average proportional yields of the different crops, sweet potato accounting for well over three-quarters of the total. The variety of crops grown, however, and their yields relative to one another, vary according to the age of gardens. The sample areas listed in Table 33 are arranged chronologically: at the top the newly-cleared areas planted for the first time, and at the bottom the oldest-established ones. The newly-cleared areas A to E have the highest yields from the widest variety of crops, as expected from the discussion of the previous chapter.[5] But they do not have the highest overall yields per unit area. Indeed they are quite low compared to some of the older-established gardens, the oldest of which gave the highest overall yields, vindicating the assertions of the Wola, noted with some incredulity in the last chapter, that some of their crops, notably sweet potato, maintain and even increase their yields as a garden is replanted time and again.

The woman responsible for this longest established area K (which abuts the oldest garden discussed in the previous chapter — see Fig. 54), said that yields from it had always been high, improving as the soil structure had developed with the breaking up and remounding of each planting. This holds so long as the soil and other site factors allow it: where they do not, gardens experience a decline with age and are soon abandoned in the fashion of classic shifting cultivation. The capability of some Wola crops, though, to

[5] See the garden depicted in Fig. 47 of Chapter 11. The five newly-cleared areas from which yields were weighed occurred within it, demarcated by the near horizontal rows of Highland *pitpit* (listed from bottom to top as A to E on Table 33).

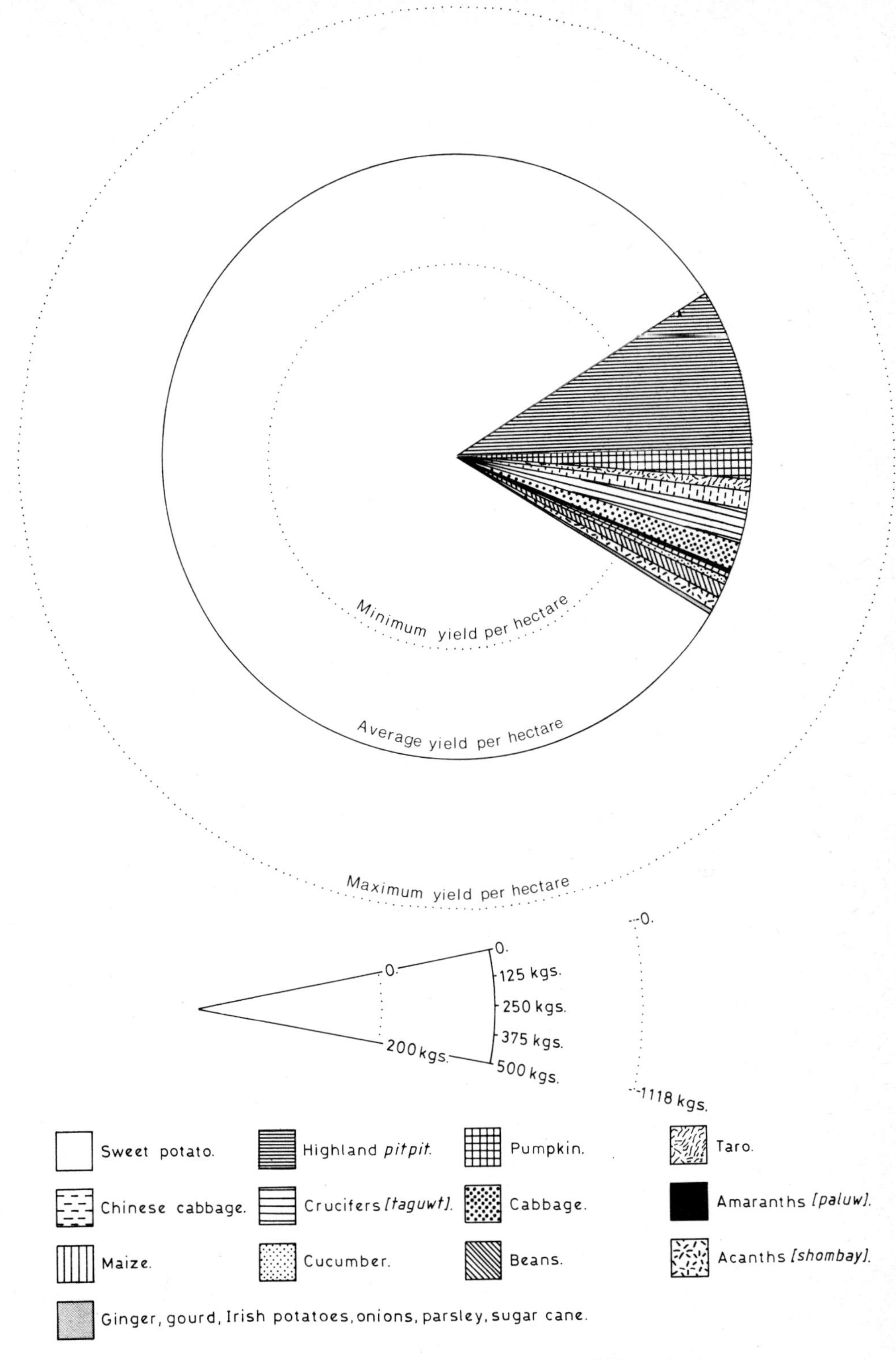

Fig. 57: Maximum, minimum and average crop yields per hectare

maintain or increase their yields with the passage of time somewhat turns the assumptions of the swidden cultivation model on their head. These people are shifting cultivators in a number of respects, but the permanence of some of their gardens gives them something of a betwixt and between status.

Some gardens planted a number of times not only produce high yields for a few plants, but also, as demonstrated in the last chapter, support a wide variety of crops. This is so with the second-oldest area J in Table 33, which produced a marked variety of crops, although only having about average yields for the sample.[6] This garden had a fold running down it, containing a deep layer of dark top-soil washed there over the years, and in this fertile pocket it produced a variety of crops again and again, over and above the staples predominating elsewhere in it. Thus illustrating a point made a number of times in earlier chapters about the exploitation of micro-environmental variations within gardens.

The yields from garden areas I and G are round about the average for the sample (area G occurred in the centre of the garden depicted in Fig. 51, in the last chapter), whereas those for area F are particularly low. This was due to pigs breaking in and rooting up the crops, not any inherent physical deficiency of the site. Table 33 notes the number of times that pigs broke into each garden area and damaged it. In the case of area F they entered only three times (as opposed to six times in areas B and G), but on two of these occasions they did extensive damage, reducing yields markedly.[7] All gardens, except the two established longest, suffered to some extent from the predations of pigs. Their break-ins are a factor affecting yields just like the soil, climate and so on, and for this reason I decided not to omit area F from this sample: it typifies the fate of some Wola gardens. The proportion so extensively damaged is difficult to judge, and it is possible that the inclusion of area F in the computation of the average yields does depress the results to some unknown extent (another element contributing to the dubiety of this figure). It is noteworthy that the number of times pigs enter gardens diminishes with their age, indicating that older ones are more effectively protected. Often this pig-proof barrier is natural and not man-made, and the Wola themselves say that they are more likely to maintain gardens on such sites for many years, thus indicating that it is not only physical conditions (such as slope, aspect, soil type and so on) that persuade them to cultivate some gardens continuously.

Pigs are not the only peril that may reduce garden yields. When soil is newly planted and exposed, it is vulnerable to damage by heavy rain. During

[6] Although I have reasons to believe that the yield results for this area are depressed because the gardener's adolescent daughter harvested from it on one or two occasions without bringing the produce for weighing. These are only suspicions. The girl and her mother denied that this was so, and I have not deleted their area from the sample because whatever crops were not weighed would have been of a small order, not enough to distort the results seriously.

[7] For more on pig damage to gardens see Sillitoe (1981).

particularly heavy storms entire sweet potato mounds are sometimes swept away. Rats also diminish garden yields. The Wola say that there are two kinds that live in and around human settlements, on occasion infesting gardens and eating their produce. They call these rats *paeket* and *kolkorkor* (they are *Rattus exulans* and *Rattus ruber*). A small garden which I had cleared and planted adjacent to our house to measure sweet potato yields (omitted from the table because they would distort it too much) illustrates how devastating rat damage can be. The plot covered 198.6 m^2 and yielded a mere 20.9 kg of sweet potato (which is only 1050.8 kg per hectare).[8] Rats also sometimes play havoc with crops propagated from seeds, eating them before they germinate; this is what happened to many of the bean and maize seeds planted in the newly-cleared Ganonkiyba areas (hence the yields shown for these crops are abnormally low).

Other factors that sometimes reduce crop returns below those that a garden would yield are aberrant climatic conditions, frosts and droughts; blights and diseases attacking the plants; and insects devastating them (as some chafers do taro for example).[9] The Wola say that the occurrence of earth tremors is an omen about sweet potato yields. If they occur in the morning then yields will be good but if they occur in the afternoon then they will be poor, the plants will develop dense foliage at the expense of tubers and there will be a hungry period. In some Wola regions, if yields are low, they stage a ritual called *iysh ponda ma honday* which they believe improves the weather and brings fertility back to the soil.

When harvesting sweet potato from gardens, the Wola distinguish two stages, which they call *waeniy* (lit: new) and *puw*. This distinction depends on the progress of the harvest. The *waeniy* stage is the first harvest, about five to six months after planting, which yields the largest returns. After this the garden enters the *puw* stage, during which yields decline following each excavation. Table 34 shows the average period of time for which each stage extended in the garden areas included in the yield survey, and the average weights and relative percentages of the four principal crops that yield produce during both stages. This table indicates that although there is no difference between newly-cleared and established gardens in the proportions of crops harvested during each stage (about 60 per cent in *waeniy* and 40 per cent in *puw*), there is a difference in the time each stage lasts and the number of days in them that people actually harvest from gardens.

The explanation for this is the larger variety of crops occurring in newly-cleared gardens. The time that elapses from planting to initial harvesting

[8] Also, throughout the harvesting period, gardens throughout the locality were, for some inexplicable reason, yielding very poor returns. So there was something climatic depressing yields in addition to the depradations of rats.

[9] No Wola has ever talked to me about legendary falls of ash leading to famine, as reported for other parts of Papua New Guinea (see Watson 1963: 152—5, Glasse 1963: 270—1, and Young 1971: 173).

Table 34: Yields during *waeniy* and *puw* stages (averages, days and kg), with average age of gardens

	FROM FIRST TO LAST HARVEST		DAYS OF HARVESTING		SWEET POTATO		HIGHLAND *PITPIT*		ACANTH GREENS		PUMPKIN	
	waeniy	*puw*	*waeniy*	*puw*	*waeniy*	*puw*	*waeniy*	*puw*	*waeniy*	*puw*	*waeniy*	*puw*
Newly-cleared garden areas ($315\ m^2$)	186	55	27	7	108.1	62.5	19.3	12.7	0.5	2.2		
	77%	23%	79%	21%	63%	37%	60%	40%	18%	82%		
Established garden areas ($151\ m^2$)	37	116	9	6	95.1	45.9	1.9	1.3			2.1	2.8
	24%	76%	60%	40%	67%	33%	59%	41%			43%	57%
Total of all garden areas ($226\ m^2$)	105	88	17	6	101	53.5	9.8	6.5				
	54%	46%	74%	26%	65%	35%	60%	40%				

varies considerably from one crop to another (see Fig. 41 in the previous chapter). Within two months a number of the green leaf crops are ready for harvesting, but the *waeniy/puw* distinction depends on the state of the sweet potato harvest, which commences some three or four months later. So in a newly-cleared garden with a variety of crops, harvesting may have been proceeding for several weeks before the digging of any sweet potato tubers (indeed some of the early ripening crops will be finished when their excavation commences). Whereas in an established garden, sweet potato tubers may be the first crop harvested, hence the duration of the *waeniy* harvest is markedly shorter — extending from the first sweet potato harvest only. Unless the garden has a variety of crops, like area J, which resembled a newly-cleared garden more than an established one in terms of the relative duration of its *waeniy* and *puw* stages.

The termination of the survey while some gardens were still in production distorts the relative times and yields given for the *waeniy* and *puw* stages. The newly-cleared areas continued yielding for a further four months, making their productive life fifteen months from planting. Although the number of harvests made during this spell would have been low, it extended the period from the first harvest to the last, such that the time ratio of *waeniy* to *puw* is approximately the same. There was considerable variation in the productive periods of the established garden areas, and hence their *waeniy/puw* ratios. Some women replant within nine months or so of last cultivating a site, while others may prolong harvesting for anything up to two years or so (although experiencing a declining yield at each subsequent harvest). In the latter events, gardens have *puw* periods in excess of 90 per cent, compared to less than 10 per cent of their time as harvested *waeniy*.

Another shortcoming arising from the premature termination of the survey was the omission of yields from plants which take a long time to mature, such as bananas, sugar cane and the tree crops. Compared with the overall yield of all crops, the returns from these are small, and their omission does not seriously distort the conclusions reached. Neither does their absence upset the *waeniy* and *puw* periods because they are often harvested following the abandonment of gardens (see Fig. 41).

Table 35 standardises the yields for each area by expressing them in kilograms per hectare. Bearing in mind both the incompleteness of the data, owing to the survey's premature conclusion, and the difficulty of comparison, given the varying proportions of different crops cultivated intermingled in the areas surveyed, the overall average yields for all crops combined, calculated from these figures, were as follows:

(*a*) Newly-cleared gardens = 7359.3 kg per ha
(*b*) Established gardens = 9991.6 kg per ha
(*c*) Both newly-cleared and established gardens = 8318.5 kg per ha

Table 35: Crop yields of sample plots (kg/ha)

AREA INDEX	SWEET POTATO	HIGHLAND *PITPIT*	PUMPKIN	ONIONS	IRISH POTATOES	TARO	CHINESE CABBAGE	CRUCIFER GREENS	CABBAGE	AMARANTH GREENS (*paluw*)	MAIZE	GOURD	CUCUMBER	BEANS	ACANTH GREENS (*shombay*)	PARSLEY	SUGAR CANE*	GINGER
A—E	5409	1014.0	23.0	4.0	8.1	87.7	125.0	238.0	188.5		15.1	2.2	52.4	93.1	86.6	0.7	5.8	6.1
F	3301																	
G	7392	72.2																
H	17892	306.0																
I	7908																	
J	3252	694.0	1711.0	26.7	26.7	26.7	3.4	10.0	26.7	267.0	107.0							
K	18324	145.6		34.4														
Average established gardens	9361	210.8	322.0	10.0	5.0	5.0	0.6	1.9	5.0	50.2	20.1							
Average established gardens	6849	721.0	131.8	6.2	7.0	57.6	79.8	152.0	121.9	18.3	16.9	1.4	33.3	59.2	55.0	0.5	3.7	3.9

* The yield figures for sugar cane are depressed because of the time it takes to grow and mature, most of the cane in these garden areas would not have been mature for cutting until several months after the conclusion of the survey.

Discounting the dubiousness of averages calculated from such widely differing examples, without a statistically significant sample to offset their randomness (revealed by comparing them with Table 35), these figures compare broadly with those reported from other regions, although Wola yields are somewhat on the low side (which follows given the poor nature of their limestome based soils, compared with those of volcanic origin elsewhere — see Chapter 1). They are close to those obtained by the Maring from their predominantly sweet potato and sugar cane gardens, as reported by Rappaport (1968: 50). With their long-maturing crops of bananas and sugar cane omitted to correspond to the Wola data, the Maring harvest 11,221 kg per hectare from these gardens, of which 6,485 kg is sweet potato, which falls very near the Wola average. This contrasts with Waddell's (1972: 117) reported sweet potato yields for the Enga of 17,400 and 20,870 kg per hectare, with which only the most productive areas in the Wola sample can compare. The range of sweet potato yields given for the Kapauku by Pospisil (1963: 444), from 8,100 to 16,900 kg per hectare, falls within the range of Wola yields.[10]

The yields discussed here come from measurements made on the produce of sweet potato gardens only, which account for by far and away the greater part of Wola cultivation (as the areas recorded on Fig. 55 show). The yields from taro gardens and small mixed vegetable gardens would be comparable, although the proportion of different crops grown would vary. The yields from the newly-cleared sweet potato garden areas, with their variety of crops, would be similar to those of taro and mixed gardens, although having a higher proportion of sweet potato. Some taro gardens, notably those on waterlogged sites which restrict the other crops that can be grown, are virtually mono-crop stands of taro — as some established sweet potato gardens consist almost entirely of sweet potato.

The newly-cleared area A in the survey (see Table 33) had about a third of its surface under taro, and its yield, converted to kilograms per hectare was only 965 kg. The site, however, was not particularly good for taro and the yield is undoubtedly a very poor one, as the higher taro yield of area E shows, which had only twenty-four plants growing on a markedly smaller but damper and more suitable patch (see Fig. 47). The Wola say that taro not only grows more slowly than sweet potato, but also yields less per unit area, that it is a difficult and unpredictable crop, which sometimes yields small returns for the effort expended cultivating it. So Rappaport's (1968: 49) figure of 3,109 kg per hectare for Maring taro—yam gardens is probably nearer that of the Wola average than the Kapauku yield of 12,800 kg per hectare (Pospisil 1963: 444), or the Kukukuku yield of 12,553 kg per hectare (Hipsley and Clements 1947: 85), or the yields given by Massal and Barrau

[10] It is not clear from Pospisil's account whether the crops were harvested Kapauku fashion over several months, or whether he harvested them all at one time when he judged them ready (if the latter was so, then his yield figures are, in all probability, somewhat depressed).

(1956: 8) of 7,532 to 15,064 kg per hectare, and up to 20,084 kg per hectare under irrigation. These latter yields are for taro growing alone, not intercropped under indigenous conditions. They illustrate again the extent to which this can influence yields and rule out direct comparison with local group returns, as argued above.

CONSUMPTION

When recording the weights of the crops harvested from the areas covered in the yield survey, a record was also kept, as mentioned, of the persons and animals who would consume the produce. Table 36 presents this information for those days on which the people concerned did not receive unweighed food from other sources, that is the produce weighed constituted their households' daily supply of food, which was the situation that pertained on 39 per cent of the days.[11] On the table 'adult male consumption days', 'child consumption days', and so on, record the number of days on which individual men and children shared the produce weighed; for example, on the ten days included in the consumption table that Leda harvested from area A, her husband received some of the produce on all ten days and her youngest son on one day, making a total of eleven 'adult male consumption days'. According to the table, 625.8 kg of vegetable produce was sufficient to feed 265 adults and 166 children for a day, which (assuming on average that children, regardless of age, consume approximately half the food adults do) averages out crudely at 1.8 kg per adult a day. The 644.8 kg harvested for pigs and cassowaries, on the other hand, was sufficient to feed 339 pigs and 27 cassowaries for a day, which (assuming that a cassowary consumes about one-third of the food of a pig) averages out crudely at 1.9 kg per pig a day.

The proportion of their produce which the Wola feed to their pigs is surprising (as Waddell 1972: 113, 118 also notes for the Enga). Throughout the survey women were asked, when they brought produce for weighing, to divide their sweet potatoes into two piles, one for human consumption and the other for pigs. They fed 48.5 per cent of all tubers weighed to their pigs[12] (which almost parallels Waddell's (1972: 118) Enga percentage of 49 per cent although it is lower than Rappaport's (1968: 280) of 55.9 per cent for the Maring).

The above per capita consumption figure for pigs might be thought reasonably accurate (these animals not receiving any more garden produce than that fed to them in the evening, as measured here — unless they broke into

[11] One area, E, is omitted entirely and others like G are hardly represented because of the confused situations in their households and their receipt of food from other sources; for example, Oliyn, who harvested area G, lived with her son's wife who returned daily from her gardens with produce which the entire household shared together with Oliyn's weighed produce.

[12] This is lower than the percentage recorded in Table 36: by chance a larger proportion of the tubers weighed on those days were fed to pigs.

Table 36: Consumption of crops harvested during yield survey (days and kg per 'consumption day' (CD))

INDEX LETTER	HARVEST DAYS	ADULT MALE CD	ADULT FEMALE CD	CHILD CD	PIG CD	CASSOWARY CD	SWEET POTATO		HIGHLAND *PITPIT*		ACANTH GREENS (*SHOMBAY*)	CABBAGE	PUMPKIN	TARO	PARSLEY	GINGER	ONIONS
							HUMANS	PIGS & CASSO-WARIES	HUMANS	PIGS							
A	10	11	12	2	33		42.2	80.1	6.6	13.1	4.7						
B	17	17	23	1	65		62.4	85.3	8.2	16.4	1.1	0.2	2.3	1.8	0.1	0.2	
C	10	15	10	20	50		48.6	84.9	4.3	9.1	0.2	0.2					
D	5	9	15	11	39	5	37.6	65.4	2.7	5.8	0.3	1.4					
E	Insufficient data on household consuming produce																
F	6	6	6	24	19		34.8	17.4									
G	1		1		2	1	1.6	1.2									
H	18	23	19	39	26	15	128.4	82.3	1.0	2.0							
I	10	15	12	8	22		51.7	56.6									
J	7	7	13	26	26		37.1	28.9	3.4	6.9			14.6				
K	17	17	34	35	57	6	127.2	88.3	0.6	1.2							0.2
Totals	101	120	145	166	339	27	571.7	590.4	26.8	54.4	6.3	1.8	16.9	1.8	0.1	0.2	0.2

gardens!). This assumption is not correct, however, because no allowance is made for the differing sizes of the pigs and hence the varying amounts they ate. Similarly, the per capita consumption figure for human beings is markedly too low. Although all the dubious days were ruthlessly omitted in the compilation of Table 36, the collection of the data was not rigorous enough to achieve a presentably accurate record. The people concerned were receiving sufficient food from elsewhere, over and above that weighed, to bias the results seriously. To off-set the shortcomings of this data, the food consumed by a larger and statistically more significant sample was surveyed in a more rigorous manner, over a period of some three months.

For this consumption survey I approached the residents of twelve homesteads and asked them if they would be willing to co-operate and allow me to weigh and document the food they ate over a number of weeks. In order to allow them to appreciate what would be required, and myself to see the work involved and adjust my approach to the survey to collect the data in as reliable a way as possible, we staged a preliminary survey of twelve days' duration. Following this I drew up the questionnaire reproduced in Appendix IV, and when each day members of the households involved brought along food for weighing, the results were entered on to such forms. This survey concerned the adult female members of the homesteads primarily because they harvested and brought home the greater part of the food consumed; it was on their willing co-operation that its success largely depended.[13] With them I worked out the following routine: each day, in the late afternoon or early evening, when they returned home with their daily load of vegetables for the family I would weigh everything before any of its was shared out and eaten. Here I used the same spring balance as employed for the yield survey.[14]

Not all the food consumed by the members of a household comes from that brought back in the latter part of the day by the womenfolk, although the greater part of it does come from this source — as Table 37 shows something like 80 per cent to 90 per cent. There are three other sources: men harvesting for themselves, men and women receiving food from the members of other homesteads, and women eating during the day in the gardens (see Table 37 for the proportions concerned). The food coming from the first two of these sources is sometimes brought home and eaten, other times consumed elsewhere. When those participating in the survey returned home with such food I was usually able to weigh it (often they brought it to me to do so). But that which they consumed elsewhere, including the food eaten by women in the gardens, was not weighed. The greater part of this unweighed food was received from hosts when visiting, or came from earth ovens

[13] I gave them the same reward as the women who co-operated in the yield survey documented earlier, and I record my gratitude to them for their ready, and on the whole happy, co-operation with me in this survey: *chay Aenda sem ten.*

[14] This spring belance could weigh accurately to 0.1 kg, hence the results recorded here are to the nearest 0.1 of a kg.

Table 37: Sources of food eaten during consumption survey

	SWEET POTATO	HIGHLAND PITPIT	PUMPKIN	GOURD	CUCUMBER	CLIMBING CUCURBIT	TARO	GINGER	IRISH POTATO	BEANS	MAIZE	ACANTHS (SHOMBAY)	AMARANTHS (KOMB)	AMARANTHS (PALUW)	HIBISCUS	PUMPKIN LEAVES	CRUCIFERS (TAGUWT)	WATERCRESS
WEIGHTS HARVESTED BY WOMEN AND EATEN AT HOMESTEAD (kgs)	5972.1 81.9%	214.8 93.4%	687.4 72.6%	7.7 34.5%	14.5 89.5%	0.2 33.3%	41.6 46.5%	1.7 70.8%	0.9 100%	40.4 73.7%	68.8 68.6%	60.8 93.5%	0.9 90%	23.0 82.7%	9.9 60.4%	38.0 86.4%	12.0 82.2%	4.8 92.3%
WEIGHTS HARVESTED BY MEN AND EATEN AT HOMESTEAD OR ELSEWHERE (kgs) [1]	205.6 2.8%	1.5 0.6%	94.5 10.0%	5.5 24.7%	0.4 2.5%	0.2 33.3%	4.7 5.3%	0.3 12.5%		5.0 9.2%	1.6 1.6%	0.7 1.1%			1.6 9.7%	1.9 4.3%	0.1 0.7%	
WEIGHTS GIVEN BY MEMBERS OF OTHER HOMESTEADS (kgs) [1]	846.2 11.6%	8.9 3.9%	132.0 13.9%	8.6 38.6%	0.1 0.6%	0.2 33.3%	26.6 29.7%	0.4 16.7%		8.8 16.0%	20.5 20.4%	2.1 3.2%	0.1 10%	2.9 10.4%	4.7 28.7%	3.9 8.9%	2.4 16.4%	0.2 3.9%
WEIGHTS HARVESTED BY WOMEN AND EATEN DURING DAY AWAY FROM HOMESTEAD (kgs) [2]	32.9 0.5%	1.3 0.6%	20.6 2.2%	0.5 2.2%	0.6 3.7%		14.0 15.7%			0.4 0.7%	1.1 1.1%	1.4 2.2%		0.1 0.4%	0.2 1.2%	0.1 0.2%	0.1 0.7%	
WEIGHT OF PRODUCE GIVEN TO OTHER HOMESTEADS (kgs) [3]	236.3 3.2%	3.4 1.5%	12.3 1.3%		0.6 3.7%		2.5 2.8%			0.2 0.4%	8.2 8.3%			1.8 6.5%		0.1 0.2%		0.2 3.8%

	PARSLEY	TARO LEAVES	SPIDERWORT	FIG LEAVES (POIZ)	CHINESE CABBAGE	CABBAGE	ONIONS	SCREW-PINE NUTS (AENK)	SCREW-PINE OIL (WABEL)	SUGAR CANE	BANANAS	BANANA PLANT HEARTS	TOMATO	CHOKO	PASSION FRUIT	BAMBOO SHOOTS	PEANUTS
WEIGHTS HARVESTED BY WOMEN AND EATEN AT HOMESTEAD (kgs)	18.7 93.0%	17.4 93.1%	0.2 100%	2.5 96.2%	0.5 55.6%	13.2 61.7%	1.8 85.7%	20.7 32.5%		37.2 32.1%	6.3 7.4%		1.5 93.7%	3.0 48.4%		0.7 100%	
WEIGHTS HARVESTED BY MEN AND EATEN AT HOMESTEAD OR ELSEWHERE (kgs) [1]	0.1 4.3%					1.6 7.5%		27.2 42.7%		14.4 12.5%	28.3 33.3%	0.5 12.5%			0.4 100%		
WEIGHTS GIVEN BY MEMBERS OF OTHER HOMESTEADS (kgs) [1]	1.1 5.5%	0.6 3.2%			0.4 44.4%	6.6 30.8%	0.3 14.3%	13.8 21.7%	2.4 60%	32.1 27.7%	43.2 50.8%	3.0 75%	0.1 6.3%	3.2 51.6%			0.1 100%
WEIGHTS HARVESTED BY WOMEN AND EATEN DURING DAY AWAY FROM HOMESTEAD (kgs) [2]	0.2 1%	0.6 3.2%		0.1 3.8%				2.0 3.1%	1.6 40%	27.5 23.8%	6.7 7.9%	0.5 12.5%					
WEIGHTS OF PRODUCE GIVEN TO OTHER HOMESTEADS (kgs) [3]		0.1 0.5%								4.4 3.8%	0.5 0.6%						

[1] Sometimes this food was eaten elsewhwre and so not weighed, but estimated.

[2] This food was not weighed; weights are estimated from informant's statements about amounts eaten.

[3] Food not consumed by members of household included in consumption survey.

prepared elsewhere in the daytime (usually in gardens), or consisted of refreshment taken in the course of the day (like chewing sugar cane).

During the survey not only were the large loads women brought back from the gardens weighed daily, but also, when possible, the food harvested by men and that received from members of other homesteads. In an attempt to account for that which slipped through the weighing net, I questioned the members of each homestead in the survey to establish what other food they had eaten each day.[15] The weights of this I estimated from informants' statements about the amounts they had eaten, the size of the tubers and so on (I calculated crude average weights for various crops against which to estimate them).

In a sense, Table 37 indicates the reliability of the data collected, showing the proportion of food weighed against that estimated. All the food harvested

[15] A liberal supply of tobacco ensured the co-operation of the male members of each homestead.

by women was weighed, plus about 50 per cent of that harvested by men and 30 per cent of that received from others. Overall 87 per cent of that consumed during the survey was weighed. This does not signify the true degree of accuracy achieved — it is too optimistic — for those participating in the survey undoubtedly ate some food which I failed to record. Although co-operating fully, it was inevitable that people forgot to inform me of some of the food they had eaten the previous day. I have no way of knowing what percentage was eaten and forgotten, although cross-checking the food that passed between those participating in the survey indicates that it was low (of the food recorded under 'given by members of other homesteads' 4.5 per cent was received from others participating in the survey but the recipients forgot to inform me of it, which as a proportion of all the food eaten, was 0.6 per cent that would have been missed if this cross-check of what others said they gave away had not been made).[16]

Table 37 also indicates how the data on different crops vary in reliability, the proportions of them weighed to those estimated. The percentage weighed is high for the staple crops, harvested largely by women. Estimation figures more with the crops eaten less often, those in the male domain given frequently to guests (and which are hence often eaten elsewhere). This is so, for example, with sugar cane, bananas, screw-pine nuts and taro.[17]

Another problem that hindered the devising of an all but accurate survey, reducing estimation to a minimum, was the passing of food between members of different homesteads. It became clear in the pilot survey that considerable amounts of food changed hands between neighbouring homesteads of relatives, and to minimise the unmeasured movement of food in and out of the sample I included in it clusters of households that interacted intensively with one another. In this way a considerable part of the unseen sharing of food was kept within the survey sample. Two homesteads in the pilot survey (H and L) were omitted from the main survey because of their considerable interaction with homesteads outside the sample, which increased estimation to unacceptable levels.[18]

The genealogy given in Fig. 60 sets out the composition of the homesteads included in the survey. Showing the relations between their members, it indicates the degree of interaction taking place between them. The circles enclose those resident in the same homestead. The individuals enclosed by

[16] The final category on the table 'produce given to other homesteads' is low in comparison to that received from others because it includes only that food harvested by women and weighed, and not all the food (unweighed) which participants in the survey gave away to others.

[17] These crops and other male-harvested ones, recorded under 'harvested by women' in Table 37 were collected by men but given to women to carry home, hence their anomalous listing.

[18] The reduction to ten was also necessary because twelve homesteads proved too many to cope with comfortably and maintain the standards of accuracy set for the survey. The couple living in homestead G dropped out of the survey about half way through when their baby son died and they left Haelaelinja for the duration of the mourning period to live elsewhere.

two circles are closely connected to both homesteads; for example, some men live together in one men's house while their female relatives occupy separate houses nearby (this is so for some of the men residing in homesteads E and K); alternatively some women link together two homesteads (like one of the wives of the man occupying homestead J who, tired of his violent treatment of her, moved with her children to live in his half-sister's homestead I). These people closely connected to two homesteads confused the definition of the groups consuming food. The amounts involved in either homestead were not always clear, making the precise allocation of produce to them somewhat inexact on occasion. But the margin of error was small, and does not affect the overall computations because the survey includes all those involved consuming the food. The dual affiliated are bracketed overall with the homestead in which they shared or obtained the greater part of their food.

Another problem was allowing for the movement of individuals in and out of the homesteads during the survey. The sample fluctuated as resident members visited elsewhere (a couple of women for extended periods) and relatives from other places came to stay as guests. These movements did not seriously embarrass the calculation of food consumed, although they introduced an element of error; those staying a night counting as a twenty-four-hour period away from the homestead, or within it in the event of a guest, when in terms of the food consumed this may not have been strictly so.

In Appendix V, the composition of the homesteads during the survey is broken down according to sex and age, and the average estimated weight and height of the individuals in each category given. This information is significant for assessing the nutritional status of the diet.

Bearing in mind the shortcomings enumerated, Tables 38 and 39 record by homestead the amounts of food consumed by both humans and animals during the survey. The categories 'human consumption days' and 'pig consumption days' are the same as those used in the earlier table, they record the total number of days on which different individuals shared the produce weighed. The pigs are divided up according to their age and size into four categories. The 'human consumption day' figures give the days people normally resident in the homestead consumed food, plus those on which guests did so (the figures in brackets give the number of individuals normally resident). Babies under one year in age and feeding at the breast are omitted from the children's column, and also from the following calculations of average amounts consumed and the nutritive value of the diet.[19] Another column on the tables records the number of days on which women harvested for the homesteads. This is generally a daily task for married women with families (except when some event, such as attendance at a funeral, precludes it), although where two or more women live together the days they harvest

[19] According to Bailey and Whiteman (1963: 379) a child under one year of age consumes 0.02 of an adult male's share.

Table 38: Food consumed by human beings, by homestead, during survey (unprepared weights)

INDEX LETTER	NUMBER OF DAYS FOR WHICH FOOD WEIGHED	NUMBER OF DAYS WOMEN HARVESTED FOR HUMAN BEINGS	NUMBER OF "HUMAN CONSUMPTION DAYS" (Brackets : number of persons resident in homestead during survey)			SWEET POTATO (kgs)		HIGHLAND PITPIT (kgs) (HUMAN EDIBLE PORTION)	PUMPKIN (kgs)	GOURD (kgs)	CUCUMBER (kgs)	CLIMBING CUCURBIT (kgs)	TARO (kgs)	GINGER (kgs)	IRISH POTATOES (kgs)	BEANS (kgs)	MAIZE (kgs)	ACANTHS (SHOMBAY) (kgs)	AMARANTHS (KOMB) (kgs)	AMARANTHS (PALUW) (kgs)	HIBISCUS (kgs)	PUMPKIN LEAVES (kgs)	CRUCIFERS (TAGUWT) (kgs)	WATERCRESS (kgs)	PARSLEY (kgs)	TARO LEAVES (kgs)	SPIDERWORT (kgs)	PIG LEAVES (POIZ) (kgs)	CHINESE CABBAGE
			ADULT MALES	ADULT FEMALES	CHILDREN (UNDER 14 YEARS APPROX)	MEN'S HOUSE	WOMEN'S HOUSE																						
A	70	125	107 (3)	167 (4)	301 (6)	181.9	343.6	3.0	64.4	0.9			2.8	0.1		0.3	0.8	1.0		0.2	1.0	0.5							
B	70	107	109 (2)	114 (2)	201 (3)	205.2	295.3	1.8	12.1							2.6		1.3		0.1	1.0		0.7	0.7					
C	70	85	135 (2)	124 (2)	209 (3)	233.9	266.3	10.3	49.4	0.7			3.2	0.2		0.9	0.4	1.9	0.1	0.2	1.2	5.0	0.6	4.1	0.6	0.1			0
D	82	146	164 (2)	180 (3)	82 (1)	414.7	395.8	13.8	90.1	8.2			3.3		0.9	4.1	3.7	4.3		0.2	1.3	2.3	0.6		3.1	1.7			
E	92	266	204 (4)	274 (4)	597 (8)	1233.0		58.2	273.9	3.0	1.6	0.2	13.5	0.2		5.3	49.1	19.2		6.1	0.4	12.6	2.7		7.3	3.2		2.0	0.
F	82	109	86 (1)	159 (2)	313 (4)	783.9		30.2	64.8	0.3	13.3		23.7	1.3		0.2	14.8	12.8	0.8	0.9	1.2	2.9	0.1		2.4	1.1	0.1		
G	42	42	42 (1)	47 (1)	2	173.7		3.2	17.1		0.6		0.5				2.1	1.7			1.0	0.6			0.2				
H	12	30	24 (2)	34 (3)	24 (2)	119.7		6.6	18.8		0.1		0.5			6.1	2.7	1.5	0.1			1.4			0.8				
I	70	113	212 (4)	125 (2)	125 (2)	621.4		27.7	125.9	5.0		0.2	5.1	0.1		6.7	0.8	7.3			0.6	8.3	0.7	0.2	4.3	0.1			0.
J	82	164	86 (1)	164 (2)	295 (4)	445.7	572.2	46.5	150.5	2.1			14.9	0.3		9.1	11.8	4.8		2.4	1.2	3.4	0.2		0.8	6.5	0.1	0.6	
K	82	94	154 (3)	101 (1)	77 (1)	296.3	354.5	24.0	54.4	2.1		0.2	19.4	0.2		19.3	5.8	6.4		15.4	7.5	5.2	9.0		0.4	5.9			0.
L	12	26	24 (2)	22 (2)	24 (2)	119.7		1.3	13.1									2.8		0.5		1.7			0.2				
	776	1307	1347 (27)	1511 (28)	2250 (36)	7056.8		226.5	934.5	22.3	15.6	0.6	86.9	2.4	0.9	54.6	92.0	65.0	1.0	26.0	16.4	43.9	14.6	5.0	20.1	18.6	0.2	2.6	0.

INDEX LETTER	CABBAGE (kgs)	ONIONS (kgs)	SCREW-PINE NUTS (AENK) (kgs)	SCREW-PINE OIL (WABEL) (kgs)	SUGAR CANE (kgs)	BANANAS (kgs)	BANANA PLANT HEARTS (kgs)	TOMATO (kgs)	CHOKO (kgs)	PASSION FRUIT (kgs)	BAMBOO SHOOTS (kgs)	PEANUTS (kgs)	WILD PLANTS			INSECTS				WILD GAME					PORK (kgs)	EUROPEAN PROCESSED FOOD				
													FUNGI (largely GRIFOLA FRONDOSA) (kgs)	TREE FERN FRONDS (largely CYATHEA SP.)	SCREW-PINE NUTS (PANDANUS ANT.*)	CRICKETS (kgs) (GYMNOSGRYLLUS ANGUSTUS)	CICADA (kgs) (POMPONIA SP.)	GRASSHOPPERS (kgs) (VALANGA SP.)	LARVAE (kgs) (OLETHRINUS TYRANNUS)	FROGS (kgs)	BIRDS (kgs)	BIRDS' EGGS (kgs)	RATS (kgs)	MARSUPIAL FLESH (kgs)		TINNED FISH (kgs)	TINNED MEAT (kgs)	RICE (kgs)	TINNED LARD (kgs)	BISCUITS (kgs)
A	0.5		3.8		11.2	12.3							26.6												6.0	0.9		0.9		
B	0.2				0.7	1.4							31.2												0.6					
C	1.0	0.2	13.0		1.8	14.4							41.9	0.2		0.012								0.1	2.6	1.5		1.5		
D	2.0	0.2	10.1	0.4	10.5	10.2		0.2					7.8											0.3	18.4	1.6		1.5		0.0
E	6.2	0.2	4.4	0.4	39.0	15.2	1.0	0.5				0.1	92.4	0.2		0.088	0.002			0.08	0.2	0.04	0.7	0.3	29.0	3.2	0.5	3.0	0.5	0.2
F	1.1	0.7	3.8	2.0	12.5	1.5	1.0	0.8	1.8		0.5		14.0	0.8		0.002				0.01				0.8	5.7	1.2		1.3		
G	1.1		0.2		2.3	1.4			0.8				10.2												0.4	0.5	0.5	0.9		
H		0.2	8.4		1.6										1.6										9.2					
I	2.0	0.1	1.1	0.4	13.7	14.1			0.9	0.1			18.9	0.1		0.272		0.002	0.007					0.1	4.7	2.7	0.5	3.2	0.5	0.2
J	0.7		9.7	0.6	7.3	5.4	1.0		2.8				35.3	1.5		0.085		0.002	0.007			0.01		0.1	0.2	0.9		0.9		
K	6.6	0.2	9.2	0.2	10.6	6.9	1.0	0.1		0.3	0.2		5.5								0.01		0.04	0.7	7.7	1.1		1.1		
L		0.3				1.7															3.2				5.2					
	21.4	2.1	63.7	4.0	111.2	84.5	4.0	1.6	6.3	0.4	0.7	0.1	283.8	2.8	1.6	0.459	0.002	0.004	0.014	0.09	3.41	0.05	0.74	2.4	89.7	13.6	1.5	14.3	1.0	0.4

* ANTARESENESIS

fluctuate somewhat as they cover for one another. The harvesting of food by young unmarried women is more variable and fluctuates from day to day, for there is not the compunction on them to harvest daily.

The weights of the food collected are given as the unprepared raw weights with no adjustments made for the proportion wasted in preparation and not consumed (with the exception of Highland *pitpit*, the waste of which is fed to pigs and accordingly recorded on their table). The amounts wasted, as later estimates show, are considerable for some crops; for example, the shells and inedible fibre of screw-pine nuts amount to 60 per cent of the weight, and this waste portion goes up to 70 percent for maize and sugar cane.

The survey covered all the food consumed, not only crops (although these made up the largest part of the diet). The amount of fungi eaten was high because of a flush of saprophytes during the early part of the survey. These

Table 39: Food consumed by pigs and cassowaries during survey (days, weights, kg). Bracketed figures = women resident sometime in household. Dogs receive scraps only

INDEX LETTER	DAYS FOR WHICH FOOD WEIGHED	DAYS WOMEN HARVESTED FOR ANIMALS	'PIG CONSUMPTION DAYS'				CON-SUMPTION DAYS	SWEET POTATO		HIGHLAND *PITPIT* (PIGS)	DOGS
			ADULT PIGS	ADO-LESCENT PIGS	SHOATS	PIGLETS		PIGS	CASSO-WARIES		
A	70	120 (3)	70	297		85	70	592.3	54.4	5.6	
B	70	90 (2)	140	21		21		504.6		3.6	
C	70	75 (2)		30	70		70	249.2	58.7	20.5	
D	82	128 (3)	279	36		673	70	816.1	44.4	26.5	1
E	92	241 (4)	361	56		333	1	1350.8	0.7	118.1	2
F	82	83 (2)	70	140				631.5		59.8	
G	42	42 (1)	39	5		97		223.7		6.1	
H	12	20 (3)		48		36		152.0		13.3	
I	70	105 (2)	171		140			516.0		56.4	1
J	82	159 (2)	210	28	133	537		1109.9		95.7	
K	82	99 (3)	226			72		616.9		49.0	
L	12	16 (3)			36			86.6		2.6	1
Totals	766	1198 (30)	1566	661	379	1854	211	6849.6	158	457.2	5

plants spring up in the forest in very large numbers on occasion (the Wola say that they are prolific when heavy rain follows a dry period). At other times they occur here and there in small numbers, so that if the survey had been held a few weeks later the amount of fungi eaten would have been negligible. Screw-pine nuts, as described earlier, occur in similar flushes, irregularly every few months, large numbers of trees producing clusters of nuts, whereas for the main part of the year just the occasional tree bears nuts. During a flush, which may last for a month or so, pandan nuts figure prominently in people's diets. There was no such flush during the course of the survey, and hence the contribution of screw-pine nuts to the diet is under-rated in Table 38. Another fluctuating source of food is the pig. The Wola tend to eat pork in gluts, rather as they do fungi and screw-pine nuts, receiving the greater part of their meat from pig kills and exchanges (see Sillitoe 1979a: 256–69), which are infrequent events. No such kill took place in the vicinity of the homesteads included in the survey (although some pork did filter through to them from kills staged in settlements some way off) and so the amount of pork consumed is also depressed in Table 38.

The periodicity of the above food sources raises again the question of the seasonality of these, and maybe other crops. The Wola, unlike the Enga (see Waddell 1972: 123–7), maintain that the yields of their crops do not fluctuate predictably according to seasons. They say, as pointed out before, that the same crops grow and produce all year round, and that the flushes of pandan nuts and fungi are unpredictable.[20] The yields of other crops fluctuate irregularly, sometimes seriously. They do so dramatically in fine periods, with cold nights accompanied by frosts, which damage crops and reduce yields to uncomfortably low levels (the last such occasion was the Highlands-wide drought of 1972). Yields can also drop less gravely at any time, for reasons which are not clear to the Wola nor to me. They call these times of privation *day biy tomb* (lit: hunger feel time). Given the random occurrence of fluctuations in crop yields, no attempt was made to accommodate them by sampling the food consumed at a number of points throughout the year. The timing was random: the pilot survey ran from 14 January through to 25 January, and the main survey from 22 March through to 12 June. There was no reason for selecting this period over any other and during it a flush of screw-pine nuts might have occurred, or for some reason crop yields might have fluctuated. As it was a glut of fungi occurred; otherwise the period was average.

Yet the situation with fungi, pandan nuts and pork raises a query about the representativeness of the survey. If extended over a longer period it would have given a somewhat different picture with regard to these foods: fungi would have diminished in significance and the other two increased. A

[20] Over a period of three years I have been unable to establish any pattern either, although there was a big flush of screw-pine nuts at the beginning of each year (in February). But there were others too at irregular intervals.

thoroughly satisfying survey accounting for all fluctuations adequately would need to run for considerably longer than the study attempted here. This survey, though, with all its qualifications regarding accuracy, documents adequately, with the exception of the foods mentioned, the average diet eaten by the Wola for the greater part of their lives. During it the following total weights of food were collected, supplying sustenance for 2858 adult and 2250 child 'consumer days' (kg):

35 kinds of cultivated crops:	9007.4
3 kinds of wild plants:	288.2
5 kinds of game animals:	6.7
1 species of domesticated animal (pig):	89.7
5 kinds of processed food:	30.7
4 species of insects:	0.5
Total:	9423.2

The graph in Fig. 58 presents these findings in another form. It shows the weights of food eaten in an average homestead over a period of three months.

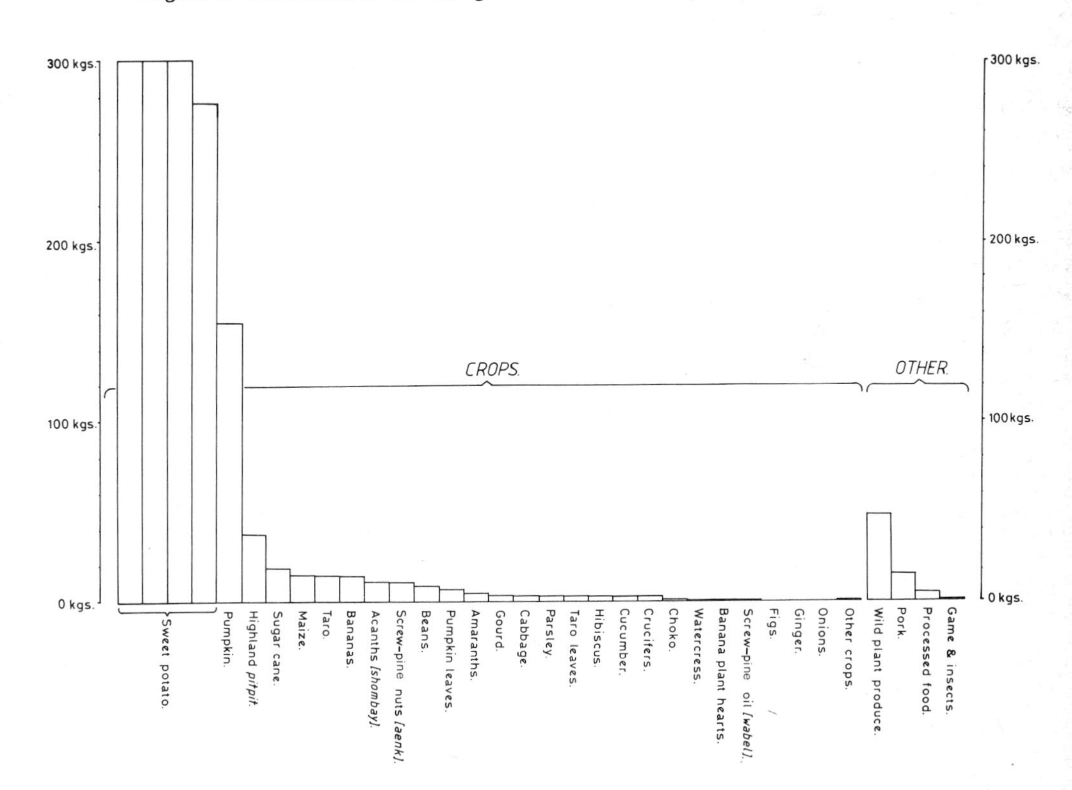

Fig 58: The food consumed in an average homestead over three months (unprepared weights)

The average homestead is calculated from the survey sample to consist of five adults and four children resident throughout the period, plus an adult guest staying for eleven nights and a child guest for three nights[21] (an adult is taken as a person over fourteen years of age). This graph shows clearly the overriding importance of the sweet potato tuber in the diet of the Wola, exceeding pumpkin, which is their next most important crop, by nearly eight times in terms of weights consumed.

A comparison of the food eaten by humans with that consumed by animals (see Tables 38 and 39) shows that 49.8 per cent of all sweet potato tubers harvested were fed to pigs and cassowaries (the latter's share was 1.1 per cent). This figure compares with that computed earlier from the produce weighed in the yield survey and consumed by pigs. The other averages calculated from the yield data, though, for human and animal consumption, are considerably lower. The source of this error is two-fold, as pointed out earlier: firstly, in the yield survey a close check was not kept on the food eaten over and above that weighed, and secondly, insufficient attention was paid to the varying amounts consumed by humans and animals of different ages and sizes.[22]

Table 38 on pig consumption divides the animals eating the produce into four age categories. Assuming that a piglet eats an estimated twenty per cent of an adult pig's share, a shoat forty per cent and an adolescent animal eighty per cent, the average weight of food given to one adult pig in a day is 2.79 kg; that fed to a cassowary is 0.75 kg.

This is more than an adult human being eats in a day. The average raw, unprepared weight of food received daily by an adult Wola male is 2.26 kg — calculated on the basis of the different proportions of food that Bailey and Whiteman (1963: 379) established males and females of varying ages consume.[23] And of this 1.69 kg consists of sweet potato tubers (pumpkin, the

[21] Giving a total of 476 adult 'consumer days' and 375 child 'consumer days'.

[22] An analysis of the figures shows that it is the first of these that introduces the margin of error into the average figure calculated for human consumption, and the second that is responsible for the error in the pig's average. Calculating human consumption crudely, as done in the analysis of the yield data, by assuming that children under fourteen years old eat half an adult's portion, reduces by 0.08 kg the average weight calculated here as eaten by an adult male per day. Whereas assuming that all pigs eat the same amounts (as done in the yield survey earlier) reduces by 1.12 kg the average weight calculated here. This demonstrates that ignoring differences in age and size accounts for the pig error, while it does not seriously bias the human figure, which is erroneous because a close check was not kept on all food eaten.

[23] This compares with the findings of Bailey and Whiteman (1963: 379), that an adult Chimbu male eats 1.99 kg of prepared raw food a day. According to their work an adult female eats 0.93 of an adult male's share (for Wola women that is 2.10 kg daily), an 11- to 15-year-old child 0.77 (1.74 kg), a 6- to 10-year-old child 0.65 (1.47 kg), a 3- to 5-year-old child 0.63 (1.42 kg), and a 1- to 3-year-old child 0.35 (0.79 kg).

See Appendix V for a breakdown of the composition of the homesteads included in the survey according to sex and age, and the average estimated weight and height of the individuals in each category. Although somewhat gross from a nutritional viewpoint (dealing with unprepared food and not measuring the proportions consumed individually), the average amounts of food eaten by each age/sex category could be calculated from this data, and their dietic values computed too.

next most important source of food by weight, constitutes only 0.22 kg of this per capita average).[24]

This figure underscores again the prominent part sweet potatoes play in the diet of the Wola, as shown by Fig. 58. Expressed in percentiles, these tubers make up 78.3 per cent of all crops eaten, and 74.9 per cent of all foodstuffs (which is nearer to Venkatachalam's (1962: 9) figure for the Chimbu of 77 per cent, than Waddell's (1972: 123) for the Enga of 63 per cent). Table 40 presents the latter statistic another way, as it is apparent to a Wola person.

Table 40: The frequency with which various crops occur in meals

	TOTAL	ADULT MALE	ADULT FEMALE	CHILD
No. of 'consumption days'	4409	1115	1232	2062
No. of individuals	76	21	22	33

	SWEET POTATO	HIGHLAND PITPIT	PUMPKIN	GOURD	CUCUMBER	CLIMBING CUCURBIT	TARO	GINGER	IRISH POTATOES	BEANS	MAIZE	ACANTHS (shombay)	AMARANTHS (komb)	AMARANTHS (paluw)	HIBISCUS	PUMPKIN LEAVES	CRUCIFERS (taguwt)	WATERCRESS
Total no. of 'consumption days' on which foodstuffs were eaten	4409	1537	1541	56	45	6	248	115	8	128	370	393	2	108	138	361	67	21
Percentage of 'consumption days' on which foodstuffs were eaten	100	34.9	35.0	1.3	1.0	0.1	5.6	2.6	0.2	2.9	8.4	8.9	0.05	2.5	3.1	8.2	1.5	0.5
Ranking order (No. foodstuffs = 44)	1	3	2	21	22	32	9	17	31	16	5	4	36	18	15	7	20	26

	PARSLEY	TARO LEAVES	SPIDERWORT	FIG LEAVES (poiz)	CHINESE CABBAGE	CABBAGE	ONIONS	SCREW-PINE NUTS (aenk)	SCREW-PINE OIL (wabel)	SUGAR CANE	BANANA	BANANA PLANT HEART	TOMATO	CHOKO	PASSION FRUIT	BAMBOO SHOOTS	PEANUTS	FUNGI	INSECTS	FROGS	BIRDS	BIRDS' EGGS	RAT	MARSUPIAL FLESH	PORK	EUROPEAN PROCESSED FOOD
Total no. of 'consumption days' on which foodstuffs were eaten	179	80	8	21	10	156	45	115	38	234	186	28	28	13	4	14	1	369	56	5	3	5	12	25	309	161
Percentage of 'consumption days' on which foodstuffs were eaten	4.1	1.8	0.2	0.5	0.2	3.5	1.0	2.6	0.9	5.3	4.2	0.6	0.6	0.3	0.1	0.3	0.02	8.4	1.3	0.1	0.07	0.1	0.3	0.6	7.0	3.7
Ranking order	12	19	31	26	30	14	22	17	23	10	11	24	24	28	34	27	37	6	21	33	35	33	29	25	8	13

It records the number of days on which various crops and other foodstuffs occurred in the meals eaten by those included in the survey sample, expressing their occurrence as frequency percentages (see Pospisil 1963: 377 for a series of similar calculations for the Kapauku). Everyone ate sweet potato on every day of the survey, this tuber scoring 100 per cent. The next two most common crops, according to the number of days on which people consumed them, are pumpkin and Highland *pitpit*, both of which score occurrences of 35 per cent. No other foodstuff exceeds 10 per cent, and the majority are well below this figure (so that, for example, a person can expect to eat

[24] According to Rappaport's (1968: 282—3) crude but workable assumption, that the calorific differences established by Venkatachalam (1962) between the diets of males and females of varying ages parallal the actual weights of food ingested, a Wola male's daily portion weighs 2.44 kg (of which 1.83 is sweet potato).

onions or cucumbers once every one hundred days, or between three or four times a year).

NUTRITIONAL VALUE OF THE DIET

Although sweet potato predominates by weight in the Wola diet, this, from a nutritional point of view, does not tell the entire story. Indeed, in some significant respects it is a distortion because other crops and foodstuffs, although small in comparison judged by weights ingested, are important for supplying essential nutritional elements in which the paramount sweet potato tuber is deficient.

Table 41 gives the nutritive value of the food eaten over three months in the one average household, consisting of five adults and four children upon which the previous graph was based. The calculations are made on the basis of the raw weights of the foods consumed and apply to the estimated edible portions only.[25] The table is unfortunately incomplete, because it omits those foods for which I have been unable to ascertain a nutritive value. These, however, make up a very small part of the Wola diet by weight (amounting to only 0.24 per cent), and their omission does not seriously distort the overall picture.[26] Another source of possible error on the table comes from the works used as authorities to calculate the nutritive values of the various foodstuffs listed.[27] These display a bewildering lack of agreement, making it difficult for the non-specialist to calculate nutritive values accurately with any certainty (see Rappaport (1968: 279) who draws attention to this problem, and McArthur (1974: 93—111) who points to the errors likely to occur when non-specialists try to interpret and use this data).[28] Following others who have attempted the kind of analysis made here, this study uses, where possible, values calculated in research on New Guinea foods.[29]

The four pie graphs in Fig. 59 summarise this data on nutrition and show the contribution of calories, protein, vitamins and minerals made to the Wola diet by the various foodstuffs which they eat (see Waddell 1972: 125 and Hipsley and Kirk 1965: 80 for similar graphs relating to the Enga and Chimbu diets).[30] These demonstrate clearly that although by weight sweet potatoes

[25] The estimated wastage rates (based on a small sample of tests) correspond more closely with Waddell's (1972: 127) for the Enga than they do with Rappaport's (1968: 280) high ones for the Maring (see Waddell for a comment on this variation).

[26] Only a rough and ready indication is given of the nutritive value of the insects eaten in the survey, calculated on the basis of the figures given for sago grubs in Hipsley & Clements (1947: 279). Similarly for the values of parsley, calculated from Platt's (1963) figures for a medium carotene leaf.

[27] The authorities referred to are listed on the table. All the values given to carbohydrates, fibre, pro-vitamin A and carotene are taken from Powell (1976: 115—17).

[28] See Oomen *et al.* (1961) who document in detail the differing chemical composition of various sweet potato tubers, demonstrating how these can vary according to region, climate, cultivar and so on.

[29] See, for example, Waddell's (1972: 232—3) admirable table of the composition of various foodstuffs occurring in the Enga diet.

[30] Crops included in the 'greens' category on these nutritive pie-graphs are pumpkin

Table 41: The nutritional value of an average homestead's diet over three months

FOOD ITEM	UNPREPARED RAW WEIGHTS (KGS)	ESTIMATED % INEDIBLE AND REMOVED IN PREPARATION	PREPARED RAW WEIGHTS (KGS) (= EDIBLE PORTION)	CALORIES (KCAL.)	PROTEIN (G.)	FAT (G.)	CARBOHYDRATES (G)	FIBRE (G)	CALCIUM (MG.)	IRON (MG.)	PRO-VITAMIN A (I.U.)	β-CAROTENE (μG)	THIAMIN (MG.)	RIBOFLAVIN (MG.)	NIACIN (MG.)	ASCORBIC ACID (MG.)	SOURCE
SWEET POTATO	1176.1	5	1117.2	1675800	10055		312816	8379	167580	8938	22344000	13406400	1452	559	7820	349684	1
PUMPKIN	155.8	10	140.2	61688	2103		7010	1822.6	28040	981			56	112	84	14020	1
HIGHLAND PITPIT	113.3	66	37.7	8671	151	75	2262	434	7917	339			68	72	527.8	12441	1
SUGAR CANE	18.5	70	5.55	3219	22	trace	777	694	555	trace			trace			trace	?
MAIZE	14.9	70	4.47	4112	152	54	925	45	224	27	15645	9387	6.7	4	76	626	2
TARO	14.4	5	13.68	19836	192		4523	253	5335	137	2394	205	13.7	4.1	55	848	1
BANANAS	14.1	30	9.87	9278	109		2695	99	790	trace	29610	15299	4.9	7.9	59	1964	1
ACANTHS (shombay)	10.8	0	10.8	3240	292	32			39852	324			10.8	22	65	6480	1
SCREW-PINE NUTS (aenk)	10.6	60	4.2	28686	500	2772	924	256	17598					15.1			2
BEANS	9.1	40	5.5	18700	1210	110	3484	377	4400	193			33	11	127		3
PUMPKIN LEAVES	7.3	0	7.3	2555	277	29	292	73	7300	2263							2
AMARANTHS	4.5	0	4.5	1350	122	13.5	252	63	16605	135	274050	162000	4.5	9	27	2700	1
GOURD	3.7	0	3.7	1036	26				740	22			1.5	1.1	22	555	4
CABBAGE	3.5	0	3.5	1225	67	3.5	245	35	1050	7			2.1			1750	2
PARSLEY	3.4	0	3.4	952	68	10.2			2720	85			2.7	6.8	17	1700	4
TARO LEAVES	3.1	0	3.1	1178	85	39	175	54	8525	31			4.7			3565	2
HIBISCUS	2.7	0	2.7	1053	153	8.1	208	49	15660	108	35100		4.1			3186	2
CUCUMBER	2.6	0	2.6	338	18.2	5.2	52	13	390	5.2			0.8			390	2
CRUCIFERS	2.4	0	2.4	960	106		163						3.6			1344	2
CHOKO	1.1	10	1.0														
WATERCRESS	0.8	0	0.8	320	35		54						0.5			400	2
BANANA PLANT HEARTS	0.7	0	0.7														
SCREW-PINE OIL (wabel)	0.7	75	0.18	1229	21	119	40	11	754					0.7			2
ONIONS	0.4	10	0.36	155	6.5	0.7	34	4.3	288	3.6	180	108	0.22	0.14	1.8	65	2
FIG LEAVES (poiz)	0.4	0	0.4														
GINGER	0.4	15	0.34	221	6.8	3.4			68	8.5			0.54	1.46	2.04	20	1
TOMATO	0.3	0	0.3	60	3				15	1.2			0.18	0.12	2.1	75	5
IRISH POTATOES	0.15	5	0.143	107	2.9				14.3	1.0			0.14	0.04	2.2	21	5
CHINESE CABBAGE	0.15	0	0.15	53	2.9	0.15	10.5	1.5	45	0.3			0.09			75	2
BAMBOO SHOOTS	0.1	25	0.075	23	2.0	0.23	3.9	0.53	9.8	0.38	15	9	0.11	0.05	0.45	3	2
CLIMBING CUCURBIT	0.1	20	0.08	35	1.2	0	7.9		16							42	2
PASSION FRUIT	0.07	50	0.035	25	0.53	0.7			3.2					0.04	0.67	10	3
SPIDERWORT	0.03	0	0.03														
PEANUTS	0.02	40	0.012	40	3.1	5.2							0.13	0.02	1.9		7
FUNGI	47.3	0	47.3	3450	180	45	720		600				900			1395	6
TREE FERN FRONDS	0.5	10	0.45														
SCREW-PINE NUTS (aendashor)	0.3	60	0.12	71	1.3	7.1			8.8	0.22			0.03		0.16		4
MARSUPIAL	0.4	20	0.32														
RAT	0.1	20	0.08														
BIRDS	0.6	5	0.57														
BIRDS' EGGS	0.008	5	0.0076														
FROGS	0.02	0	0.02														
INSECTS	0.08	0	0.08	145	4.9	10.5	7.2		6.4							0	6
PORK	15	0	15	43800	2250	3795			1500				150				6
PROCESSED FOOD	5.2	0	5.2	16191	629	618			4543	87			1.4	1.7	24		5,8
TOTAL (ALL FOODS)	1645.728		1456.123	1909802	18858.33	7756.48	337680.5	12663.93	333152.5	13697.4	23016894	13593408	2722.44	828.27	8915.12	403359	

SOURCES: 1. Hipsley and Kirk (1965); 2. Powell (1975); 3. Peters (1957); 4. Peters (1958); 5. Platt (1963); 6. Hipsley and Clements (1947); 7. Massal and Barrau (1956); 8. McCance and Widdowson (1960).
The source for all carbohydrate, fibre, pro-vitamin A and β-carotene values is Powell (1975).

---

leaves, amaranths, cabbage, parsley, hibiscus, taro leaves, cricifers, watercress and Chinese cabbage; and under the 'other' category are sugar cane, gourd, cucumber, onions, ginger, tomato, Irish potato, bamboo shoots, climbing cucurbits, passion fruit, peanuts and insects.

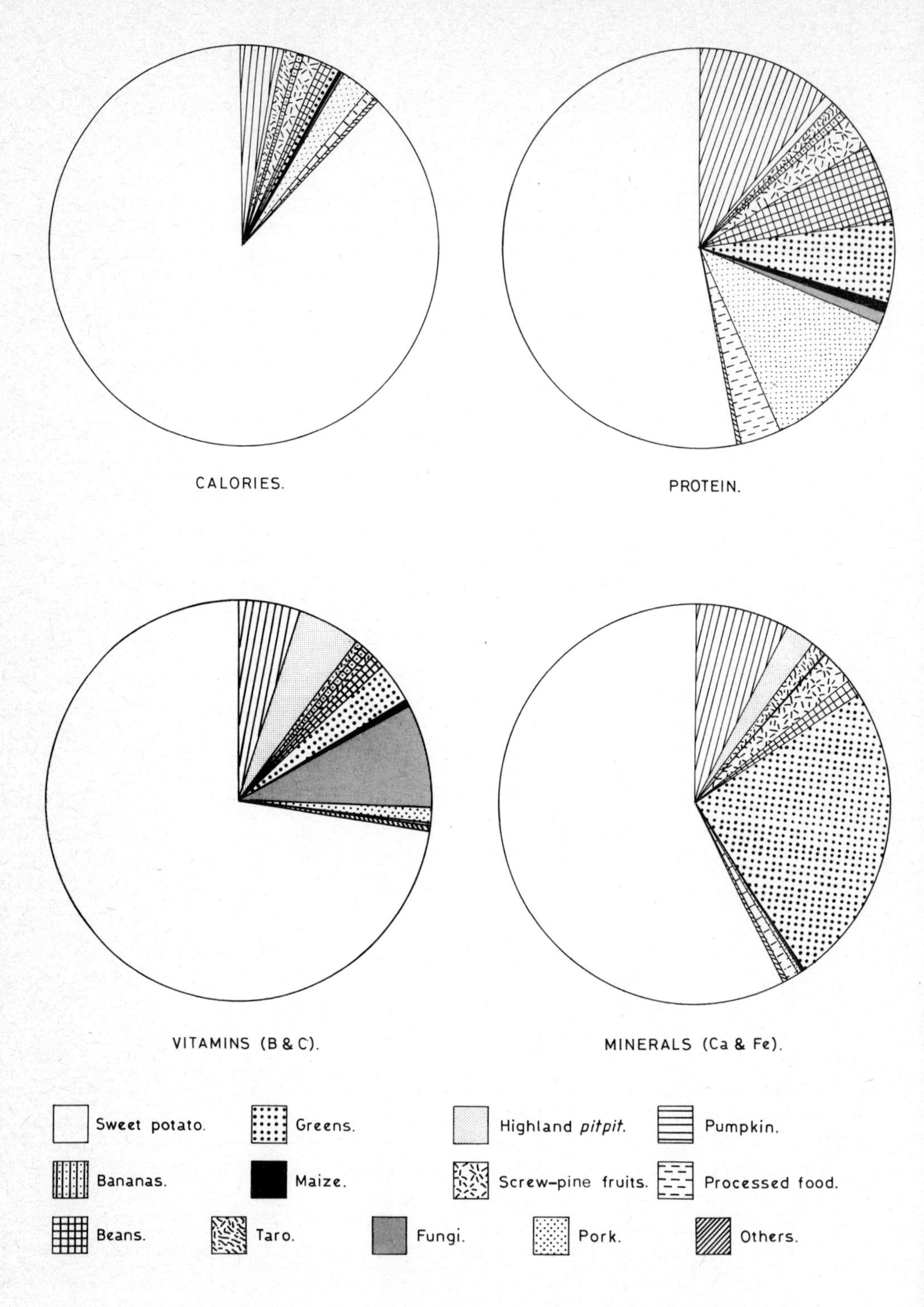

Fig. 59: The nutritional contribution of various footstuffs to the Wola diet

comprise the larger part of the Wola diet, their overall nutritional contribution is not of the same order. In the case of calories and vitamins (B and C) these tubers do contribute amounts comparable with the weights eaten. Indeed they contribute more to the diet in calories than they do by weight: supplying 87 per cent of all calories, 12.6 per cent more than they contribute by weight, whereas the difference between vitamins contributed and weights eaten is only 1 per cent. It is however, the percentages of protein and minerals (calcium and iron) coming from sweet potatoes that are markedly low compared to the weights consumed. They supply 53.3 per cent of the protein in the diet and 57.6 per cent of the minerals.

Given the inordinately high calorific contribution of sweet potato to the diet, it follows that no other foodstuff is significant in this regard; the next largest contributor of calories is pumpkin, supplying a mere 3.2 per cent. This is not so with the remaining nutritive elements, of which certain other crops make significant contributions to the diet. Pumpkin and Highland *pitpit*, the next two most significant crops by weight, each contribute 5 per cent of the vitamins consumed. It is fungi though, supplying 8.4 per cent of the total vitamins ingested, which contribute the largest percentage after sweet potato. This raises again the point made earlier about the periodicity of the consumption of certain foodstuffs biasing the data collected. The protein and mineral contributions of screw-pine nuts and pork, the two other principal foods eaten periodically, underlines this. Pandan nuts supply considerable amounts of protein and minerals to the diet, and pork a significant proportion of the protein ingested. Their contribution, as indicated, is considerably depressed in the data upon which this discussion rests. Both contribute significantly more nutritionally to the Wola diet than the pie-graphs show, particularly protein and to a lesser extent minerals. Of the other crops documented in the survey, the various greens are an important source of minerals, supplying 24.6 per cent of the total. They are also, to a lesser extent, a significant source of protein, contributing 6.5 per cent of that ingested. Other notable sources of protein are pumpkin and beans, supplying 11.1 per cent and 6.4 per cent respectively. All the above footstuffs make disproportionate contributions of protein and minerals to the Wola diet, compared to the weights eaten, so compensating for the nutritional deficiencies of sweet potato.

The nutritional status of the Wola clearly varies considerably over time. These people do not regularly consume a number of foods, some of which (such as screw-pine nuts, greens and pork) contribute a significant proportion of the nutrients contained in their diet. They subsist largely on sweet potato, supplemented on most days with one or two other vegetables. Their nutritional situation is modest, every so often reaching a high spot (for example, the week following a pig kill). Although, on occasion, actual intakes vary considerably, they correspond for the majority of the time to the daily average presented here.

The data on food consumption, considered both by weight and nutritive composition, indicate that the Wola diet has changed relatively little overall since contact with the outside world. Pumpkin stands out as the most significant of the crops introduced since this time. Maize, cabbage and kidney beans now make noticeable contributions to the diet too. The amounts of processed food eaten are small. They make a noticeable protein contribution, though, given the deficiency of the predominantly vegetable diet in this regard.[31] It is probable that the consumption of processed food will increase with the passage of time and the integration of the Wola into the wider Papua New Guinean economy through the cultivation of cash crops (as mentioned in Chapter 7; see Waddell (1972: 126) whose study shows that the Enga consume more introduced and processed foods, indicating the probable next stage for the Wola).

There are some disconcerting differences between the nutritional data presented here and those contained in some other studies, as Table 42 indicates. The Wola diet compares generally with that of the nearby Enga (Waddell 1972), and that of the distant Maring (Rappaport 1968) and Chimbu (Venkatachalam 1962; Bailey and Whiteman 1963; Hipsley and Kirk 1965). But the results recorded for four settlements in Hipsley and Clements (1947) are notably different in some regards.[32] A possible explanation, suggested by Rappaport (1968: 284) is that the surveys on which these studies stand were not rigorous enough to document all the food consumed (their surprisingly low Calorie scores suggest some omissions). The two separate surveys documented here support this suggestion. They show clearly how overlooking casual food consumption can seriously bias results. In the survey of crop yields, insufficient attention was paid to food consumed over and above that weighed daily as the women's harvest. While these loads constituted the largest single source of food to the members of the homesteads concerned, they obtained and ate significant additional amounts, as the later more rigorous consumption survey demonstrated.

The results of this nutritional excursus compare broadly with those of most studies conducted in other regions of the Highlands of Papua New

[31] Some of this contribution came from the anthropologist. During the survey I tried to restrict the amount of such food which I gave away. Other sources were trade stores on Nipa station and at a mission in the Nembi valley.

[32] The use of different authorities, giving foods discrepant nutritional compositions, to calculate dietary intakes, introduces an element of distortion into this comparison. So does the differing conduct of surveys to collect nutritional data. Dieticians work intensively with small numbers for short periods and measure prepared food, whereas anthropologists, geographers and others do the reverse, and pay less attention to the physical condition of their subjects. To amalgamate both approaches presents considerable difficulties, a serious one in my experience would be persuading people to put up with the disruption to their daily lives and co-operate for any period with the painstaking measurements of a thorough nutritional survey without cheating (see McArthur 1974: 93—111 for a discussion of the issues involved). On balance though, the results obtained by either approach are broadly comparable (as Table 42 shows), indicating that the strengths and weaknesses of each cancel one another out overall.

Table 42: The Wola's nutritional intake compared with those elsewhere in Papua New Guinea

SOURCE		WEIGHT (g) OF ALL PREPARED FOODSTUFFS	CALORIES (Kcal)	PROTEIN (g)	FAT (g)	CARBOHYDRATE (g)	FIBRE (g)	CALCIUM (mg)	IRON (mg)	PRO VITAMIN A (μg)	β CAROTENE (μg)	THIAMIN (mg)	RIBOFLAVIN (mg)	NIACIN (mg)	ASCORBIC ACID (mg)
WOLA (This study)	daily per capita	1710	2243	22.2	9.1	396.6	14.9	391.3	16.1	2050	1795	3.2	0.97	10.5	473.8
ENGA (Waddell)	"	1911–1772	2364–2415	29.5–34.7	7.3–26.8			372–519.8	15.4–15.7		1433–4126	2.1–2.0	1.1–1.1	12–13.9	382.6–490.7
MARING (Rappaport)	"	2287	2015	34.7–46.8				1200							
CHIMBU (Bailey & Whiteman)	"	1686	2351	25.6	4.3			653.3	10.7	1430		2.3	1.2	12.4	649.9
CHIMBU (Hipsley & Kirk)	daily adult male	1748	2360	19.9	5.5		10.2	811	15.8		8430	2.1	1.2	12.1	542.7
CHIMBU (Venkatachalam)	"	2365	2883	30				476				2.7	1.8		740
BUSAMA (Hipsley & Clements)	daily per capita	794	1223	19.1				500				0.7			137
KAIAPIT (Hipsley & Clements)	"	1013	1609	24.8				600				1.05			216
PATEP (Hipsley & Clements)	"	1387	1904	24.4				600				1.7			244
KAVATARIA (Hipsley & Clements)	"	1256	1600	41.3				300				1.25			142
RECOMMENDED NUTRIENT INTAKES															
(Platt)	daily per capita		2500	60				800	20		3000	1.6	1.8	12	30
(Hipsley)	daily adult male		2350	30				600	9		4050	1.0	1.4	10	30

Guinea, and support their conclusions regarding shortfalls from the intakes recommended by nutritional scientists to ensure a healthy population; see Table 42. Their reliance on the single root crop of sweet potato makes these people particularly vulnerable, both to events that reduce its productivity (notably of climatic origin), and from a nutritional viewpoint, in failing to supply adequate amounts of all the nutrients considered necessary for a healthy diet (see Oomen and Malcolm 1958, Oomen *et al.* 1961). It is especially deficient in protein, resulting overall in an unbalanced diet, particularly in this direction (see Bailey and Whiteman 1963; Venkatachalam 1962). The daily per capita intake of protein for the Wola is 22.2 g, which is not far short of the 25 to 30 g advocated by Oomen and Malcolm (1958: 134) and Hipsley (1961: 118) but well below that recommended by other authorities.[33]

[33] For example, Langley (in Hipsley and Clements 1947: 134), the F.A.O. (1957), Hipsley (1961: 106—11), and Platt (1947: 379), who recommend something like two, or even three, times the protein intake of the Wola. However, the apparent health and fitness of Highlanders on their low protein diet suggests that these recommended levels may be too high (Bailey and Whiteman 1963: 384; Rappaport 1968: 78). See also Oomen and Corden (1970) on the puzzling nitrogen metabolism of New Guinea Highlanders, some of whom excrete more nitrogen than they ingest, a further indication that the protein recommendations of different authorities are awry for these people (Oomen and Corden suggest that they have bacteria in their guts capable of fixing gaseous nitrogen to compensate).

The other glaring deficiency in the Wola diet is the low intake of calcium, the daily *per capita* consumption of which is only a third or half that recommended (Venkatachalam 1962: 11 notes a similar short-fall in the Chimbu diet). Otherwise, as the figures in Table 42 show, their diet more or less meets the intakes recommended for other nutrients. The limestone on which the Wola live may account for the apparent calcium shortfall, for they display no signs of an acute insufficiency of this mineral in their diet. Although they obtain little additional calcium from their drinking water,[34] between 0.4 and 8 mg daily from ground water and a mere 0.2 mg daily from rain-water (which they catch in gourds and which supplies considerable amounts of drinking water),[35] it is possible that, growing in limestone soils, their crops (notably green leafy vegetables, which supply appreciable amounts of calcium in Highland diets according to Venkatachalam 1962: 11, and Hipsley and Kirk 1965: 81) have calcium contents above those recorded by the authorities used here to compute nutritional composition.

These dietary deficiencies do not appear to handicap Wola adults in any noticeable way. They are strong and fit, and lead strenuous lives. Their population is a vigorous and expanding one, reproducing at a rate which easily exceeds the death rate.[36] It is their children who suffer from these nutritional deficiencies. Since Venkatachalam's and Ivinskis's (1957) report on kwashiorkor (a condition of protein—calorie malnutrition to which infants are primarily prone) in New Guinea, authorities have agreed that it is this segment of the population that suffers most from the dietary deficiencies outlined. Up to about six months of age Wola infants appear well nourished, but after this their mother's milk (on which they usually continue to feed to some extent for over two years) does not supply all their calorie and protein needs. Their low protein intake from then until they reach about five years of age renders them vulnerable to under-nourishment and susceptible to sickness.[37] This is one of the factors which contributes to the observed high rate of infant mortality (see Oomen and Malcolm 1958; McKay 1960).

THE GASTROLOGY OF CROPS

The way in which people prepare food prior to eating it affects its nutritional value. The Wola waste relatively little in preparation, benefiting nutritionally from nearly everything their food has to offer (see Waddell

[34] An estimated 300 ml *per capita* daily (the water contained in vegetables eaten supplying something like five times this amount); I think Hipsley and Kirk's (1965) estimate of 71 ml per head daily for adult Chimbu males is too low for the Wola.

[35] The calcium content of rain-water is 0.6 mg/l, and of water from two streams flowing through the Haelaelinja region (the Tagiy and Bombok) are 1.4 and 26.5 mg/l; I am grateful to Philip Day of the Chemistry Department in Manchester for analysing water samples on my behalf.

[36] Unlike some unfortunate populations on the Highland fringe, for example at Bosavi (Kelly 1977: 28—31).

[37] See Allen *et al.* (1978: 16—19) on child malnutrition in the Nembi plateau region.

1972: 127, who makes the same point for the Enga). Indeed their protein-deficient diet prompts these people to reject hardly any edible part and they put a wasteful European to shame (for example, after plucking a bird, removing its guts and roasting it over hot embers, they eat it all, except for talons and beak). The Wola prepare and cook their crops in a number of ways, as Table 43 shows.

The commonest method of cooking, which they use daily, is to bake food in the ashes of a fire. Tubers they firstly scrape with a bamboo sliver to remove their outside skin, together with any soil adhering to it, an action called *siybiyay* (lit: scrape). Following this they stand them near the hot embers of a fire to dry and seal their outside surface so that ash will not adhere to it, an action called *shor biy* (lit: dry do). Then they excavate a hole in the ashes of the fire with tongs made from split saplings, into which they put the tubers, covering them over with hot embers. They leave them about half an hour to cook, called *kwishay* (lit: bake). When they retrieve the cooked tubers they slap them with their fingers to knock off any ash that is on them before eating, called *taenk bay* (lit: ash do). When they bake greens buried in the ashes of a fire, the Wola usually wrap them up in a parcel with one or two large leaves. Other crops, such as cucurbits, pulses, maize and various shoots, they simply push into the embers with no preparation, and then pick off their burnt skin or outer leaves before eating (see Pospisil 1963: 367 on the similar practices of the Kapauku).

The Wola do not cook many crops by roasting over hot embers (although they commonly cook some game in this way). They call this method of cooking *haeray* (lit: light or roast), and prepare foods for it in the same manner as for baking. They frequently roast balls of screw-pine nuts, chopping them into two or four pieces (depending on size) and trimming off their soft green skin down to the tops of the shells. Another way they prepare pandan nuts for roasting, described earlier, is to tie individual nuts around a stick in a spiral. The slow roasting of nuts over hot embers they call *ombagay*.

The majority of crops grown by the Wola, and meat too, are suitable for cooking in leaf-lined pits with hot stones. This method, which in effect steams food, is called *saway*. It requires the heating of stones on a large fire, and their transference with split sapling tongs to a cordyline leaf-, or banana frond-lined pit that will accomodate the food to be cooked (see plate XII in Sillitoe 1979a). Again food preparation is minimal: tubers are scraped and large cucurbits and cabbages cut into sections, while other crops are put in the oven as harvested. A few hot stones, wrapped in leaves to prevent them burning the food, are placed among the vegetables, and sometimes water is sprinkled in the pit to produce additional steam. The leaves lining the pit are then folded over the contents and a layer of hot stones placed across the top. The oven is left for one or two hours, depending on the amount of food in it (see Sillitoe 1977 for a more detailed account of this method of cooking).

Table 43: The culinary preparation of crops

CROPS	BAKED IN ASHES OF FIRE	ROASTED OVER HOT EMBERS	STEAMED IN EARTH OVEN	STEAMED IN BAMBOO TUBE	BOILED	EATEN RAW	STEAMED IN HOT STONE PARCEL	DRIED OVER FIRE
Sweet potato	*	*	*		*	*		
Taro	*		*					
Tannia	*		*					
Irish potato	*		*		*			
Yam	*		*		*			
Kudzu	*		*					
Ginger						*		
Amaranth greens	*[1]		*	*	*	*	*	
Crucifer greens	*[1]		*	*	*		*	
Acanth greens	*[1]		*	*	*	*		
Hibiscus greens	*[1]		*	*	*			
Parsley	*[1]		*	*		*	*	
Watercress	*[1]		*	*	*		*	
Chinese cabbage	*[1]		*	*	*		*	
Cabbage			*	*	*		*	
Onions	*	*	*	*	*	*		
Spiderwort			*	*			*	
Taro leaves			*	*				
Pumpkin leaves			*	*	*		*	
Figs	*[2]		*	*	*	*[2]		
Highland breadfruit			*	*	*			
Pumpkin	*		*		*			
Gourd	*		*		*			
Cucumber			*			*		
Climbing cucurbit	*		*		*			
Choko	*		*		*			
Beans	*		*	*				
Maize	*	*	*					
Highland *pitpit*	*		*	*	*	*		
Bamboo shoots	*		*	*		*		
Sugar cane	*[3]					*		
Bananas — fruit	*		*		*	*		
Banana plant heart						*		
Passion fruit						*		
Tomatoes						*		
Screw-pine (*aenk*)	*	*	*	*[4]		*		*[5]
Screw-pine (*wabel*)			*					

Notes

[1] In leaf parcel. [2] Fruit only. [3] Inflorescence. [4] Pith only. [5] Nuts only.

A quicker method of cooking using the earth-oven principle consists of wrapping one or two hot stones in a leaf parcel together with some greens. Unlike cooking in an oven, this can be done indoors. The stones are simply heated on a domestic fire, and transferred with tongs when hot to a prepared bundle of spinach, which is parcelled up. The contents may take up to an hour to cook. The Wola call this method of cookery *ombuwgiy* (see Pospisil 1963: 367 for a description of a similar Kapauku practice).

The Wola also cook greens, and pulses and various shoots (plus some small game and fungi too) by simmering in lengths of bamboo[38] tube called *pay kuwla*. They fill these tubes by poking food into them and shaking it down by knocking their nodal enclosed ends on the ground, an action called *pay bombay* (lit: container knock-fill; see plate X in Sillitoe 1979a). They close the open end with a plug of leaves and cook the contents by placing the tubes on hot embers, turning them round every so often. The contents settle down in the tube after they have simmered a while and are usually topped up during cooking, an act called *meninyay* (lit: top up). The cooked contents are shaken out on to a leaf, called *nay goiyay* (lit: food pour out; see Pospisil 1963: 368 for an account of similar Kapauku practices). The boiling of vegetables in a container with water is a recent culinary innovation, somewhat akin to that of the traditional bamboo tube.[39] But it reduces the nutritional value of food more than any of the above traditional methods of cooking: more is wasted in paring and peeling and some goodness is lost into the water which, except for greens, is usually thrown away.

Just as the culinary practices outlines pass on most of the nutritive goodness contained in food, so do the mensal mores surrounding its consumption. There are few conventions restricting the eating of food, few taboos hampering the effective exploitation of the diet's potential to the full.[40] The totemic-like ideas subscribed to by the Wola, described in Chapter 9, demand no more of people than a token restriction on the consumption of any foodstuffs identified in the origin myths of their communities (which contrasts with the Kapauku, according to Pospisil 1963: 364). The taboo forbidding men to take any food from the hands of female relatives during their menstrual periods (see Sillitoe 1979b: 78) likewise does not restrict their diet, it only interferes temporarily with the source of their food. Other conventions

[38] *Nastus elatus*.

[39] Our large empty food tins were in great demand as containers, although they were a headache to give away without provoking disputes by appearing to favour some households over others.

[40] This contrasts with some other people in Papua New Guinea, such as those occupying islands in the D'Entrecasteaux Archipelago; see for example Young (1971: 159–66), who describes ways the Goodenough Islanders try to suppress their hunger, so conserving their stocks of tubers for use in exchanges. When food is naturally short the Wola try to quell their hunger too, but not when it is in adequate supply, which they would consider lunacy. One way men try to suppress hunger, is to stuff pads of vegetation down inside their bark belts to press against their stomachs, as their stomachs press against their belts when full, and, as they say, 'make us think they are full'. They call these pads *day baen* (lit: hunger *baen*).

similarly restrict the source but not the range of people's food; as does the belief, for example, that eating a mixture of sweet potato tubers harvested from different gardens gives severe heartburn and pain (the easing of which requires the eating of some powdery wood ash).

The few taboos on the consumption of certain foods observed from time to time by the Wola are neither irksome nor extensive, as Table 44 shows.[41] If someone had to observe ritual taboos six times in the course of his or her life this would be an exception, and the restrictions would cover less than six months in total. People observe the *injiy* sickness taboos more frequently, for reasons of comfort and not fear of supernatural sanctions. The Wola avoid chewing sugar cane when suffering from a cold, because they think its copious juice stimulates the production of mucus and they abstain from eating hot flavoured foodstuffs that would burn and worsen a sore throat. They stick to soft and bland food.

The few food taboos observed spasmodically by the Wola support Morris's (1976) negative response to Douglas's (1966) hypothesis relating to dietary prohibitions. There is, as argued in Chapter 9, no apparent overarching classificatory schema, encompassing both the concrete and symbolic, to which they relate. Certain crops, like other prohibited foods, are forbidden on occasion not because in some sense they are taxonomic misfits which, occupying anomalous categories, are apt symbols for the focusing of ambiguous ideas and contradictory emotions in esoteric contexts. The most important taboo observed by Wola men demonstrates this: it forbids the consumption of *all* food associated with menstruating women. They believe that breaking this prohibition leads to serious illness, demanding the performance of a purificatory rite to prevent death (Sillitoe 1979b: 85–7).

Other conventions preventing people from eating certain foods are a matter of etiquette. Men, for example, rarely eat insects, other than wood-boring larvae: these are things belonging to women and children. Similarly, propriety demands that adults should give small birds and rodents to children to eat as a treat. They would be embarrassed and ashamed to consume them themselves; it would be ill-mannered. In a small way these rules redress the protein deficiency of the children's normal diet, which makes them particularly vulnerable to illness and death, by passing protein-rich perks to them.

These conventions, obliging certain persons to pass food on to others, relate to the sharing of food in general. This is something that pervades Wola commensality. The acts of giving and receiving are all-important in their society (see Sillitoe 1979a), and they extend to food. Throughout Melanesia people value food as more than a simple necessity sustaining life: in giving it and receiving it they make, maintain and manipulate social relationships.[42]

[41] For comparison see Whiteman's (1965: 308–11) account of Chimbu beliefs and taboos relating to different foods.

[42] See Young (1971: 146–227) who describes in detail how the Goodenough Islanders do this with their yams. Refer also to Bell (1948) and Lea (1969).

Table 44: Prohibitions on food consumption

RITUAL/ ILLNESS	NUMBER INVOLVED	FREQUENCY	FOOD TABOOED	FOOD RECOMMENDED	DURATION OF TABOO	CONSEQUENCES OF BREAKING TABOO
Iriy iyb kwalay ritual (see Sillitoe 1979: 83—5)	Any men who wish	Once in a lifetime	Highland *pitpit* Marsupials Birds	Sweet potato Bananas Sugar cane	Several weeks	Poor head of hair and weakness
Iysh ponda ma honday ritual	14 men	Once a generation	Sweet potato	Sugar cane Taro Ginger Marsupials Birds	2 months	Drought and sweet potato shortage
Porot ritual (see Sillitoe 1979: 160—4)	Any men who wish	Once or twice a lifetime		Bananas Marsupial Birds Pork	5 days	Fail to obtain wealth in exchanges
Shor kem ritual	Entire community	Once a generation	Highland *pitpit* Sugar cane Marsupials Insects Birds Rats		1 month	Leprosy
Injiy sickness (= cold, influenza)	Everyone sometime	Variable	Highland *pitpit* Ginger Sugar cane	Sweet potato Taro leaves Acanth greens Greens generally	A week or so	Discomfort and continuing sickness

They underline the significance of food in this regard by centering occasional large ceremonial events around its distribution, the most important of which for the Wola are pig kills (see Sillitoe 1979a: 256—69), and to a lesser extent the sharing out of taro and pandanus nuts. Food is not exchanged only on such formal occasions: it passes between people daily in a multitude of small, social transactions. There is an ethos that no one should eat without sharing some of his or her food with others present.[43] For the Wola, not to do so would be rude and insulting (as it is for the Kapauku too: see Pospisil 1963: 365).

The Wola, like other Highlanders (for example, see Waddell 1972: 121; Bailey and Whiteman 1963: 378), eat two meals a day. The largest is in the late afternoon/early evening and the arrangements for cooking it vary from household to household and from day to day. Sometimes men and women residing in the same homestead cook and consume their food separately, at others women bake vegetables for their husbands and other male relatives. Sometimes one member of a household cooks and eats his or her meal, and then shares in another cooked later by a relative. The usual method of cooking employed late in the day is baking in ashes, although occasionally members of a homestead co-ordinate their food preparation in the late afternoon, combining their efforts and preparing an earth oven in which to cook a joint meal. Sometimes a family does not wait for its female members to return with the day's food but joins them in the gardens during the afternoon where they co-operate to cook a meal in an earth oven (someone will put aside a share of the cooked food for those not present and carry it home for them to eat in the evening).

The other, smaller meal of the day often consists of leftovers from the previous evening heated up in the ashes of a fire: this is called *day biy* (lit: cooked do). People eat this meal first thing in the morning soon after waking. They may then go through to the late afternoon without eating anything. Other times they may help themselves to something for refreshment, such as some sugar cane, or eat some food offered to them by friends during the day (small children usually carry a couple of cold sweet potato tubers, given to them by their mothers in the morning, to nibble when hungry).

Food constantly passes between the members of any homestead. When the womenfolk return from the gardens in the afternoon they share out their load, either raw into three piles (one each for the men's house, themselves and the pigs), or later after they have cooked it. Similarly, when someone returns home with food from some other source (for example, with some bananas given by a relative elsewhere, or a marsupial caught while hunting, or pork received in an exchange) they share it out. A person returning with a more rarely-eaten food feels a sense of satisfaction at sharing a treat with the

[43] Whenever I stopped at a house where people were eating, for instance, I was invariably offered a share.

family. This is so much so that if someone else tried to usurp them and share out the food there would be an argument. This happened, for example, when Pundiya cut down some bananas and put them in his house to share out later. On this occasion his younger half-brother Pes found the fruits and divided them between Pundiya's wife, children and himself, leaving a pile for his elder half-brother.[44] Pundiya was furious when he returned and threw away his share: the gift was his to bestow, not Pes's, and he would have hit him in his anger had he not been ten minutes' walk away in my house (instead he satisfied himself by bawling disparaging remarks).

The composition of Wola homesteads, within which people constantly give and receive food, varies from nuclear families of one man and his wife or wives and their children, to extended family groupings consisting of varying combinations of relatives. The genealogy in Fig. 60 shows the composition of the homesteads involved in the survey of food consumption discussed earlier in this chapter.[45] Each circle encloses a homestead. Those individuals included in two overlapping circles spent considerable periods in the two linked homesteads, obtaining and sharing food within both; this made difficult, as pointed out ealier, the precise calculation of the food consumed in each homestead.[46] Food sharing takes place constantly and most intensely within such homestead groupings. They constitute the fundamental food producing and consuming units of Wola society; the sexual division of labour, as exemplified in the conventions hedging the cultivation of crops, making them inter-dependent groups.

The composition of homesteads fluctuates. Occasionally people visit relatives and friends, staying as guests for varying periods of time. Table 45 lists the guests staying in the homesteads depicted in Fig. 60 during the consumption survey and the number of nights they stayed. Some stayed only a night, others a week or longer. Sometimes they came for no particular reason, at others because some affair brought them there, and at others because of some issue at home — some women, for example, to escape from their husbands after a violent quarrel. They include both close and distant relatives. Visitors rarely bring their own food, depending on their relatives' hospitality, which is certain to be forthcoming. They know their hosts will treat them as members of the homestead during their stay, including them in all the informal sharing of food. Indeed they often favour them with any choice morsels available, and sometimes take their visit as an excuse to harvest some delicacy like bananas, knowing that in return at some future date they will receive the same cordial treatment. Guests who stay for a number of days fit into their hosts' homestead in a residential role; for

[44] These people constitute homestead D on Fig. 60.

[45] I have manipulated some of the birth orders on this genealogy to simplify it and facilitate the connection of closely-related homesteads.

[46] For the purposes of calculation these people were put in the homestead from which they said they obtained, and with which they shared, the most food.

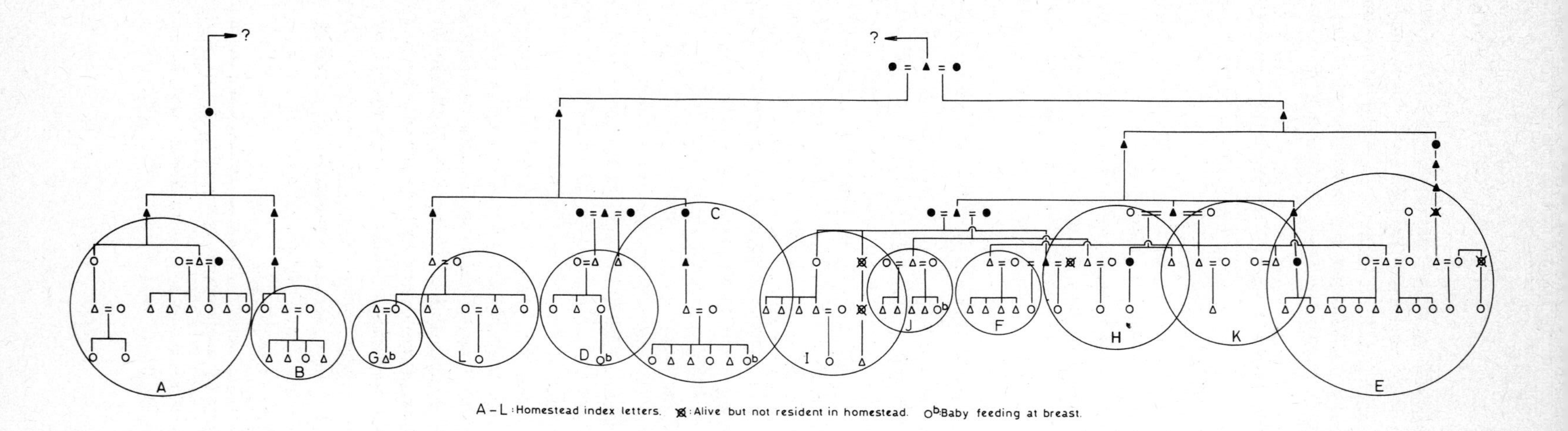

Fig. 60: The composition of the homesteads that co-operated in the consumption survey

Table 45: Guests staying in homesteads during consumption survey. No guests stayed in homesteads H, I and L

INDEX LETTER	NIGHTS STAYED MALE	FEMALE	CHILD	RELATIONSHIP OF GUESTS TO SENIOR MALE OF HOMESTEAD MALE ADULTS	FEMALE ADULTS	MALE CHILDREN	FEMALE CHILDREN
A	1	6	3	MBS	D FBSD		DD ZSD
B	—	1	1		ZHZ		WBD
C	—	5	—		WZD		
D	2	2	2	DH	D		DD
E	13	7	10	WBS W½B ½B FFZSSSS W's relative Friend	WZD FFZSSSW FFFFZSSSD FFFFZSSDD W's relative		WBD WBD
F	4	6	—	Friend	BWD FFFFZSSSD W's relative		
G	12	5	2	FB FBS WFFFFZSSSS Friend	Z FBD FBW WFFF½BDSSSDD		W½Z WBD
J	4	—	—	B FFFFZSSS Friend			
K	4	1	—	WBS Friend	WM		

example, a woman will accompany the womenfolk to their gardens and help with the work there.

The Wola share food not only with their families and guests but also with anyone who drops by during the course of a meal. Sometimes they share too, with people met during the day when carrying food home; it is quite usual to see women stopping in the afternoon and giving others produce from their net bags, especially if it is relatively uncommon and they have considerable amounts of it (for example, when they are harvesting from a new garden yielding its first flush of many various crops, or returning from the forest with a load of fungi they have collected).

The ideal, however, that they should always share food readily, can sometimes prove awkward and embarrassing for the Wola. A family occasionally comes by something special to eat which it would resent having to share. This is so, for example, when men return from a successful hunting trip with a marsupial, bird, or some other delicacy. To avoid meeting anyone, families sometimes hide themselves to consume such food, either cooking and eating it in a remote place away from their houses or waiting until night when visitors are unlikely.[47] They resort to such subterfuges if they have a little delicacy, to avoid those shameless individuals who would almost certainly turn up otherwise hoping for a share. This is poor behaviour, as rude as not offering to share food with others. If someone does come though, during the course of such a secret meal, they are inevitably offered a share, although, if their arrival is a genuine accident, they may be too embarrassed to accept it.

On the whole people willingly give and happily accept food. It is largely fortuitous whom they share with outside the homestead: this depends on chance meetings during the day. Fig. 61 illustrates this point. It shows the general flow of food between those living in different homesteads during the course of the food consumption survey.[48] The flow lines reflect the actual daily interaction between individuals, showing the categories of relatives with whom they came in contact and shared food. This distribution is presented according to three territorial zones, becoming progressively more distant from ego: (1) to members of the same *semg^{e}nk* or small localised land-holding group, (2) to members of the same *semonda* or local community, and (3) to people living elsewhere. The largest volume of food flows to members of the same *semg^{e}nk* and *semonda*. These people live close to one another and interact frequently; they are more likely to drop by at meal times or meet on a path, in a garden or somewhere else, and share food informally.

The flow lines in Fig. 61 combine the sharing of men and women. Table 46 distinguishes between them. The sexual variation this reveals in the sharing of food with different categories of relative is largely a function of the mainly virilocal residence pattern. Women usually living in their husbands' settlements received considerably more food from their spouses' relatives resident in the same *semg^{e}nk* and *semonda* (69 per cent of that recorded), and men more from their spouses' relatives living elsewhere (79 per cent). Also, residing with their own consanguines, men received the greater part of the food handed on by paternal relatives (99 per cent) and affinal relatives of both paternal and maternal relatives (84 per cent). Women received more from their children and their affines (76 per cent), men more from friends (87 per cent).

The table also shows that men share food more frequently (3.6 times more often than women). This is an interesting reflection of the more solitary life led by women compared to the social one of their menfolk. While they are working alone in their gardens, men are likely to be congregated somewhere

[47] People also go to similar lengths when they have fish and rice; they sometimes adopted laughable subterfuges at my house to receive and steal off with this food without others knowing!

[48] Both homesteads included in the survey sample, and others outside it.

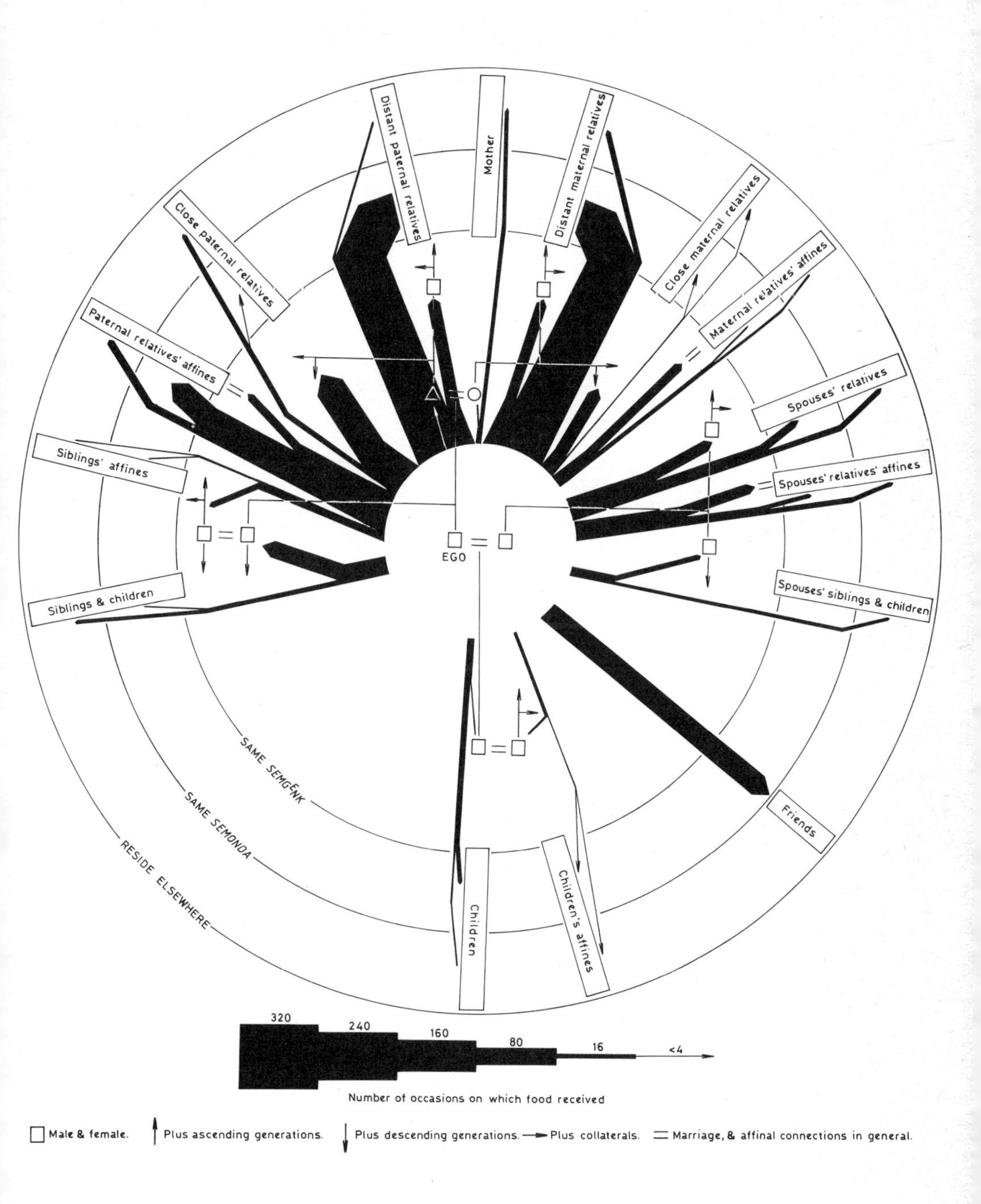

Fig. 61: The flow of food between members of different homesteads during the survey

Table 46: Food received during survey from different categories of relative living in different homesteads

RELATIONSHIPS (ego as recipient)	NUMBER OF DIFFERENT KIN RELATIONSHIPS IN EACH CATEGORY		MEN						WOMEN					
			Same semg^{e}nk		Same semonda		Resident elsewhere		Same semg^{e}nk		Same semonda		Resident elsewhere	
	Male	Female	NUMBER OF TIMES	AVERAGE WEIGHT (kgs)	NUMBER OF TIMES	AVERAGE WEIGHT (kgs)	NUMBER OF TIMES	AVERAGE WEIGHT (kgs)	NUMBER OF TIMES	AVERAGE WEIGHT (kgs)	NUMBER OF TIMES	AVERAGE WEIGHT (kgs)	NUMBER OF TIMES	AVERAGE WEIGHT (kgs)
CONSANGUINES														
Children	1	2			4 (25%)	1.60	5 (40%)	0.67	4 (0%)	1.42	17 (23.5%)	0.53	5 (20%)	4.12
Siblings	4	3	32 (9.4%)	0.63			7 (0%)	1.06	33 (9.1%)	1.26			6 (50%)	2.0
Siblings' children	4	1	7 (14.3%)	1.23	6 (0%)	0.56	2 (0%)	0.54	4 (25%)	0.43				
Father	0	1							1 (0%)	1.35				
Mother	1	1					3 (0%)	3.96	10 (0%)	1.43			15 (6.7%)	2.51
F's relatives	12	2	145 (4.8%)	0.71	1 (0%)	0.54	24 (8.3%)	0.84	3 (66.7%)	0.3				
FF's relatives	12	6	19 (10.5%)	0.74	68 (0%)	0.56	2 (0%)	0.95	14 (0%)	0.86				
FM's relatives	3	1	22 (22.7%)	0.79					1 (0%)	0.54				
FFF's relatives	8	2	1 (0%)	0,81	54 (3.7%)	0.59	1 (0%)	0.81			6 (16.7%)	0.77	1 (0%)	0.1
FMF's relatives	3	0			13 (0%)	0.63								
FFM's relatives	1	0			1 (0%)	1.89								
More distant paternal relatives	24	10	4 (25%)	0.27	84 (7.1%)	0.79	6 (0%)	0.58	5 (0%)	0.69	13 (30.8%)	0.77	1 (0%)	0.1
M's relatives	4	1	8 (25%)	1.62			1 (0%)	1.08			2 (0%)	0.75		
MM's relatives	2	0	4 (0%)	0.63			1 (0%)	0.54						
MF's relatives	6	0	72 (5.6%)	0.63			2 (0%)	0.95						
MFM's relatives	0	1									1 (0%)	0.2		
MFF's relatives	1	0					1 (0%)	0.54						
MMF's relatives	4	0	38 (2.6%)	0.65			1 (0%)	0.1						
More distant maternal relatives	23	0			230 (4.3%)	0.69	17 (11.8%)	0.53						
AFFINES														
Spouses' siblings	2	2	19 (5.3%)	1.02			16 (6.3%)	1.18	3 (33.3%)	0.38			1 (100%)	0.2
Spouses' siblings' children	2	0	3 (0%)	1.06			1 (0%)	0.81						
Spouses' parents	2	1	15 (0%)	2.38			10 (0%)	2.37	1 (0%)	1.89				
Spouses' F's relatives	3	7	1 (100%)	0.2			5 (0%)	1.35	13 (46.2%)	0.32	1 (0%)	0.2	1 (100%)	0.2
Spouses' M's relatives	0	1							1 (0%)	0.54				
Spouses' FF's relatives	4	5	4 (0%)	1.49					7 (14.3%)	0.64	1 (0%)	0.81	2 (0%)	1.47
Spouses' FM's relatives	0	2							8 (12.5%)	0.94				
Spouses' more distant paternal relatives	9	16	4 (0%)	0.6	5 (0%)	0.73	3 (0%)	1.26			29 (48.3%)	0.74		
Spouses' more distant maternal relatives	2	1					2 (0%)	0.24			1 (0%)	3.83		
Children's affines	2	4			3 (33.3%)	0.67	1 (0%)	3.0	9 (11.1%)	0.99	1 (100%)	0.2	4 (0%)	1.07
Sibling's affines	3	3	20 (5%)	0.87			1 (0%)	0.81	6 (0%)	0.53	3 (66.6%)	0.23	2 (0%)	2.57
Sibling's children's affines	6	0	1 (0%)	0.81	4 (50%)	0.37	3 (0%)	0.98						
F's relatives' affines	9	2	24 (8.3%)	0.96			17 (5.9%)	0.99	6 (0%)	0.75			1 (0%)	1.08
FF's relatives' affines	7	5	4 (0%)	0.75	36 (0%)	1.11	3 (33.3%)	0.24	6 (16.7%)	0.91			4 (0%)	0.37
FM's relatives' affines	3	0	1 (0%)	2.7	1 (0%)	0.81	3 (0%)	0.84						
More distant paternal relatives' affines	25	13			56 (1.8%)	0.98	9 (0%)	0.79			22 (9.1%)	1.05	3 (0%)	0.57
M's relatives affines	2	0	18 (0%)	0.65			2 (0%)	0.56						
MM's relatives' affines	1	0	4 (0%)	0.58										
MF's relatives' affines	3	0	23 (4.3%)	0.89										
More distant maternal relatives' affines	9	1			6 (0%)	1.1	12 (16.7%)	0.77					1 (0%)	0.2
Spouses siblings affines	4	2	3 (0%)	0.42			8 (0%)	1.02	16 (18.8%)	1.14				
Spouses' F's relatives' affines	0	3							11 (18.2%)	1.09			1 (0%)	1.62
Spouses' M's relatives' affines	0	1							15 (0%)	0.37				
Spouses' more distant relatives' affines	3	19	1 (0%)	1.62	3 (0%)	0.84	2 (0%)	1.0	6 (16.7%)	0.53	18 (16.7%)	0.87	8 (37.5%)	0.79
Friends	3	2					87 (3.4%)	0.93					13 (15.4%)	0.95

discussing matters of interest (often those relating to ceremonial exchange). Interacting with others more, they give and receive food with people from different homesteads more frequently. The table records too, the average weight of food passing between different categories of relatives (the figures in brackets show the percentage number of times it was meat or something other than garden produce). These indicate in general that, as might be expected, people tend to receive larger shares from closer relatives than they do from distant ones, that a brother will give more than a distant agnate.

The bilateral pattern of interaction revealed in the food shared with different categories of relatives accords with the bilaterality found in an examination of the flow of valuables in ceremonial exchanges (see Sillitoe 1979a: 284), and the cognatic nature generally of the kin-founded obligations that underpin Wola social life.

CONSIDERING CROPS

The constant handing round of food that characterises Wola daily life predictably colours not only their feelings and behaviour towards one another, but also their thoughts and attitudes about the plants concerned. Some of the less common crops are luxuries and more valued than others, but all are passed about: this influences their image of the ubiquitous sweet potato, shared most frequently of all, just as it does more rare and relished foods like bananas. Indeed the frequency with which plants occur in the diet, as argued at several junctures throughout this book, impinges both directly and indirectly, in a number of interconnected and ramifying ways, on Wola thoughts and behaviour regarding them.

This is no more than an aspect of the acknowledged complex reticulation comprising human intellectual and social life, of its interconnectedness, each link conditioning others in varying degrees. Nothing exists in cultural isolation. The social sciences, though, are forced to simplify and isolate elements from this complex reality for analytical purposes, a necessity that leads them ineluctably to distort in some measure. A laudable attempt in anthropology to keep this inevitable distortion within ascribed limits, to accord to reality its complexity and yet simplify it for analytic purposes, was Malinowski's early functional approach, subsequently refined and modified in many ways. Yet paradoxically, while acknowledging and even arguing vociferously for the consideration of everything within its cultural context, this functional approach invariably slips into a mechanistic and artificial frame, looking for the simple function of isolated aspects of custom and behaviour, thus failing to account for them in a convincing manner according to its own initially all-embracing ground rules.

This has prompted some to reject functionalism, or try and disguise its use under some other label. This may be counter-productive, as I hope this study demonstrates in some small measure in an ethno-botanical context. The tenets of the functional approach are not dead or out-dated, though it is time they were more widely employed in a sophisticated manner, superseding the simple-minded mechanistic search for singular functions (a stage that was perhaps necessary in hindsight for the development of certain aspects of sociological theory). Any culture is a connected whole, and many facets of it serve functions that promote its ordered maintenance — while few would seriously dispute this, the difficulty is in accounting for this complex reality without simplifying it to a point where only lip-service is paid to this self-evident fact, resulting in artificial and distorting accounts.

The pretended wholesale dumping of functionalism in its many guises, because of the inevitable distortions arising from its attempts to cope with social reality, is deceptive. Trying to avoid its pitfalls, some concentrate almost exclusively on what people say they think about things, or what aliens believe they express about them (not always one and the same issue).

Yet unavoidably, similar difficulties plague this approach too, some of its advocates occasionally resorting to an apparent bare-faced disregard of reality when it embarrasses their models. Nothing, this book argues, is more distorting than taking what a few informants say and treating their comments apart from other aspects of their experience. People do not think in isolation, but in relation to what else goes on in their lives — how else could it be? The Wola do not simply name and classify crops, they cultivate and consume them too.

A problem is that people may not express verbally all their thoughts relating to an issue, they may simply live them. The mode in which the Wola cultivate their crops enters into their overall conception of these plants, for instance impinging on their classification, both in a fashion that the investigator can extrapolate and in others that are covert. But the Wola may not customarily express either to themselves: for they know. They pass on this knowledge from generation to generation by both word of mouth and practical example, individuals supplementing it from experience, filling in gaps not learnt, so that each holds a slightly different picture in the mind. This raises again the problem of appreciating and dealing sympathetically with preliterate knowledge and ideas. Asking people to classify phenomena in isolation is to invite distortion, for they will probably be unable to articulate some of their ideas and a considerable part of their related knowledge in this manner. One difficulty is gaining access to this frequently unexpressed thought, and another is to present it when apprehended in a way that does justice to the subtlety of the ideas concerned. A helpful pointer here may be their relation to expressed concepts, as for example the consumption of crops impinges on Wola thoughts about and their ordering of these plants.

This advocacy of the old functional dictum, that sufficient attention must be paid to the complex and mutually influencing network of connections between the numerous elements that make up any cultural field (albeit inevitably simplified for excogitation), is quite different to the more recent argument that people's thinking, both on the mundane and symbolic planes, constitutes an integrated whole. Whatever may be argued for orderly (and surprisingly simple) structures on subconscious planes, life is too messy and bitty to allow for them in reality. This inexorably comes back again to the problem of reducing what goes on out there to manageable proportions for intellectual manipulation, while allowing for the extreme complexity that actually exists and impinges on the elements so isolated (including the occasional misfit between what is said and what is done).

This book, I hope, shows that the cultivation of related ideas with this in mind is profitable, that, although digging over only a small area of Wola experience, the return may be a worthwhile intellectual crop.

APPENDICES

APPENDIX I
List of Pidgin terms for crops

Common English name	*Pidgin name (Neo-Melanesian)*	*Botanical name*
Acanth greens	*kumu*	*Dicliptera papuana*
Acanth greens	*kumu*	*Rungia klossii*
Amaranth greens	*aupa; kumu*	*Amaranthus tricolor*
Amaranth greens	*kumu*	*Amaranthus caudatus*
Amaranth greens	*kumu*	*Amaranthus cruentus*
Bamboo	*mambu*	*Nastus elatus*
Bananas	*banana*	*Musa* hort var.
Beans: common	*bin*	*Phaseolus vulgaris*
Beans: hyacinth	*bin*	*Lablab niger*
Beans: winged	*bin*	*Psophocarpus tetragonolobus*
Cabbage	*kabis*	*Brassica oleracea*
Carrot	*wanpela yelopela sayor*	*Daucus carota*
Chinese cabbage	*kumu*	*Brassica chinensis*
Climbing cucurbit	*pikinini bilong rop*	*Trichosanthes pulleana*
Crucifer greens	*kumu*	*Rorippa* sp.
Cucumber	*kukamba*	*Cucumis sativus*
Dye plant	*gras bilong penim*	*Plectranthus scutellarioides*
Fig	*fikus*	*Ficus wassa*
Ginger	*kawawar*	*Zingiber officinale*
Gourd	*kambang*	*Langenaria siceraria*
Hibiscus greens	*aipika*	*Hibiscus manihot*
Highland breadfruit	*liyp bilong diwai*	*Ficus dammaropsis*

Highland *pitpit*	*pitpit*[1]	*Setaria palmifolia*
Irish potato	*poteto; patete*	*Solanum tuberosum*
Kudzu	*sayor bilong rop*	*Pueraria lobata*
Lemon	*muli*	*Citrus aurantifolia*
Maize	*kon; mais*	*Zea mays*
Onion	*anian*	*Allium cepa*
Palm lily	*tanket*	*Cordyline fruticosa*
Paper mulberry	*diwai bilong wokim laplap bilong tumbuna*	*Broussonetia papyrifera*
Parsley	*kumu*	*Oenanthe javanica*
Passion fruit	*prut*	*Passiflora edulis*
Pea	*hebsen*	*Pisum sativum*
Peanut	*bilinat; kasang; galip*	*Arachis hypogea*
Pineapple	*painap; ananas*	*Ananas salivas*
Pumpkin	*pamken*	*Cucurbita maxima*
Sago	*saksak*	*Metroxylum rumphii*
Screw-pine	*karuga*	*Pandanus brosimos* & *Pandanus julianetti*
Screw-pine	*marita*	*Pandanus conoideus*
Sedge	*putput*	*Eleocharis cf. dubia*
She-oak	*yal*	*Casuarina oligodon*
Spiderwort	*gras bilong kaikai*	*Commelina diffusa*
Sugar cane	*suga*	*Saccharum officinarum*
Sweet potato	*kaukau*	*Ipomoea batatas*
Tannia	*kongkong*	*Xanthosoma sagittifolium*
Taro	*bara*	*Colocasia esculenta*
Tobacco	*tabak; brus*	*Nicotiana tabacum*
Tomato	*tomato*	*Lycopersicon esculentum*
Watercress	*kango; kumu*	*Nasturtium officinale*
Yam	*yam*	*Dioscorea alata*
Yam	*mami*	*Dioscorea esculenta*

APPENDIX II
Glossary of Wola botanical terms

aegolaegol pay: growing round and round in a spiral.
aend say: to grow, increase in size.
aenk el: literally the 'eye', the tip of a single nut in a screw-pine cluster.
aenk hobor: the soft, moist kernel of a screw-pine nut (*P. brosimos* & *P. julianetti*).

[1] Not to be confused with the *pitpit* — *Saccharum robustum* — which grows at lower altitudes.

aenk iriy: the bristles on the tips of screw-pine nuts, exposed when the *kingilma* are trimmed off.
aenk iy: fibrous pith composing the inside of the trunk of a screw-pine.
aenk kor: the elongated, flowering organ of a male screw-pine.
aenk shombor: the small, sterile crowns growing out from the trunk of a screw-pine.
aenk togo: the entire nut cluster of a screw-pine.
ay: the stalk attaching a nut cluster to the crown of a screw-pine.
ay: the fruit bearing organ of a banana plant.
bor day: ripe (of a fruit).
buk: the tough bark of the screw-pine.
buriy: hard, tough, strong.
dekay: fine hairs growing on any part of a plant.
dend biy: sweet or appetizing taste.
dimbuw: a watery, inedible sweet potato tuber.
dink dink bay: serrated edge of leaf: also small lobes on leaf.
doliymba: sprout or lateral bud growing from side of main stem of plant, usually develop when growing tip is plucked off main stem.
duwn pay: growing straight.
dwem bay: leaves growing in a dense, compact mass (for example, the head of a cabbage).
el: internodes of sugar cane.
el: an old plant which is no longer very productive.
habuwp: a taro tuber with more than one cluster of stalks, a multiple topped corm.
hae: erect, free standing.
hokay hobor: the flesh of a sweet potato tuber.
honday: a single nut from a screw-pine cluster.
hondba: a sucker of a plant.
horok: black, flaky outer skin of an old tuber.
huguwp huguwp pay: a spindly plant which has bolted upwards in search of sunlight.
humbiy: bark.
huw: leaf rib.
huwniy: a young plant at the height of its productive life.
iyl: fruit and/or seeds.
iyla: a bulb or bulb-like protuberance; also the fibrous base of some shoots.
iyl biyay: ripe fruits fallen from a plant.
iysh: tree.
iysh: wood.
iysha: the stem or stalk of a plant, or the trunk of a tree.
iysh hul: literally a ‘tree bone’, a twig.
jimb bay: the node where a leaf stalk joins a stem.
ka biy: sweet scent of any plant.

kaendow: the inflorescence of a banana plant enclosed by bracts; sterile male flowers enclosed by bracts at end of fruiting stem.
kas: fissures in the surface of a tuber.
kat: a hand of banana fruits.
kelen bay: greasy sap.
kingilma: the same as *aenk el.*
kirikiykirikay pay: growing crooked.
koma: aerial prop roots.
koray: the soft heart of the growing tip of certain canes.
krai biy: bitter or hot taste.
kunguwp: the thin, stringy roots of a sweet potato plant which fail to develop into tubers; also the tail-like growing tip of some sweet potato tubers.
kwaliyl iriy: silks of maize cob.
lor: the nodes of a cane stem.
lor: a knot in wood.
ma: the new, growing part of a plant together with its new leaves.
maend bay: a young plant or sapling which has withered and died.
magow: the swollen, fibrous stalk at the heart of a screw-pine nut cluster.
ma injiy: literally the 'taro mother', parent tuber.
ma sinjbiyay: a diseased taro tuber.
momborlom: a screw-pine nut cluster containing few kernels.
momuw: round, spherical.
mongba: the soft pith holding the nuts of a screw-pine in a cluster on the *magow* stem.
naip say: ripe (of a fruit).
nay hae: spines or prickles on a plant.
pabkim lay: to yield; to have the potential to yield.
paenjay: sticky sap.
paepuwliym: a flower.
pak: an immature plant not yet yielding a crop.
pat henget: the soft flesh inside the pod of the winged bean in which seeds are embedded.
paziy: leaf stalk which attaches itself to the stem of the parent plant by becoming a tubular sheath that wraps around it.
pebshow: smooth and shiny, glabrous.
pil: small roots.
piy: fine, hair-like roots.
piyla: large roots.
pona: the bracts enclosing the spherical nut cluster of a screw-pine.
pondil: the spike shaped inflorescence of aroids.
pora: the set developing from a parent tuber.
port kay: a plant shoot just breaking the surface of the soil.
p^{e}ga: a branch.

punduw: the fibrous stalk by which a sweet potato tuber is attached to its parent vine.
puwla: the dry, flaking outer skin on the stem of a banana plant.
shoba: the fork formed where a branch joins a tree, a lateral stem joins a plant, and a branching vine joins a trailing or climbing plant.
shoba: an old branching sweet potato vine.
shor: a leaf.
shor baebray: a yellow leaf.
shor baeka kaeruw tibay: when a plant or tree falls over under the heavy weight of its fruit.
shor dekel: a split or tear in a leaf.
shor hungabalay: a new, furled leaf.
shor kabiy: a dry, dead leaf.
sol: a young unbranching sweet potato vine.
suwkunumb: the fluffy inflorescence of a cane grass.
suwpuw: the pod of legumes.
suwpuw: overlapping leaf sheaths.
suwpuw: the hard shell of a nut.
taenday: same as *doliymba.*
taeng biy: the white powder which occurs on the leaves of some plants.
tay: the base of a herb or tree where it enters the ground.
tibiy: the base of a screw-pine spherical nut cluster.
tiy diywael bay: the simultaneous ripening of many fruits on a plant or tree.
tiyt wiy: a hollow plant stem.
tomiy: soft, weak, pliable.
tuwmay: climbing.
uwga: the tuberous connection between a parent corm and a cormel.
way: a cutting from a plant that propagates vegetatively, which it is intended to use as planting material.
wiy: recumbent, creeping.
ya: a vine or tendrils.

APPENDIX III

The occurrence of crops in different kinds of garden by area and number of cultivations. The upper figure is the area (m^2), the lower figure the number of gardens

Type of garden	No. times planted	Total no. of gardens	Total area of gardens (m^2)	Sweet potato	Highland pitpit	Sugar cane	Pumpkin	Palm lily	Acanthus (shombey)	Taro	Bananas	Cabbage	Maize	Screw-pine (aenk)	Beans	Onions	Irish potato	Hibiscus	Amaranthus (paluw)	Crucifer (tagwt)	Amaranthus (komb)	Ginger	Cucumber	Gourd	Tomato	Bamboo	Chinese cabbage	Sedge	Tobacco	Watercress	Parsley	Passion fruit
TARO GARDENS (Dry)	1	19	11465	4160 5	11204 17	8434 14	7744 12	4160 6	9375 15	11465 19	4379 9	8455 13	8069 11	3992 5	7526 11	3930 5	1254 3	7818 11	6355 10	6731 10	6355 9	6678 9	6815 8	6564 7	1714 1			2153 2				
TARO GARDENS (Wet)	1	15	5382	1421 3	2979 7	251 1	3481 7		1170 2	5382 15	52 1	1672 4	4149 9	376 1	1985 5			2007 3	2810 8	4170 9		2811 7	1986 3	2215 4				2812 6				
MIXED GARDENS	1	37	3415	1083 14	2237 18	2266 22	2338 22	1727 15	1538 18	2619 30	1930 18	1465 17	1737 19	299 2	1145 12	639 7	629 9	732 3	799 7	595 5	498 8	178 2	439 1	146 1	95 2	345 3				73 1		
MIXED GARDENS old house site	1	3	94				84 2		42 1	42 1	42 1	10 1	52 2			42 1	52 2				10 1				10 1							
SWEET POTATO GARDENS (New)	1	163	167947	167947 163	151848 145	137007 124	115396 107	113965 96	106144 93	102199 85	75164 68	107651 96	67504 60	43307 37	67425 58	98659 54	44518 31	46755 37	43663 30	51189 32	37661 26	20089 17	31411 23	32121 25	28882 18	13790 7	15595 14		2241 5	1756 1	5644 2	690 1
SWEET POTATO GARDENS (Established)	2	91	118377	118377 91	106107 79	99406 73	85347 60	89977 61	58604 45	41762 31	66254 43	36245 21	25890 18	62257 39	33737 19	27487 17	23096 12	6209 6	4849 3	1568 1	9720 7	12334 7	4661 2	732 1	5079 3	3815 2				1777 1		
	3	29	37144	37144 29	30695 24	22554 18	8570 7	24132 20	8758 7	9647 9	22073 17	3261 3	10891 6	26107 17	2320 2	4223 4	1066 1	376 1	2006 3	1254 1	1254 1	1254 1				982 1						
	4	29	36862	36862 29	25667 20	21257 16	13859 11	23932 15	14651 9	5957 6	22418 12	2947 3	10955 9	22333 12	564 1	293 1	564 1	293 1		293 1	293 1	293 1				5653 1	293 1					
	5	15	14195	14195 15	8697 9	7380 6	7233 8	5812 6	5958 6	188 1	6815 6	1212 1	2509 3	7107 7	1212 1										1212 1							
	6	19	19832	19832 19	16153 14	11200 10	11868 10	7896 6	13290 10	5392 6	7374 5	6370 4	6120 4	7838 7	5618 3	3156 2	2174 1									4636 2						
	7	4	3428	3428 4	3428 4	3428 4	2090 2	1672 3	2090 2	2090 2	3114 3	314 1	334 1	3428 4		648 2	648 2	334 1							334 1							
	8	6	5124	5124 6	3526 4	2152 2	3536 4	2925 3	773 1	543 1	2925 4	1609 1		1927 3												611 1						
	9	3	4953	4953 3		1045 1		3135 1			3135 1																					
	10	5	9887	9887 5	9887 5	6355 3	4996 2	9887 5	6355 3	6355 3	3575 2	4139 2	4996 2	7797 4	2780 1	4996 2			2780 1													
	12	5	5581	5581 5	5581 5	2822 2	5581 5	3303 2	1819 1	1819 1	3303 2	1484 1	1484 1	3303 2	1484 1	1484 1			1484 1							1819 1						
	20	7	11684	11684 7	10472 6	6521 3	8946 4	1066 1	1066 1	1066 1	3135 1	9322 5	4034 2	4034 2		2195 1	1839 1															
	Many	13	12395	12395 13	7985 7	5413 4	8674 7	7002 6	1735 1	1839 1	4515 3	4598 3	293 1	3784 3	3574 2	2863 2	2759 2															

APPENDIX IV

Consumption survey questionnaire

The following is a copy of the form used in the consumption survey. The produce brought along each day by those participating was entered up daily on such a form, each day having a separate form.

Ref/No: *CONSUMPTION SURVEY*

Name of respondent: *Semg^{e}nk:* *Sex:* *Date:*

Crop	*Number of tubers, fruits etc.*	*Weight*	*Remarks*
Sweet potato (human)			
Sweet potato (pig)			
Highland *pitpit*			
Pumpkin			
Maize			
Sugar cane			
Beans			
Amaranths & Acanths			
Peas			
Irish potatoes			
Bananas			
Cabbage			
Tomatoes			
Cucumbers			
Taro			
Onions			
Bamboo shoots			
Gourds			
Pork			
Pandanus (*aenk*)			
Fungi			
Birds			
Marsupials			
Insects			
Water			
Firewood			
Other			

Names of people consuming:

Number (& owners) of pigs consuming:

Garden harvested from:

Other food consumed by members of this group during the day:

Remarks:

APPENDIX V

The age and sex composition of the homesteads that co-operated in the consumption survey

MALES

HOMESTEAD INDEX LETTER	Over 50yrs.		20 - 50yrs.		14 - 20yrs.		10 - 14yrs.		5 - 10yrs.		3 - 5yrs.		1 - 3yrs.		<1yr.	
	No. individuals	No. 'consump-tion days'	No. individuals	No. 'consump-tion days'	No. individuals	No. 'consump-tion days'	No. individuals	No. 'consump-tion days'	No. individuals	No. 'consump-tion days'	No. individuals	No. 'consump-tion days'	No. individuals	No. 'consump-tion days'	No. individuals	No, 'consump-tion days'
A	1	70	1+1G	18	1	19			2	124	1	70	1	17		
B			1	66	1	43			1	68			1	70		
C			1	70	1	65			1	69	1	70				
D	1	82	1+1G	82			1	82								
E			3+6G	126	1	78			2	184			1	70		
F			1+1G	86			1	82	1	80	1	82	1	69		
G			1+4G	42											1	42
H			1	12	1	12										
I			2	120	2	92	1	70								
J			1+3G	86			2	131	2	164						
K			2+1G	80	2+1G	74					1	77				
L			1	12	1	12										
Average estimated weight of individuals in survey (kgs.)	54						36		23		14		9		5	
Average estimated height of individuals in survey (cm.)	157				152		137		107		84					

FEMALES

HOMESTEAD INDEX LETTER	Over 50yrs.		20 - 50yrs.		14 - 20yrs.		10 - 14yrs.		5 - 10yrs.		3 - 5yrs.		1 - 3 yrs.		<1yr.	
	No. individuals	No. 'consump-tion days'	No. individuals	No. 'consump-tion days'	No. individuals	No. 'consump-tion days'	No. individuals	No. 'consump-tion days'	No. individuals	No. 'consump-tion days'	No. individuals	No. 'consump-tion days'	No. individuals	No. 'consump-tion days'	No. individuals	No. 'consump-tion days'
A	1	17	2+2G	93	1	57			1+1G	18	1	70	1G	2		
B			1+1G	63	1	51			1G	1	1	62				
C			1L	70	1+1G	54			1	70					1	70
D	1	81	1L+1GL	27	1	72									1+1G	27
E	1	39	3+5G	223	1	12	2	125	1+2G	92	1	44	1	82		
F			1+2G	85	1+1G	74										
G			1L+4G	47					1G	1	1G	1				
H	1	12	2	22							1	12	1	12		
I	1	70	1P	55							1	55				
J			1+1L	164											1	70
K	1+1G	12	2	89												
L	1	10	1P	12			1	12					1	12		
Average estimated weight of individuals in survey (kgs.)	47						32		23		14		9		5	
Average estimated height of individuals in survey (cm.)	147				145		130		107		84					

G = guest ; L = lactating (breast feeding children less than 1 yr. old) ; P = pregnant.

REFERENCES

Akehurst, B. C. 1968. *Tobacco.* London: Longmans.

Allen, B. J. *et al.* 1978. A preliminary report on a child nutrition and agricultural survey of the Nembi Plateau, Southern Highlands Province. Port Moresby: Mimeo — Office of Environment and Conversation.

Aufenanger, H. 1961. 'The cordyline plant in the central Highlands of New Guinea', *Anthropos*, 56: 393–408.

Bailey, K. V. and Whiteman, J. 1963. 'Dietary studies in the Chimbu', *Tropical and Geographical Medicine*, 15: 377–88.

Barnes, A. C. 1964. *The Sugarcane.* London: Leonard Hill (World Crop Series).

Barrau, J. 1962. *Les plantes alimentaires de L'Oceania.* Marseille: Annals du Musée coloniale de Marseille.

—, 1965. 'Witnesses of the past: notes on some food plants of Oceania', *Ethnology*, 4: 282–94.

—, 1979. 'Coping with exotic plants in folk taxonomies'. In *Classifications in their social context* (pp. 139–49), edited by R. F. Ellen and D. Reason. London: Academic Press.

Bell, F. L. S. 1948. 'The place of food in the social life of Tanga', *Oceania*, 19: 51–74.

Berlin, B. 1972. 'Speculations on the growth of ethnobotanical nomenclature', *Language in Society*, 1: 51–86.

Berlin, B., Breedlove, D. E. and Raven, P. H. 1966. 'Folk taxonomies and biological classification', *Science*, 154: 273–5.

—, 1968. 'Covert categories and folk taxonomies', *American Anthropologist*, 70: 290–9.

—, 1973. 'General principles of classification and nomenclature in folk biology', *American Anthropologist*, 75: 214–42.

—, 1974. *Principles of Tzeltal plant classification.* New York: Academic Press.

Berlin, B. and Kay, P. 1969. *Basic color terms: their universality and evolution.* Berkeley; Calif.: University of California Press.

Black, M. B. 1967. 'An ethnoscience investigation of Ojibwa ontology and world view', Ph.D. dissertation, Stanford University.

Bloch, M. 1977. 'The past and the present in the present', *Man*, 12: 278–92.

Bond, G. 1957. 'The development and significance of root nodules of *Casuarina*', *Annals of Botany*, 21: 373–80.

—, 1976. 'The results of the IBP survey of root-nodule formation in non-leguminous angiosperms'. In *Symbiotic nitrogen fixation in plants* (pp. 443–74), edited by P. S. Nutman. Cambridge: University Press (I.B.P. 7).

Bourke, R. M. 1975. *Know your bananas.* L.A.E.S. Information Bulletin No. 6. Keravat, New Britain: D.A.S.F.

Bowers, N. 1964. 'A further note on a recently reported root crop from the New Guinea Highlands', *Journal of the Polynesian Society*, 73: 333–5.

Brennan, P.W. 1977. *Let sleeping snakes lie: central Enga traditional religious belief and ritual.* Adelaide: The Australian Assoc. for the Study of Religions, Special Studies in Religions No. 1.

Brookfield, H. C. and White, J. P. 1968. 'Revolution for evolution in the prehistory of the New Guinea Highlands,' *Ethnology*, 7: 43–52.

Brown, C. H. 1977. 'Folk botanical life-forms: their universality and growth', *American Anthropologist*, 79: 317–42.

— *et al.*, 1976. 'Some general principles of biological and nonbiological folk classification', *American Anthropologist*, 3: 73–85.

Bulmer, R. N. H. 1965. 'Beliefs concerning the propagation of new varieties of sweet potato in two New Guinea Highlands societies', *Journal of the Polynesian Society*, 74: 237–9.

—, 1966. 'Birds as possible agents in the propagation of the sweet potato', *The Emu*, 65: 165–82.

—, 1967. 'Why is the cassowary not a bird? A problem of zoological taxonomy among the Karam of the New Guinea Highlands', *Man*, 2: 5–25.

—, 1970. 'Which came first, the chicken or the egg-head?'. In *Échanges et communications* (pp. 1069–91) edited by J. Pouillon and P. Maranda. The Hague: Mouton.

—, 1978. 'Totems and taxonomy', In *Australian Aboriginal Concepts* edited by L. Hiatt. Canberra: Australian Institute of Aboriginal Studies.

—, 1979. 'Mystical and mundane in Kalam classification of birds'. In *Classifications in their social context* (pp. 57–79) edited by R. F. Ellen and D. Reason. London: Academic Press.

— and Bulmer, S. 1964. 'The prehistory of the Australian New Guinea Highlands', *American Anthropologist* (Special Publication), 64: 39–76.

— and Menzies, J. 1972/73. 'Karam classification of marsupials and rodents. Parts 1 and 2', *Journal of the Polynesian Society*, 81: 472–99, 82: 86–107.

— and Tyler, M. 1968. 'Karam classification of frogs', *Journal of the Polynesian Society*, 77: 333–85.

Burkhill, I. H. 1951. 'Dioscoreaceae', *Flora Malesiana* (Series 1), 4: 293–335.

Clarke, W. C. 1971. *Place and People: an ecology of a New Guinean community.* Berkeley, Calif.: University of California Press.

– and Street, J. M. 1967. 'Soil fertility and cultivation practices in New Guinea', *Journal of Tropical Geography*, 24: 7–11.

Cobley, L. S. 1976. *The botany of tropical crops* (revised by W. M. Steele). London: Longmans.

Conklin, H. C. 1954. 'The relation of Hanunóo culture to the plant world', Ph.D. dissertation, Yale University.

Corner, E. J. H. 1967. 'Ficus in the Soloman Islands and its bearing on the post-Jurassic history of Melanesia', *Philosophical Transactions of the Royal Society of London*, 253: 23–159.

Coursey, D. G. 1967. *Yams.* London: Longmans.

Diamond, J. M. 1979. 'Review of Majnep and Bulmer 1977: Birds of my Kalam country', *Journal of the Polynesian Society*, 88: 116–17.

Douglas, M. 1966. *Purity and danger: an analysis of concepts of pollution and taboo.* London: Routledge and Kegen Paul.

Durkheim, E. and Mauss, M. 1963. *Primitive classification* (translated by R. Needham). London: Cohen and West.

Ellen, R. F. 1975. 'Variable constructs in Nuaulu zoological classification', *Social Science Information*, 14 (3/4): 201–28.

–, 1979a. 'Introductory essay'. In *Classifications in their social context* edited by R. F. Ellen and D. Reason. London: Academic Press.

–, 1979b. 'Omniscience and ignorance: variation in Nuaulu knowledge, identification and classification of animals', *Language in Society*, 8 (2): 337–59.

Evans-Pritchard, E. E. 1940. *The Nuer: a description of the modes of livelihood and political institutions of a Nilotic people.* Oxford: University Press.

F.A.O. 1957. *Protein Requirements.* Rome: F.A.O. Nutritional Studies Series 16.

Fischer, H. 1968. *Negwa: Eine Papua Gruppe im Wandel.* München: Klaus Renner Verlag.

Fitzpatrick, E. A. 1965. 'Climate of the Wabag-Tari area'. In *General report on lands of the Wabag-Tari area, Territory of Papua New Guinea, 1960–61* (pp. 56–69). Melbourne: C.S.I.R.O. Land Research Series No. 15.

Flenley, J. R. 1967. 'The present and former vegetation of the Wabag region of New Guinea', Canberra: Ph.D. dissertation, Australien National University.

Fowler, C. S. and Leland, J. 1967. 'Some Northern Paiute native categories', *Ethnology*, 6: 381–404.

Frake, C. O. 1961. 'The diagnosis of disease among the Subanum of Mindanao', *American Anthropologist*, 63: 113–32.

French, B. and Bridle, C. 1978. *Food crops of Papua New Guinea.* East New Britain: Vudal College of Agriculture.

Friedberg, C. 1968. 'Les méthodes d'enquête en ethnobotanique. Comment mettre en évidence les taxonomies indigènes?', *Journal d'Agriculture Tropicale et de Botanique Appliquée*, 15: 297–324.

—, 1970. 'Analyse de quelques groupements de végétaux comme introduction à l'étude de la classification botanique Bunaq'. In *Échanges et Communications* (pp. 1092–131), edited by J. Pouillon and P. Maranda. The Hague: Mouton.

—, 1974. 'Les processus classificatoires appliqués aux objets naturels et leur mise en évidence. Quelques principes methodologiques'. *Journal d'Agriculture Tropicale et de Botanique Appliquée*, 21: 313–43.

—, 1979. 'Socially significant plant species and their taxonomic position among the Bunaq of Central Timor'. In *Classifications in their social context* (pp. 81–101), edited by R. F. Ellen and D. Reason. London: Academic Press.

Gal, S. 1973. 'Inter-informant variability in an ethno-zoological taxonomy', *Anthropological Linguistics*, 15: 203–19.

Gardener, P. M. 1976. 'Birds, words and a requiem for the omniscient informant', *American Ethnologist*, 3: 446–68.

Geertz, C. 1963. *Agricultural involution: the process of ecological change in Indonesia.* Berkeley, Calif.: University of California Press.

Gell, A. 1975. *Metamorphosis of the cassowaries.* London: Athlone Press. L.S.E. Monographs in Social Anthropology No. 51.

—, 1979. 'The Umeda language poem', *Canberra Anthropology*, 2: 44–62.

Girard, F. 1957. 'Quelque plantes alimentaires et rituelles en usage chez les Buang', *Journal d'agriculture tropicale et de botanique appliquée*, 4: 212–27.

Glasse, R. M. 1963. 'Bingi at Tari', *Journal of the Polynesian Society*, 72: 270–1.

Glick, L. B. 1964. 'Categories and relations in Gimi natural science', *American Anthropologist* (Special Publication), 66: 273–80.

Golson, J. 1976. 'Archaeology and agricultural history in the New Guinea Highlands'. In *Problems in economic and social archaeology* (pp. 201–20), edited by G. de G. Sieveking, I. H. Longworth and K. E. Wilson. London: Duckworth.

Goody, J. R. 1977. *The domestication of the savage mind.* Cambridge: University Press.

Grassl, C. O. 1969. '*Saccharum* names and their interpretation', *Proceedings of the International Society of Sugarcane Technology* 868–75.

Haddon, A. C. 1947. 'Smoking and tobacco pipes in New Guinea', *Philosophical Transactions of the Royal Society of London* (Series B) 232: 1–278.

Hardesty, D. L. 1977. *Ecological Anthropology.* New York: John Wiley and Sons.

Hays, T. E. 1974. 'Mauna: explorations in Ndumba ethnobotany'. Seattle, Wash.: Ph.D. dissertation, University of Washington.

—, 1976. 'An empirical method for the identification of covert categories in ethnobiology', *American Ethnologist*, 3: 489—507.

—, 1979. 'Plant classification and nomenclature in Ndumba, Papua New Guines Highlands', *Ethnology*, 18: 253—70.

—, 1981. 'Some cultivated plants in Ndumba, Eastern Highlands Province', *Science in New Guinea*, 8 (2): 122—31.

Heider, E. R. 1972. 'Probabilities, sampling and ethnographic methods: the case of Dani colour names', *Man*, 7 (3): 418—66.

Heider, K. G. 1969. 'Sweet potato notes and lexical queries, or, the problem of all those names for sweet potatoes in the New Guinea Highlands', *Kroeber Anthropological Society Papers*, 41: 78—86.

Henty, E. E. 1969. *A manual of the grasses of New Guinea.* Lae: T.P.N.G. Department of Forests. Botany Bulletin No. 1.

Herklots, G. A. C. 1972. *Vegetables in South-East Asia.* London: George Allen and Unwin.

Hipsley, E. H. 1961. *Food and nutrition notes and reviews.* Canberra: Commonwealth Department of Health Vol. 18: 97—132.

— and Clements, F. W. (eds.) 1947. *Report of the New Guinea nutrition survey expedition.* Canberra: Department of External Territories.

— and Kirk, N. 1965. *Studies of dietary intake and the expenditure of energy by New Guineans.* Noumea, New Caledonia: South Pacific Commission. Technical Paper No. 147.

Holttum, R. E. 1967. 'The bamboos of New Guinea', *Kew Bulletin*, 21: 263—92.

Hunn, E. 1976. 'Toward a perceptual model of folk biological classification', *American Ethnologist*, 3: 508—24.

—, 1977, *Tzeltal folk zoology: the classification of discontinuities in nature.* New York: Academic Press.

Johnson, A. 1974. 'Ethnoecology and planting practices in a swidden agricultural system', *American Ethnologist*, 1: 87—101.

Kay, P. 1970. 'Some theoretical implications of ethnographic semantics', *Current directions in anthropology*, 3: 19—31.

Keleny, G. P. 1962. 'Notes on the origin and introduction of the basic food crops of the New Guinea people'. In *Symposium on the impact of man on humid tropics vegetation (Goroka T.P.N.G.)* (pp. 76—85). Djakarta: UNESCO.

Kelly, R. C. 1977. *Etoro Social Structure.* Ann Arbor, Mich.: University of Michigan Press.

La Barre, W. 1947. 'Potato taxonomy among the Aymara Indians of Bolivia', *Acta Americana*, 5: 83—103.

Lea, D. 1969. 'Some non-nutritive functions of food in New Guinea'. In *Settlement and encounter* (pp. 173—84) edited by F. Gale and G. H. Lawton. Melbourne: Oxford University Press.

Leenhardt, M. 1946. 'Le Ti', *Journal de la Societe des Oceanistes*, 2: 192–3.

Lévi-Strauss, C. 1966. *The savage mind.* London: Weidenfeld and Nicolson.

—, 1969. *Totemism* (trans. R. Needham). Harmondsworth: Penguin.

Lewis, G. A. 1980. *Day of shining red: an essay on understanding ritual.* Cambridge: University Press.

McAlpine, J. R., Keig, G. and Short, K. 1975. *Climatic tables for Papua New Guinea.* Melbourne: C.S.I.R.O. Land Use Research Technical Paper No. 37.

McArthur, M. 1974. 'Pigs for the ancestors: a review article', *Oceania*, 45: 87–123.

McCance, R. A. and Widdowson, E. M. 1960. *The composition of foods.* London: H.M.S.O. Medical Research Council Special Report No. 297.

McKay, S. R. 1960. 'Growth and nutrition of infants in the Western Highlands of New Guinea', *Medical Journal of Australia*, 1: 452–9.

Majnep, I. S. and Bulmer, R. N. H. 1977. *Birds of my Kalam country.* Oxford: University Press.

Malinowski, B. 1935. *Coral Gardens and their Magic* (2 vols.). London: Geroge Allen and Unwin.

Martin, F. W. and Ruperte, R. M. 1975. *Edible leaves of the tropics.* Mayagüey: Antillian College Press.

Massal, E. and Barrau, J. 1956. *Food plants of the South Sea Islands.* Noumea, New Caledonia: South Pacific Commission. Technical Paper No. 94.

Morris, B. 1976. 'Whither the savage mind? Notes on the natural taxonomies of a hunting and gathering people', *Man*, 11: 542–57.

—, 1979. 'Symbolism as ideology: thoughts around Navaho taxonomy and symbolism'. In *Classifications in their social context* (pp. 117–38) edited by R. F. Ellen and D. Reason. London: Academic Press.

Mowry, H. 1933. 'Symbiotic nitrogen fixation in the genus *Casuarina*', *Soil Science*, 36: 409–21.

Needham, R. 1963. Introduction to E. Durkheim and M. Mauss, *Primitive classification* (pp. vii–xlviii). London: Cohen and West.

—, 1971. 'Introduction'. In *Rethinking kinship and marriage.* A.S.A. Monograph No. 11. (pp. xiii–cxvii), edited by R. Needham. London: Tavistock.

—, 1975. 'Polythetic classification: convergence and consequences', *Man*, 10: 349–69.

—, 1979. *Symbolic classification.* Santa Monica, Calif.: Goodyear Publishing Company.

Nishiyama, I. 1963. 'The origin of the sweet potato plant'. In *Plants and the migration of Pacific peoples* edited by J. Barrau. Honolulu: B. P. Bishop Museum.

Oomen, H. A. P. C. and Corden, M. W. 1970. *Metabolic studies in New Guineans: Nitrogen metabolism in sweet potato eaters.* Noumea, New Caledonia: South Pacific Commission. Technical Paper No. 163.

Oomen, H. A. P. C. and Malcolm, S. H. 1958. *Nutrition and the Papuan Child.* Noumea, New Caledonia: South Pacific Commission. Technical Paper No. 118.

Oomen, H. A. P. C. *et al.* 1961. 'The sweet potato as the staff of life of the highland Papuan', *Tropical and Geographical Medicine*, 13: 55–66.

Panoff, F. 1969. 'Some facets of Maenge horticulture', *Oceania*, 40 (1): 20–31.

—, 1970. 'Food and faeces: a Melanesian rite', *Man*, 5 (2): 237–52.

—, 1972. 'Maenge taro and cordyline: elements of a Melanesian key', *Journal of the Polynesian Society*, 81: 375–90.

Pelto, P. J. and Pelto, G. H. 1975. 'Intra-cultural diversity: some theoretical issues', *American Ethnologist*, 2: 1–18.

Perry, R. A. 1965. 'Outline of the geology and geomorphology of the Wabag-Tari area'. In *General report on lands of the Wabag-Tari area, Territory of Papua New Guinea, 1960–61* (pp. 70–84). Melbourne: C.S.I.R.O. Land Research Series No. 15.

Peters, F. E. 1957. *Chemical composition of South Pacific foods: an annotated bibliography.* Noumea: South Pacific Commission Technical Paper No. 100.

—, 1958. *The chemical composition of South Pacific foods.* Noumea, New Caledonia: South Pacific Commission Technical Paper No. 115.

Platt, B. S. 1947. 'Colonial nutrition and its problems', *Transactions of the Royal Society of Tropical Medicine and Hygiene*, 40: 379–98.

—, 1962. *Tables of representative values of foods commonly used in tropical countries.* London: H.M.S.O. Medical Research Council Special Report No. 302.

Pospisil, L. 1963. *Kapauku Papuan economy.* New Haven, Conn.: Yale University Press. Yale University Publications in Anthropology No. 67.

Powell, J. M. 1970. 'The impact of man on the Mount Hagen region, New Guinea'. Canberra: Ph.D. dissertation Australian National University.

—, 1976. 'Ethnobotany'. In *New Guinea vegetation* (Part 3) (pp. 106–83), edited by K. Paijmans. Amsterdam: Elsevier Sci. Pub. Co.

— *et al.* 1975. *Agricultural traditions of the Mount Hagen area.* Port Moresby: University of Papua New Guinea. U.P.N.G. Geography Department Occasional Paper No. 2.

Purseglove, J. W. 1968. *Tropical crops: dicotyledons.* London: Longmans.

—, 1972. *Tropical crops: monocotyledons.* London: Longmans.

Randall, R. 1976. 'How tall is a taxonomic tree? Some evidence for dwarfism', *American Anthropologist*, 3: 229–42.

Rappaport, R. A. 1968. *Pigs for the ancestors: ritual in the ecology of a New Guinea people.* New Haven, Conn.: Yale University Press.

—, 1972. 'The flow of energy in an agricultural society'. In *Biology and culture in modern perspective* (Readings from Scientific American) (pp. 345–56), edited by J. G. Jorgensen. San Francisco, Calif.: W. H. Freeman.

Richards, A. I. 1932. *Hunger and work in a savage tribe: a functional study of nutrition among the Southern Bantu.* London: Routledge.

—, 1939. *Land, labour and diet in northern Rhodesia: an economic study of the Bemba tribe.* Oxford: University Press.

Riesenfeld, A. 1952. 'Tobacco in New Guinea and the other areas of Melanesia', *Journal of the Royal Anthropological Institue of Great Britain and Ireland*, 81: 69–102.

Roys, R. L. 1931. *The ethnobotany of the Mayas.* New Orleans: Tulane University Press. Middle American Research Series No. 2.

Rutherford, G. K. and Haantjens, H. A. 1965. 'Soils of the Wabag-Tari area'. In *General report on lands of the Wabag-Tari area Territory of Papua New Guinea 1960–61* (pp. 85–99). Melbourne: C.S.I.R.O. Land Research Series No. 15.

Ryan, D. J. 1958. 'Names and naming in Mendi', *Oceania*, 29: 109–16.

Sankoff, G. 1971. 'Quantitative analysis of sharing and variability in a cognitive model', *Ethnology*, 10: 389–408.

Sauer, J. D. 1950. 'The grain amaranths: a survey of their history and classification', *Annals of the Missouri Botanical Garden*, 37: 561–632.

Sillitoe, P. 1977. 'The stone oven: an alternative to the barbecue from the Highlands of Papua New Guinea'. In *The anthropologists' cookbook.* London: Routledge and Kegan Paul.

—, 1979a. *Give and take: exchange in Wola society.* Canberra: Australian National University Press.

—, 1979b. 'Man-eating women: fears of sexual pollution in the Papua New Guinea Highlands', *Journal of the Polynesian Society*, 88: 77–97.

—, 1979c. 'Stone versus steel', *Mankind*, 12: 151–61.

—, 1979d. 'Cosmetics from trees: an underrated trade in Papua New Guinea', *Australian Natural History*, 19: 292–7.

—, 1981. 'Pigs in disputes', *Oceania*, 51: 256–65.

Simmonds, N. W. 1959. *Bananas.* London: Longmans.

Smartt, J. 1976. *Tropical pulses.* London: Longmans.

Smith, H. H. 1923. *Ethnobotany of the Menomini Indians.* Milwaukee, Wis.: Public Museum. Museum Bulletin Vol. 4 No. 1.

—, 1928. *Ethnobotany of the Meskwaki Indians.* Milwaukee, Wis.: Public Museum. Museum Bulletin Vol. 4 No. 2.

Sokal, R. R. and Sneath, P. H. 1963. *Principles of numerical classification.* San Francisco, Calif.: Freeman.

Spencer, J. E. 1966. *Shifting cultivation in south-eastern Asia.* Berkeley, Calif.: University of California Press.

Stevenson, G. C. 1965. *Genetics and breeding of sugar cane.* London: Longmans.

Stone, B. 1974. 'Studies in Malesian Pandanaceae, XIII. New noteworthy Pandanaceae from Papuasia', *Contrib. Herb. Aust.*, 4: 7–40.

—, 1982. 'New Guinea Pandanaceae: first approach to ecology and biogeography' in *Monographiae Biologicae*, Vol. 42 (pp. 401–36), edited by J. L. Gressitt. The Hauge: W. Junk.

Strathern, A. J. 1976. 'Some notes on the cultivation of winged beans in two Highlands areas of Papua New Guinea', *Science in New Guinea*, 4: 145–52.

Strathern, A. M. 1969. 'Why is the pueraria a sweet potato?', *Ethnology*, 8 (2): 189–98.

—, 1972. *Women in between.* London: Seminar Press.

Treide, B. 1967. *Wildpflanzen in der Ernährung der Grundbevölkerung Melanesians.* Berlin: Akademie Verlag.

Turton, D. 1980. 'There's no such beast: cattle and colour naming among the Mursi', *Man*, 15: 320–38.

Venkatachalam, P. S. 1962. *A study of the diet, nutrition and health of the people of the Chimbu area (New Guinea Highlands).* Port Moresby: Department of Public Health. D.P.H. Monograph No. 4.

— and Ivinskis, V. 1957. 'Kwashiorkor in New Guinea', *Medical Journal of Australia*, 1: 275–7.

Waddell, E. 1972. *The mound builders: agricultural practices, environment and society in the Central Highlands of New Guinea.* Seattle, Wash.: University of Washington Press.

Wagner, R. 1972. *Habu: the innovation of meaning in Daribi religion.* Chicago, Ill.: University Press.

Warner, J. N. 1962. 'Sugar cane: an indigenous Papuan cultigen', *Ethnology*, 1: 405–11.

Watson, J. B. 1963. 'Krakatoa's echo', *Journal of the Polynesian Society*, 72: 152–5.

—, 1964. 'A previously unreported root crop from the New Guinea highlands', *Ethnology*, 3: 1–5.

—, 1965. 'From hunting to horticulture in the New Guinea Highlands', *Ethnology*, 4: 295–309.

—, 1968. 'Pueraria: names and traditions of a lesser crop of the central highlands, New Guinea', *Ethnology*, 7: 268–79.

Werner, O. 1969. 'The basic assumptions of ethnoscience', *Semiotica*, 1: 329–38.

Whitaker, T. W. 1971. 'Endemism and pre-Columbian migration of the bottle gourd, *Lagenaria siceraria* (Moi) Stand 1'. In *Man across the sea* edited by C. L. Riley *et al.* Austin, Tex.: University of Texas Press.

Whiteman, J. 1965. 'Customs and beliefs relating to food, nutrition and health in the Chimbu area', *Tropical and Geographical Medicine*, 17: 301–16.

Whitney, L. D. *et al.* 1939. *Taro varieties in Hawaii.* Hawaii: Agricultural Station Bulletin No. 84.

Whorf, B. L. 1956. *Language, thought and reality.* Boston: M.I.T. Press.

Womersley, J. S. 1972a. 'Crop plants'. In *Encyclopaedia of Papua and New Guinea* (pp. 222–32). Melbourne: University Press.

—, 1972b. 'Malesia'. In *Encyclopaedia of Papua and New Guinea* (p. 684), Melbourne: University Press.

—, 1972c. 'Plants, indigenous uses'. In *Encyclopaedia of Papua and New Guinea* (pp. 908—12). Melbourne: University Press.

—, (ed.) 1978. *Handbooks of the flora of Papua New Guinea* (Vol. 1) Melbourne: University Press.

Worsley, P. 1967. 'Groote Eylandt totemism and *Le Totemisme aujourd'hui'.* In *The structural study of myth and totemism* (pp. 141—59), edited by E. R. Leach. London: Tavistock.

Wyman, L. C. and Harris, S. K. 1941. *Navajo Indian medical ethnobotany.* Albuquerque, N. Mex.: University of New Mexico. University of New Mexico Bulletin 336.

Yen, D. E. 1963. 'Sweet potato variation and its relation to human migration in the Pacific'. In *Plants and the migration of Pacific peoples* (pp. 93—117) edited by J. Barrau. Honolulu: B. P. Bishop Museum.

—, 1971. 'Construction of the hypothesis for distribution of sweet potato'. In *Man across the sea* (pp. 328—42) edited by C. L. Riley *et al.* Austin: University of Texas Press.

—, 1974. *The sweet potato and Oceania: an essay in ethnobotany.* Honolulu: B. P. Bishop Museum. Museum Bulletin 236.

—, and Wheeler, J. M. 1968. 'Introduction of taro into the Pacific: the indications of the chromosome numbers', *Ethnology*, 7: 259—67.

Young, M. W. 1971. *Fighting with food.* Cambridge: University Press.

INDEX